WORLD HEALTH ORGANIZATION

INTERNATIONAL AGENCY FOR RESEARCH ON CANCER

IARC MONOGRAPHS
ON THE
EVALUATION OF THE CARCINOGENIC RISK OF CHEMICALS TO HUMANS

Allyl Compounds, Aldehydes, Epoxides and Peroxides

VOLUME 36

This publication represents the views and expert opinions
of an IARC Working Group on the
Evaluation of the Carcinogenic Risk of Chemicals to Humans
which met in Lyon,

19-26 June 1984

February 1985

IARC MONOGRAPHS

In 1969, the International Agency for Research on Cancer (IARC) initiated a programme on the evaluation of the carcinogenic risk of chemicals to humans involving the production of critically evaluated monographs on individual chemicals. In 1980, the programme was expanded to include the evaluation of the carcinogenic risk associated with employment in specific occupations.

The objective of the programme is to elaborate and publish in the form of monographs critical reviews of data on carcinogenicity for chemicals and complex mixtures to which humans are known to be exposed, and on specific occupational exposures, to evaluate these data in terms of human risk with the help of international working groups of experts in chemical carcinogenesis and related fields, and to indicate where additional research efforts are needed.

This project was supported by PHS Grant No. 1 UO1 CA33193-02 awarded by the US National Cancer Institute, Department of Health and Human Services.

© International Agency for Research on Cancer 1985

ISBN 92 832 1236 3 (soft-cover edition)

ISBN 92 832 1536 2 (hard-cover edition)

Distributed for the International Agency for Research on Cancer by the Secretariat of the World Health Organization

PRINTED IN FRANCE

CONTENTS

NOTE TO THE READER 5

LIST OF PARTICIPANTS 7

PREAMBLE 11
   Background 11
   Objective and Scope 11
   Selection of Chemicals and Complex Exposures for Monographs 12
   Working Procedures 12
   Data for Evaluations 13
   The Working Group 13
   General Principles 13
   Explanatory Notes on the Monograph Contents 20

GENERAL REMARKS ON THE SUBSTANCES CONSIDERED 31

THE MONOGRAPHS
  Allyl compounds
    Allyl chloride 39
    Allyl isothiocyanate 55
    Allyl isovalerate 69
    Eugenol 75

  Aldehydes
    Acetaldehyde 101
    Acrolein 133
    Malonaldehyde 163

  Epoxides
    Diglycidyl resorcinol ether 181
    Ethylene oxide 189
    Propylene oxide 227
    Styrene oxide 245

  Peroxides
    Benzoyl peroxide 267
    Hydrogen peroxide 285
    Lauroyl peroxide 315

APPENDIX: Activity profiles for short-term tests 325

SUPPLEMENTARY CORRIGENDA TO VOLUMES 1-35 347

CUMULATIVE INDEX TO THE MONOGRAPH SERIES 349

NOTE TO THE READER

The term 'carcinogenic risk' in the *IARC Monographs* series is taken to mean the probability that exposure to the chemical will lead to cancer in humans.

Inclusion of a chemical in the monographs does not imply that it is a carcinogen, only that the published data have been examined. Equally, the fact that a chemical has not yet been evaluated in a monograph does not mean that it is not carcinogenic.

Anyone who is aware of published data that may alter the evaluation of the carcinogenic risk of a chemical to humans is encouraged to make this information available to the Unit of Carcinogen Identification and Evaluation, Division of Environmental Carcinogenesis, International Agency for Research on Cancer, 150 cours Albert Thomas, 69372 Lyon Cedex 08, France, in order that the chemical may be considered for re-evaluation by a future Working Group.

Although every effort is made to prepare the monographs as accurately as possible, mistakes may occur. Readers are requested to communicate any errors to the Unit of Carcinogen Identification and Evaluation, so that corrections can be reported in future volumes.

IARC WORKING GROUP ON THE EVALUATION OF THE CARCINOGENIC RISK OF CHEMICALS TO HUMANS: ALLYL COMPOUNDS, ALDEHYDES, EPOXIDES AND PEROXIDES

Lyon, 19-26 June 1984

Members[1]

P. Bannasch, Abteilung für Cytopathologie, Institut für Experimentelle Pathologie, Deutsches Krebsforchungszentrum, Postfach, 6900 Heidelberg 1, Federal Republic of Germany

G. Boorman, Head, Tumor Pathology Section, National Toxicology Program, National Institute of Environmental Health Sciences, PO Box 12233, Research Triangle Park, NC 27709, USA

V.J. Feron, Head, Department of Biological Toxicology, Institute of Toxicology and Nutrition TNO, PO Box 360, 3700 AJ Zeist, The Netherlands

L. Fishbein, Deputy Director for Scientific Coordination, National Center for Toxicological Research, Jefferson, AR 72079, USA

M.J. Gardner, MRC Environmental Epidemiology Unit, Southampton General Hospital, Southampton SO9 4XY, UK

J. Hayashi, Head, Pathology Division, National Institute of Hygienic Sciences, 1-18-1 Kamiyooga, Setagaya-ku, Tokyo 158, Japan

K. Hemminki, Institute of Occupational Health, Haartmaninkatu 1, 00290 Helsinki 29, Finland

D. Henschler, Director, Institut für Pharmakologie und Toxikologie der Universität Würzburg, Versbacher Landstrasse 9, 87 Würzburg, Federal Republic of Germany (*Vice-Chairman*)

---

[1]Unable to attend: N. Loprieno, Istituto di Biochimica, Biofisica e Genetica, Universita' di Pisa, Via San Maria 53, 56100 Pisa, Italy

J.C.M. van der Hoeven, Notox Pathobiology Research, Hambakerwetering 31, 5231 DD's Hertogenbosch, The Netherlands

C. Hogstedt, National Board of Occupational Safety and Health, Occupational Health Department, 17184 Solna, Sweden

R.J. Kavlock, Chief, Perinatal Toxicology Branch, Developmental Biology Division, Health Effects Research Laboratory, US Environmental Protection Agency, Research Triangle Park, NC 27711, USA

T. Meinhardt, National Institute for Occupational Safety and Health, Robert A. Taft Laboratories, 4676 Columbia Parkway, Cincinnati, OH 45226, USA

G. Obe, Freie Universität Berlin, Institut für Allgemeine Genetik, Arnimallee 5-7, 1000 Berlin 33, Federal Republic of Germany

P.J. O'Brien, Department of Biochemistry, University of Newfoundland, Memorial University of Newfoundland, St John's, Newfoundland, A1B 3X9, Canada

J.M. Patel, University of Florida, Division of Pulmonary Medicine, Health Center, Box J-225, Gainesville, FL 32610, USA

H.S. Rosenkranz, Director, Case Western Reserve University, Center for the Environmental Health Sciences, School of Medicine, Cleveland, OH 44106, USA (*Chairman*)

Y.N. Solovyev, Director, Institute of Carcinogenesis, All-Union Cancer Research Centre of the USSR, Academy of Medical Sciences, Karshirskoye Shosse 24, 115478 Moscow, USSR

M.D. Waters, Director, Genetic Toxicology Division, Mail Drop 68, US Environmental Protection Agency, Health Effects Research Laboratory, Research Triangle Park, NC 27711, USA

Representative of the National Cancer Institute

S.M. Sieber, Deputy Director, Division of Cancer Etiology, National Cancer Institute, Building 31, Room 11A03, Bethesda, MD 20205, USA

Representative of SRI International

K.E. McCaleb, Director, Chemical-Environmental Department, SRI International, 333 Ravenswood Avenue, Menlo Park, CA 94025, USA

PARTICIPANTS

Observers[1]

Chemical Manufacturers Association

J. Clary, Toxicology Director, Celanese Corporation, 1211 Avenue of the Americas, New York, NY 10036, USA

European Chemical Industry, Ecology and Toxicology Centre

P. Grasso, 131 Old Lodge Lane, Purley, London CR2 4AU, UK

Secretariat

    H. Bartsch, Division of Environmental Carcinogenesis
    J.R.P. Cabral, Division of Environmental Carcinogenesis
    M. Friesen, Division of Environmental Carcinogenesis
    L. Haroun, Division of Environmental Carcinogenesis (*Co-Secretary*)
    E. Heseltine, Editorial and Publications Services
    E. Hietanen, Division of Environmental Carcinogenesis
    M. Hollstein, Division of Environmental Carcinogenesis
    J. Kaldor, Division of Epidemiology and Biostatistics
    P. Kalliokoski, Division of Environmental Carcinogenesis
    A. Likhachev, Division of Environmental Carcinogenesis
    D. Mietton, Division of Environmental Carcinogenesis
    R. Montesano, Division of Environmental Carcinogenesis
    I. O'Neill, Division of Environmental Carcinogenesis
    C. Partensky, Division of Environmental Carcinogenesis
    I. Peterschmitt, Division of Environmental Carcinogenesis, Geneva, Switzerland
    S. Poole, Birmingham, UK
    R. Saracci, Division of Epidemiology and Biostatistics
    L. Simonato, Division of Epidemiology and Biostatistics
    L. Tomatis, Director
    H. Vainio, Division of Environmental Carcinogenesis (*Head of the Programme*)
    J. Wahrendorf, Division of Epidemiology and Biostatistics
    J. Wilbourn, Division of Environmental Carcinogenesis (*Co-Secretary*)
    H. Yamasaki, Division of Environmental Carcinogenesis

Secretarial assistance

    J. Cazeaux
    M.-J. Ghess
    M. Lézère
    S. Reynaud
    J. Valles

---

[1]Unable to attend: M.-Th. van der Venne, Commission of the European Communities, Health and Safety Directorate, Bâtiment Jean Monnet, A2/115, Avenue Alcide-de-Gasperi, Kirchberg, Grand Duchy of Luxembourg

IARC MONOGRAPHS PROGRAMME ON THE EVALUATION OF THE CARCINOGENIC RISK OF CHEMICALS TO HUMANS[1]

PREAMBLE

1. BACKGROUND

In 1969, the International Agency for Research on Cancer (IARC) initiated a programme to evaluate the carcinogenic risk of chemicals to humans and to produce monographs on individual chemicals. Following the recommendations of an ad-hoc Working Group, which met in Lyon in 1979 to prepare criteria to select chemicals for *IARC Monographs* (1), the *Monographs* programme was expanded to include consideration of exposures to complex mixtures which occur, for example, in many occupations.

The criteria established in 1971 to evaluate carcinogenic risk to humans were adopted by all the working groups whose deliberations resulted in the first 16 volumes of the *IARC Monographs* series. This preamble reflects subsequent re-evaluation of those criteria by working groups which met in 1977(2), 1978(3), 1982(4) and 1983(5).

2. OBJECTIVE AND SCOPE

The objective of the programme is to elaborate and publish in the form of monographs critical reviews of data on carcinogenicity for chemicals, groups of chemicals and industrial processes to which humans are known to be exposed, to evaluate the data in terms of human risk with the help of international working groups of experts, and to indicate where additional research efforts are needed. These evaluations are intended to assist national and international authorities in formulating decisions concerning preventive measures. No recommendation is given concerning legislation, since this depends on risk-benefit evaluations, which seem best made by individual governments and/or other international agencies.

---

[1] This project is supported by PHS Grant No. 1 U01 CA33193-02 awarded by the US National Cancer Institute, Department of Health and Human Services.

The *IARC Monographs* are recognized as an authoritative source of information on the carcinogenicity of environmental and other chemicals. A users' survey, made in 1984, indicated that the monographs are consulted by various agencies in 45 countries. As of February 1985, 36 volumes of the *Monographs* had been published or were in press. Four supplements have been published: two summaries of evaluations of chemicals associated with human cancer, an evaluation of screening assays for carcinogens, and a cross index of synonyms and trade names of chemicals evaluated in the *Monographs* series(6).

3. SELECTION OF CHEMICALS AND COMPLEX EXPOSURES FOR MONOGRAPHS

The chemicals (natural and synthetic, including those which occur as mixtures and in manufacturing processes) and complex exposures are selected for evaluation on the basis of two main criteria: (a) there is evidence of human exposure, and (b) there is some experimental evidence of carcinogenicity and/or there is some evidence or suspicion of a risk to humans. In certain instances, chemical analogues are also considered. The scientific literature is surveyed for published data relevant to the *Monographs* programme; and the IARC *Survey of Chemicals Being Tested for Carcinogenicity*(7) often indicates those chemicals that may be scheduled for future meetings.

As new data on chemicals for which monographs have already been prepared become available, re-evaluations are made at subsequent meetings, and revised monographs are published.

4. WORKING PROCEDURES

Approximately one year in advance of a meeting of a working group, a list of the substances or complex exposures to be considered is prepared by IARC staff in consultation with other experts. Subsequently, all relevant biological data are collected by IARC; recognized sources of information on chemical carcinogenesis and systems such as CANCERLINE, MEDLINE and TOXLINE are used in conjunction with US Public Health Service Publication No. 149(8). The major collection of data and the preparation of first drafts for the sections on chemical and physical properties, on production and use, on occurrence, and on analysis are carried out by SRI International, Menlo Park, CA, USA, under a separate contract with the US National Cancer Institute. Most of the data so obtained refer to the USA and Japan; IARC supplements this information with that from other sources in Europe. Representatives from industrial associations may assist in the preparation of sections describing industrial processes. Bibliographical sources for data on mutagenicity and teratogenicity are the Environmental Mutagen Information Center and the Environmental Teratology Information Center, both located at the Oak Ridge National Laboratory, TN, USA.

Six months before the meeting, reprints of articles containing relevant biological data are sent to an expert(s), or are used by IARC staff, to prepare first drafts of monographs. These drafts are then compiled by IARC staff and sent, prior to the meeting, to all participants of the Working Group for their comments.

The Working Group then meets in Lyon for seven to eight days to discuss and finalize the texts of the monographs and to formulate the evaluations. After the meeting, the master copy of each monograph is verified by consulting the original literature, then edited by a professional editor before publication. The aim is to publish monographs within nine months of the Working Group meeting. Each volume of monographs is printed in 4000 copies for distribution to governments, regulatory agencies and interested scientists. The monographs are also available *via* the WHO Distribution and Sales Service.

5. DATA FOR EVALUATIONS

With regard to biological data, only reports that have been published or accepted for publication are reviewed by the working groups, although a few exceptions have been made: in certain instances, reports from government agencies that have undergone peer review and are widely available are considered. The monographs do not cite all of the literature on a particular chemical or complex exposure: only those data considered by the Working Group to be relevant to the evaluation of carcinogenic risk to humans are included.

Anyone who is aware of data that have been published or are in press which are relevant to the evaluations of the carcinogenic risk to humans of chemicals or complex exposures for which monographs have appeared is asked to make them available to the Unit of Carcinogen Identification and Evaluation, Division of Environmental Carcinogenesis, International Agency for Research on Cancer, Lyon, France.

6. THE WORKING GROUP

The tasks of the Working Group are five-fold: (a) to ascertain that all data have been collected; (b) to select the data relevant for evaluation; (c) to ensure that the summaries of the data enable the reader to follow the reasoning of the Working Group; (d) to judge the significance of the results of experimental and epidemiological studies; and (e) to make an evaluation of the carcinogenicity of the chemical or complex exposure.

Working Group participants who contributed to the consideration and evaluation of chemicals or complex exposures within a particular volume are listed, with their addresses, at the beginning of each publication. Each member serves as an individual scientist and not as a representative of any organization or government. In addition, observers are often invited from national and international agencies and industrial associations.

7. GENERAL PRINCIPLES APPLIED BY THE WORKING GROUP IN EVALUATING CARCINOGENIC RISK OF CHEMICALS OR COMPLEX MIXTURES

The widely accepted meaning of the term 'chemical carcinogenesis', and that used in these monographs, is the induction by chemicals (or complex mixtures of chemicals) of neoplasms that are not usually observed, the earlier induction of neoplasms that are commonly observed, and/or the induction of more neoplasms than are usually found - although fundamentally different mechanisms may be involved in these three situations. Etymologically, the term 'carcinogenesis' means the induction of cancer, that is, of malignant neoplasms; however, the commonly accepted meaning is the induction of various types of neoplasms or of a combination of malignant and benign tumours. In the monographs, the words 'tumour' and 'neoplasm' are used interchangeably. (In the scientific literature, the terms 'tumorigen', 'oncogen' and 'blastomogen' have all been used synonymously with 'carcinogen', although occasionally 'tumorigen' has been used specifically to denote a substance that induces benign tumours.)

(a) Experimental Evidence

(i) *Evidence for carcinogenicity in experimental animals*

The Working Group considers various aspects of the experimental evidence reported in the literature and formulates an evaluation of that evidence.

Qualitative aspects: Both the interpretation and evaluation of a particular study as well as the overall assessment of the carcinogenic activity of a chemical (or complex mixture) involve several considerations of qualitative importance, including: (a) the experimental parameters under which the chemical was tested, including route of administration and exposure, species, strain, sex, age, etc.; (b) the consistency with which the chemical has been shown to be carcinogenic, e.g., in how many species and at which target organ(s); (c) the spectrum of neoplastic response, from benign neoplasm to multiple malignant tumours; (d) the stage of tumour formation in which a chemical may be involved: some chemicals act as complete carcinogens and have initiating and promoting activity, while others may have promoting activity only; and (e) the possible role of modifying factors.

There are problems not only of differential survival but of differential toxicity, which may be manifested by unequal growth and weight gain in treated and control animals. These complexities are also considered in the interpretation of data.

Many chemicals induce both benign and malignant tumours. Among chemicals that have been studied extensively, there are few instances in which the neoplasms induced are only benign. Benign tumours may represent a stage in the evolution of a malignant neoplasm or they may be 'end-points' that do not readily undergo transition to malignancy. If a substance is found to induce only benign tumours in experimental animals, it should nevertheless be suspected of being a carcinogen, and it requires further investigation.

Hormonal carcinogenesis: Hormonal carcinogenesis presents certain distinctive features: the chemicals involved occur both endogenously and exogenously; in many instances, long exposure is required; and tumours occur in the target tissue in association with a stimulation of non-neoplastic growth, although in some cases hormones promote the proliferation of tumour cells in a target organ. For hormones that occur in excessive amounts, for hormone-mimetic agents and for agents that cause hyperactivity or imbalance in the endocrine system, evaluative methods comparable with those used to identify chemical carcinogens may be required; particular emphasis must be laid on quantitative aspects and duration of exposure. Some chemical carcinogens have significant side effects on the endocrine system, which may also result in hormonal carcinogenesis. Synthetic hormones and anti-hormones can be expected to possess other pharmacological and toxicological actions in addition to those on the endocrine system, and in this respect they must be treated like any other chemical with regard to intrinsic carcinogenic potential.

Complex mixtures: There is an increasing amount of data from long-term carcinogenicity studies on complex mixtures and on crude materials obtained by sampling in an occupational environment. The representativity of such samples must be considered carefully.

Quantitative aspects: Dose-response studies are important in the evaluation of carcinogenesis: the confidence with which a carcinogenic effect can be established is strengthened by the observation of an increasing incidence of neoplasms with increasing exposure.

The assessment of carcinogenicity in animals is frequently complicated by recognized differences among the test animals (species, strain, sex, age) and route and schedule of administration; often, the target organs at which a cancer occurs and its histological type may vary with these parameters. Nevertheless, indices of carcinogenic potency in particular experimental systems (for instance, the dose-rate required under continuous exposure to halve the probability of the animals remaining tumourless(9)) have been formulated in the hope that, at least among categories of fairly similar agents, such indices may be of some predictive value in other species, including humans.

Chemical carcinogens share many common biological properties, which include metabolism to reactive (electrophilic(10-11)) intermediates capable of interacting with DNA. However, they may differ widely in the dose required to produce a given level of tumour induction. The reason for this variation in dose-response is not understood, but it may be due to differences in metabolic activation and detoxification processes, in different DNA repair capacities among various organs and species or to the operation of qualitatively distinct mechanisms.

Statistical analysis of animal studies: It is possible that an animal may die prematurely from unrelated causes, so that tumours that would have arisen had the animal lived longer may not be observed; this possibility must be allowed for. Various analytical techniques have been developed which use the assumption of independence of competing risks to allow for the effects of intercurrent mortality on the final numbers of tumour-bearing animals in particular treatment groups.

For externally visible tumours and for neoplasms that cause death, methods such as Kaplan-Meier (i.e., 'life-table', 'product-limit' or 'actuarial') estimates(9), with associated significance tests(12,13), have been recommended. For internal neoplasms that are discovered 'incidentally'(12) at autopsy but that did not cause the death of the host, different estimates(14) and significance tests(12,13) may be necessary for the unbiased study of the numbers of tumour-bearing animals.

The design and statistical analysis of long-term carcinogenicity experiments were reviewed in Supplement 2 to the *Monographs* series(15). That review outlined the way in which the context of observation of a given tumour (fatal or incidental) could be included in an analysis yielding a single combined result. This method requires information on time to death for each animal and is therefore comparable to only a limited extent with analyses which include global proportions of tumour-bearing animals.

Evaluation of carcinogenicity studies in experimental animals: The evidence of carcinogenicity in experimental animals is assessed by the Working Group and judged to fall into one of four groups, defined as follows:

(1) *Sufficient evidence* of carcinogenicity is provided when there is an increased incidence of malignant tumours: (a) in multiple species or strains; or (b) in multiple experiments (preferably with different routes of administration or using different dose levels); or (c) to an unusual degree with regard to incidence, site or type of tumour,

or age at onset. Additional evidence may be provided by data on dose-response effects.

(2) *Limited evidence* of carcinogenicity is available when the data suggest a carcinogenic effect but are limited because: (a) the studies involve a single species, strain or experiment; or (b) the experiments are restricted by inadequate dosage levels, inadequate duration of exposure to the agent, inadequate period of follow-up, poor survival, too few animals, or inadequate reporting; or (c) the neoplasms produced often occur spontaneously and, in the past, have been difficult to classify as malignant by histological criteria alone (e.g., lung adenomas and adenocarcinomas and liver tumours in certain strains of mice).

(3) *Inadequate evidence* is available when, because of major qualitative or quantitative limitations, the studies cannot be interpreted as showing either the presence or absence of a carcinogenic effect.

(4) *No evidence* applies when several adequate studies are available which show that, within the limits of the tests used, the chemical or complex mixture is not carcinogenic.

It should be noted that the categories *sufficient evidence* and *limited evidence* refer only to the strength of the experimental evidence that these chemicals or complex mixtures are carcinogenic and not to the extent of their carcinogenic activity nor to the mechanism involved. The classification of any chemical may change as new information becomes available.

(ii) *Evidence for activity in short-term tests*[1]

Many short-term tests bearing on postulated mechanisms of carcinogenesis or on the properties of known carcinogens have been developed in recent years. The induction of cancer is thought to proceed by a series of steps, some of which have been distinguished experimentally(16-20). The first step - initiation - is thought to involve damage to DNA, resulting in heritable alterations in or rearrangements of genetic information. Most short-term tests in common use today are designed to evaluate the genetic activity of a substance. Data from these assays are useful for identifying potential carcinogenic hazards, in identifying active metabolites of known carcinogens in human or animal body fluids, and in helping to elucidate mechanisms of carcinogenesis. Short-term tests to detect agents with tumour-promoting activity are, at this time, insufficiently developed.

Because of the large number of short-term tests, it is difficult to establish rigid criteria for adequacy that would be applicable to all studies. General considerations relevant to all tests, however, include (a) that the test system be valid with respect to known animal carcinogens and noncarcinogens; (b) that the experimental parameters under which the chemical (or complex mixture) is tested include a sufficiently wide dose range and duration of exposure to the agent and an appropriate metabolic system; (c) that appropriate controls be used; and (d) that the purity of the compound or, in the case of complex mixtures, that the source and representativity of the sample being tested be specified. Confidence in positive results is increased if a dose-response relationship is demonstrated and if this effect has been reported in two or more independent studies.

---

[1]Based on the recommendations of a working group which met in 1983(5)

Most established short-term tests employ as end-points well-defined genetic markers in prokaryotes and lower eukaryotes and in mammalian cell lines. The tests can be grouped according to the end-point detected:

Tests of *DNA damage*. These include tests for covalent binding to DNA, induction of DNA breakage or repair, induction of prophage in bacteria and differential survival of DNA repair-proficient/-deficient strains of bacteria.

Tests of *mutation* (measurement of heritable alterations in phenotype and/or genotype). These include tests for detection of the loss or alteration of a gene product, and change of function through forward or reverse mutation, recombination and gene conversion; they may involve the nuclear genome, the mitochondrial genome and resident viral or plasmid genomes.

Tests of *chromosomal effects*. These include tests for detection of changes in chromosome number (aneuploidy), structural chromosomal aberrations, sister chromatid exchanges, micronuclei and dominant-lethal events. This classification does not imply that some chromosomal effects are not mutational events.

Tests for *cell transformation*, which monitor the production of preneoplastic or neoplastic cells in culture, are also of importance because they attempt to simulate essential steps in cellular carcinogenesis. These assays are not grouped with those listed above since the mechanisms by which chemicals induce cell transformation may not necessarily be the result of genetic change.

The selection of specific tests and end-points for consideration remains flexible and should reflect the most advanced state of knowledge in this field.

The data from short-term tests are summarized by the Working Group and the test results tabulated according to the end-points detected and the biological complexities of the test systems. The format of the table used is shown below. In these tables, a '+' indicates that

Overall assessment of data from short-term tests

	Genetic activity			Cell transformation
	DNA damage	Mutation	Chromosomal effects	
Prokaryotes				
Fungi/green plants				
Insects				
Mammalian cells (*in vitro*)				
Mammals (*in vivo*)				
Humans (*in vivo*)				

the compound was judged by the Working Group to be significantly positive in one or more assays for the specific end-point and level of biological complexity; '-' indicates that it was judged to be negative in one or more assays; and '?' indicates that there were contradictory results from different laboratories or in different biological systems, or that the result was judged to be equivocal. These judgements reflect the assessment by the Working Group of the quality of the data (including such factors as the purity of the test compound, problems of metabolic activation and appropriateness of the test system) and the relative significance of the component tests.

An overall assessment of the evidence for *genetic activity* is then made on the basis of the entries in the table, and the evidence is judged to fall into one of four categories, defined as follows:

(i) *Sufficient evidence* is provided by at least three positive entries, one of which must involve mammalian cells *in vitro* or *in vivo* and which must include at least two of three end-points - DNA damage, mutation and chromosomal effects.

(ii) *Limited evidence* is provided by at least two positive entries.

(iii) *Inadequate evidence* is available when there is only one positive entry or when there are too few data to permit an evaluation of an absence of genetic activity or when there are unexplained, inconsistent findings in different test systems.

(iv) *No evidence* applies when there are only negative entries; these must include entries for at least two end-points and two levels of biological complexity, one of which must involve mammalian cells *in vitro* or *in vivo*.

It is emphasized that the above definitions are operational, and that the assignment of a chemical or complex mixture into one of these categories is thus arbitrary.

In general, emphasis is placed on positive results; however, in view of the limitations of current knowledge about mechanisms of carcinogenesis, certain cautions should be respected: (i) At present, short-term tests should not be used by themselves to conclude whether or not an agent is carcinogenic, nor can they predict reliably the relative potencies of compounds as carcinogens in intact animals. (ii) Since the currently available tests do not detect all classes of agents that are active in the carcinogenic process (e.g., hormones), one must be cautious in utilizing these tests as the sole criterion for setting priorities in carcinogenesis research and in selecting compounds for animal bioassays. (iii) Negative results from short-term tests cannot be considered as evidence to rule out carcinogenicity, nor does lack of demonstrable genetic activity attribute an epigenetic or any other property to a substance(5).

(b) Evaluation of Carcinogenicity in Humans

Evidence of carcinogenicity can be derived from case reports, descriptive epidemiological studies and analytical epidemiological studies.

An analytical study that shows a positive association between an exposure and a cancer may be interpreted as implying causality to a greater or lesser extent, on the basis of the following criteria: (a) There is no identifiable positive bias. (By 'positive bias' is meant the operation of factors in study design or execution that lead erroneously to a more strongly positive association between an exposure and disease than in fact exists. Examples of positive bias include, in case-control studies, better documentation of the exposure for cases than for controls, and, in cohort studies, the use of better means of detecting cancer in exposed individuals than in individuals not exposed.) (b) The possibility of positive confounding has been considered. (By 'positive confounding' is meant a situation in which the relationship between an exposure and a disease is rendered more strongly positive than it truly is as a result of an association between that exposure and another exposure which either causes or prevents the disease. An example of positive confounding is the association between coffee consumption and lung cancer, which results from their joint association with cigarette smoking.) (c) The association is unlikely to be due to chance alone. (d) The association is strong. (e) There is a dose-response relationship.

In some instances, a single epidemiological study may be strongly indicative of a cause-effect relationship; however, the most convincing evidence of causality comes when several independent studies done under different circumstances result in 'positive' findings.

Analytical epidemiological studies that show no association between an exposure and a cancer ('negative' studies) should be interpreted according to criteria analogous to those listed above: (a) there is no identifiable negative bias; (b) the possibility of negative confounding has been considered; and (c) the possible effects of misclassification of exposure or outcome have been weighed. In addition, it must be recognized that the probability that a given study can detect a certain effect is limited by its size. This can be perceived from the confidence limits around the estimate of association or relative risk. In a study regarded as 'negative', the upper confidence limit may indicate a relative risk substantially greater than unity; in that case, the study excludes only relative risks that are above the upper limit. This usually means that a 'negative' study must be large to be convincing. Confidence in a 'negative' result is increased when several independent studies carried out under different circumstances are in agreement. Finally, a 'negative' study may be considered to be relevant only to dose levels within or below the range of those observed in the study and is pertinent only if sufficient time has elapsed since first human exposure to the agent. Experience with human cancers of known etiology suggests that the period from first exposure to a chemical carcinogen to development of clinically observed cancer is usually measured in decades and may be in excess of 30 years.

The evidence for carcinogenicity from studies in humans is assessed by the Working Group and judged to fall into one of four groups, defined as follows:

1. *Sufficient evidence* of carcinogenicity indicates that there is a causal relationship between the exposure and human cancer.

2. *Limited evidence* of carcinogenicity indicates that a causal interpretation is credible, but that alternative explanations, such as chance, bias or confounding, could not adequately be excluded.

3. *Inadequate evidence*, which applies to both positive and negative evidence, indicates that one of two conditions prevailed: (a) there are few pertinent data; or (b) the available studies, while showing evidence of association, do not exclude chance, bias or confounding.

4. *No evidence* applies when several adequate studies are available which do not show evidence of carcinogenicity.

(c) Relevance of Experimental Data to the Evaluation of Carcinogenic Risk to Humans

Information compiled from the first 29 volumes of the *IARC Monographs*(4,21,22) shows that, of the chemicals or groups of chemicals now generally accepted to cause or probably to cause cancer in humans, all (with the possible exception of arsenic) of those that have been tested appropriately produce cancer in at least one animal species. For several of the chemicals (e.g., aflatoxins, 4-aminobiphenyl, diethylstilboestrol, melphalan, mustard gas and vinyl chloride), evidence of carcinogenicity in experimental animals preceded evidence obtained from epidemiological studies or case reports.

For many of the chemicals (or complex mixtures) evaluated in the *IARC Monographs* for which there is *sufficient evidence* of carcinogenicity in animals, data relating to carcinogenicity for humans are either insufficient or nonexistent. **In the absence of adequate data on humans, it is reasonable, for practical purposes, to regard chemicals for which there is sufficient evidence of carcinogenicity in animals as if they presented a carcinogenic risk to humans.** The use of the expressions 'for practical purposes' and 'as if they presented a carcinogenic risk' indicates that, at the present time, a correlation between carcinogenicity in animals and possible human risk cannot be made on a purely scientific basis, but only pragmatically. Such a pragmatical correlation may be useful to regulatory agencies in making decisions related to the primary prevention of cancer.

In the present state of knowledge, it would be difficult to define a predictable relationship between the dose (mg/kg bw per day) of a particular chemical required to produce cancer in test animals and the dose that would produce a similar incidence of cancer in humans. Some data, however, suggest that such a relationship may exist(23,24), at least for certain classes of carcinogenic chemicals, although no acceptable method is currently available for quantifying the possible errors that may be involved in such an extrapolation procedure.

8. EXPLANATORY NOTES ON THE CONTENTS OF MONOGRAPHS ON CHEMICALS AND COMPLEX MIXTURES

The sections 1 and 2, as outlined below, are those used in monographs on individual chemicals. When relevant, similar information is included in monographs on complex mixtures; additional information is provided as considered necessary.

(a) Chemical and Physical Data (Section 1)

The Chemical Abstracts Services Registry Number, the latest Chemical Abstracts Primary Name (9th Collective Index)(25) and the IUPAC Systematic Name(26) are recorded in section 1. Other synonyms and trade names are given, but no comprehensive list is provided. Some of the trade names are those of mixtures in which the compound being evaluated is only one of the ingredients.

The structural and molecular formulae, molecular weight and chemical and physical properties are given. The properties listed refer to the pure substance, unless otherwise specified, and include, in particular, data that might be relevant to carcinogenicity (e.g., lipid solubility) and those that concern identification.

A separate description of the composition of technical products includes available information on impurities and formulated products.

(b) Production, Use, Occurrence and Analysis (Section 2)

The purpose of section 2 is to provide indications of the extent of past and present human exposure to the chemical.

(i) Synthesis

Since cancer is a delayed toxic effect, the dates of first synthesis and of first commercial production of the chemical are provided. This information allows a reasonable estimate to be made of the date before which no human exposure could have occurred. In addition, methods of synthesis used in past and present commercial production are described.

(ii) Production

Since Europe, Japan and the USA are reasonably representative industrialized areas of the world, most data on production, foreign trade and uses are obtained from those countries. It should not, however, be inferred that those areas or nations are the sole or even the major sources or users of any individual chemical.

Production and foreign trade data are obtained from both governmental and trade publications by chemical economists in the three geographical areas. In some cases, separate production data on organic chemicals manufactured in the USA are not available because their publication could disclose confidential information. In such cases, an indication of the minimum quantity produced can be inferred from the number of companies reporting commercial production. Each company is required to report on individual chemicals if the sales value or the weight of the annual production exceeds a specified minimum level. These levels vary for chemicals classified for different uses, e.g., medicinals and plastics; in fact, the minimal annual sales value is between $1000 and $50 000, and the minimal annual weight of production is between 450 and 22 700 kg. Data on production in some European countries are obtained by means of general questionnaires sent to companies thought to produce the compounds being evaluated. Information from the completed questionnaires is compiled, by country, and the resulting estimates of production are included in the individual monographs.

(iii) Use

Information on uses is meant to serve as a guide only and is not complete. It is usually obtained from published data but is often complemented by direct contact with manufacturers of the chemical. In the case of drugs, mention of their therapeutic uses does not necessarily represent current practice nor does it imply judgement as to their clinical efficacy.

Statements concerning regulations and standards (e.g., pesticide registrations, maximum levels permitted in foods, occupational standards and allowable limits) in specific countries are mentioned as examples only. They may not reflect the most recent situation, since such legislation is in a constant state of change; nor should it be taken to imply that other countries do not have similar regulations.

(iv) *Occurrence*

Information on the occurrence of a chemical in the environment is obtained from published data, including that derived from the monitoring and surveillance of levels of the chemical in occupational environments, air, water, soil, foods and tissues of animals and humans. When no published data are available to the Working Group, unpublished reports, deemed appropriate, may be considered. When available, data on the generation, persistence and bioaccumulation of a chemical are also included.

(v) *Analysis*

The purpose of the section on analysis is to give the reader an indication, rather than a complete review, of methods cited in the literature. No attempt is made to evaluate critically or to recommend any of the methods.

(c) *Biological Data Relevant to the Evaluation of Carcinogenic Risk to Humans (Section 3)*

In general, the data recorded in section 3 are summarized as given by the author; however, comments made by the Working Group on certain shortcomings of reporting, of statistical analysis or of experimental design are given in square brackets. The nature and extent of impurities/contaminants in the chemicals being tested are given when available.

(i) *Carcinogenicity studies in animals*

The monographs are not intended to cover all reported studies. Some studies are purposely omitted (a) because they are inadequate, as judged from previously described criteria(27-30) (e.g., too short a duration, too few animals, poor survival); (b) because they only confirm findings that have already been fully described; or (c) because they are judged irrelevant for the purpose of the evaluation. In certain cases, however, such studies are mentioned briefly, particularly when the information is considered to be a useful supplement to other reports or when it is the only data available. Their inclusion does not, however, imply acceptance of the adequacy of their experimental design or of the analysis and interpretation of their results.

Mention is made of all routes of administration by which the test material has been adequately tested and of all species in which relevant tests have been done(30). In most cases, animal strains are given. Quantitative data are given to indicate the order of magnitude of the effective carcinogenic doses. In general, the doses and schedules are indicated as they appear in the original; sometimes units have been converted for easier comparison. Experiments in which the compound was administered in conjunction with known carcinogens and experiments on factors that modify the carcinogenic effect are also reported. Experiments on the carcinogenicity of known metabolites and derivatives are also included.

(ii) *Other relevant biological data*

LD_{50} data are given when available, and other data on toxicity are included when considered relevant.

Data on effects on reproduction, on teratogenicity and embryo- and fetotoxicity and on placental transfer, from studies in experimental animals and from observations in humans, are included when considered relevant.

Information is given on absorption, distribution and excretion. Data on metabolism are usually restricted to studies that show the metabolic fate of the chemical in experimental animals and humans, and comparisons of data from animals and humans are made when possible.

Data from short-term tests are also included. In addition to the tests for genetic activity and cell transformation described previously (see pages 16-18), data from studies of related effects, but for which the relevance to the carcinogenic process is less well established, may also be mentioned.

The criteria used for considering short-term tests and for evaluating their results have been described (see page 18). In general, the authors' results are given as reported. An assessment of the data by the Working Group which differs from that of the authors, and comments concerning aspects of the study that might affect its interpretation are given in square brackets. Reports of studies in which few or no experimental details are given, or in which the data on which a reported positive or negative result is based are not available for examination, are cited, but are identified as 'abstract' or 'details not given' and are not considered in the summary tables or in making the overall assessment of genetic activity.

For several recent reviews on short-term tests, see IARC(30), Montesano et al.(31), de Serres and Ashby(32), Sugimura et al.(33), Bartsch et al.(34) and Hollstein et al.(35).

(iii) *Case reports and epidemiological studies of carcinogenicity to humans*

Observations in humans are summarized in this section. These include case reports, descriptive epidemiological studies (which correlate cancer incidence in space or time to an exposure) and analytical epidemiological studies of the case-control or cohort type. In principle, a comprehensive coverage is made of observations in humans; however, reports are excluded when judged to be clearly not pertinent. This applies in particular to case reports, in which either the clinico-pathological description of the tumours or the exposure history, or both, are poorly described; and to published routine statistics, for example, of cancer mortality by occupational category, when the categories are so broadly defined as to contribute virtually no specific information on the possible relation between cancer occurrence and a given exposure. Results of studies are assessed on the basis of the data and analyses that are presented in the published papers. Some additional analyses of the published data may be performed by the Working Group to gain better insight into the relation between cancer occurrence and the exposure under consideration. The Working Group may use these analyses in its assessment of the evidence or may actually include them in the text to summarize a study; in such cases, the results of the supplementary analyses are given in square brackets. Any comments by the Working Group are also reported in square brackets; however, these are kept to a minimum, being restricted to those instances in which it is felt that an important aspect of a study, directly impinging on its interpretation, should be brought to the attention of the reader.

(d) Summary of Data Reported and Evaluation (Section 4)

Section 4 summarizes the relevant data from animals and humans and gives the critical views of the Working Group on those data.

(i) *Exposures*

Human exposure to the chemical or complex mixture is summarized on the basis of data on production, use and occurrence.

(ii) *Experimental data*

Data relevant to the evaluation of the carcinogenicity of the test material in animals are summarized in this section. The animal species mentioned are those in which the carcinogenicity of the substance was clearly demonstrated. Tumour sites are also indicated. If the substance has produced tumours after prenatal exposure or in single-dose experiments, this is indicated. Dose-response data are given when available.

Significant findings on effects on reproduction and prenatal toxicity, and results from short-term tests for genetic activity and cell transformation assays are summarized, and the latter are presented in tables. An overall assessment is made of the degree of evidence for genetic activity in short-term tests.

(iii) *Human data*

Case reports and epidemiological studies that are considered to be pertinent to an assessment of human carcinogenicity are described. Other biological data that are considered to be relevant are also mentioned.

(iv) *Evaluation*

This section comprises evaluations by the Working Group of the degrees of evidence for carcinogenicity of the exposure to experimental animals and to humans. An overall evaluation is then made of the carcinogenic risk of the chemical or complex mixture to humans. This section should be read in conjunction with pages 00 and 00 of this Preamble for definitions of degrees of evidence.

When no data are available from epidemiological studies but there is *sufficient evidence* that the exposure is carcinogenic to animals, a footnote is included, reading: 'In the absence of adequate data on humans, it is reasonable, for practical purposes, to regard chemicals for which there is *sufficient evidence* of carcinogenicity in animals as if they presented a carcinogenic risk to humans.'

References

1. IARC (1979) Criteria to select chemicals for *IARC Monographs*. *IARC intern. tech. Rep. No. 79/003*

2. IARC (1977) IARC Monograph Programme on the Evaluation of the Carcinogenic Risk of Chemicals to Humans. Preamble. *IARC intern. tech. Rep. No. 77/002*

3. IARC (1978) Chemicals with *sufficient evidence* of carcinogenicity in experimental animals - *IARC Monographs* volumes 1-17. *IARC intern. tech. Rep. No. 78/003*

4. IARC (1982) *IARC Monographs on the Evaluation of the Carcinogenic Risk of Chemicals to Humans*, Supplement 4, *Chemicals, Industrial Processes and Industries Associated with Cancer in Humans* (IARC Monographs Volumes 1 to 29)

5. IARC (1983) Approaches to classifying chemical carcinogens according to mechanism of action. *IARC intern. tech. Rep. No. 83/001*

6. IARC (1972-1985) *IARC Monographs on the Evaluation of the Carcinogenic Risk of Chemicals to Humans*, Volumes 1-36, Lyon, France

   Volume 1 (1972) Some Inorganic Substances, Chlorinated Hydrocarbons, Aromatic Amines, *N*-Nitroso Compounds and Natural Products (19 monographs), 184 pages

   Volume 2 (1973) Some Inorganic and Organometallic Compounds (7 monographs), 181 pages

   Volume 3 (1973) Certain Polycyclic Aromatic Hydrocarbons and Heterocyclic Compounds (17 monographs), 271 pages

   Volume 4 (1974) Some Aromatic Amines, Hydrazine and Related Substances, *N*-Nitroso Compounds and Miscellaneous Alkylating Agents (28 monographs), 286 pages

   Volume 5 (1974) Some Organochlorine Pesticides (12 monographs), 241 pages

   Volume 6 (1974) Sex Hormones (15 monographs), 243 pages

   Volume 7 (1974) Some Anti-thyroid and Related Substances, Nitrofurans and Industrial Chemicals (23 monographs), 326 pages

   Volume 8 (1975) Some Aromatic Azo Compounds (32 monographs), 357 pages

   Volume 9 (1975) Some Aziridines, *N*-, *S*- and *O*-Mustards and Selenium (24 monographs), 268 pages

   Volume 10 (1976) Some Naturally Occurring Substances (22 monographs), 353 pages

   Volume 11 (1976) Cadmium, Nickel, Some Epoxides, Miscellaneous Industrial Chemicals and General Considerations on Volatile Anaesthetics (24 monographs), 306 pages

   Volume 12 (1976) Some Carbamates, Thiocarbamates and Carbazides (24 monographs), 282 pages

   Volume 13 (1977) Some Miscellaneous Pharmaceutical Substances (17 monographs), 255 pages

   Volume 14 (1977) Asbestos (1 monograph), 106 pages

   Volume 15 (1977) Some Fumigants, the Herbicides, 2,4-D and 2,4,5-T, Chlorinated Dibenzodioxins and Miscellaneous Industrial Chemicals (18 monographs), 354 pages

   Volume 16 (1978) Some Aromatic Amines and Related Nitro Compounds - Hair Dyes, Colouring Agents, and Miscellaneous Industrial Chemicals (32 monographs), 400 pages

   Volume 17 (1978) Some *N*-Nitroso Compounds (17 monographs), 365 pages

Volume 18 (1978) Polychlorinated Biphenyls and Polybrominated Biphenyls (2 monographs), 140 pages

Volume 19 (1979) Some Monomers, Plastics and Synthetic Elastomers, and Acrolein (17 monographs), 513 pages

Volume 20 (1979) Some Halogenated Hydrocarbons (25 monographs), 609 pages

Volume 21 (1979) Sex Hormones (II) (22 monographs), 583 pages

Volume 22 (1980) Some Non-Nutritive Sweetening Agents (2 monographs), 208 pages

Volume 23 (1980) Some Metals and Metallic Compounds (4 monographs), 438 pages

Volume 24 (1980) Some Pharmaceutical Drugs (16 monographs), 337 pages

Volume 25 (1981) Wood, Leather and Some Associated Industries (7 monographs), 412 pages

Volume 26 (1981) Some Antineoplastic and Immunosuppressive Agents (18 monographs), 411 pages

Volume 27 (1981) Some Aromatic Amines, Anthraquinones and Nitroso Compounds, and Inorganic Fluorides Used in Drinking-Water and Dental Preparations (18 monographs), 344 pages

Volume 28 (1982) The Rubber Manufacturing Industry (1 monograph), 486 pages

Volume 29 (1982) Some Industrial Chemicals (18 monographs), 416 pages

Volume 30 (1982) Miscellaneous Pesticides (18 monographs), 424 pages

Volume 31 (1983) Some Food Additives, Feed Additives and Naturally Occurring Substances (21 monographs), 314 pages

Volume 32 (1983) Polynuclear Aromatic Compounds, Part 1, Chemical, Environmental and Experimental Data (42 monographs), 477 pages

Volume 33 (1984) Polynuclear Aromatic Compounds, Part 2, Carbon Blacks, Mineral Oils and Some Nitroarenes (8 monographs), 245 pages

Volume 34 (1984) Polynuclear Aromatic Compounds, Part 3, Industrial Exposures in Aluminium Production, Coal Gasification, Coke Production and Iron and Steel Founding (4 monographs), 219 pages

Volume 35 (1984) Polynuclear Aromatic Compounds, Part 4, Bitumens, Coal-Tars and Derived Products, Shale-Oils and Soots (4 monographs), 271 pages

Volume 36 (1985) Allyl Compounds, Aldehydes, Epoxides and Peroxides (14 monographs), 369 pages

Supplement No. 1 (1979) Chemicals and Industrial Processes Associated with Cancer in Humans (IARC Monographs, Volumes 1 to 20), 71 pages

PREAMBLE 27

Supplement No. 2 (1980) Long-term and Short-term Screening Assays for Carcinogens: A Critical Appraisal, 426 pages

Supplement No. 3 (1982) Cross Index of Synonyms and Trade Names in Volumes 1 to 26, 199 pages

Supplement No. 4 (1982) Chemicals, Industrial Processes and Industries Associated with Cancer in Humans (IARC Monographs, Volumes 1 to 29), 292 pages

7. IARC (1973-1984) *Information Bulletin on the Survey of Chemicals Being Tested for Carcinogenicity*, Numbers 1-11, Lyon, France

Number 1 (1973) 52 pages
Number 2 (1973) 77 pages
Number 3 (1974) 67 pages
Number 4 (1974) 97 pages
Number 5 (1975) 88 pages
Number 6 (1976) 360 pages
Number 7 (1978) 460 pages
Number 8 (1979) 604 pages
Number 9 (1981) 294 pages
Number 10 (1982) 326 pages
Number 11 (1984) 370 pages

8. PHS 149 (1951-1983) Public Health Service Publication No. 149, *Survey of Compounds which have been Tested for Carcinogenic Activity*, Washington DC, US Government Printing Office

1951 Hartwell, J.L., 2nd ed., Literature up to 1947 on 1329 compounds, 583 pages
1957 Shubik, P. & Hartwell, J.L., Supplement 1, Literature for the years 1948-1953 on 981 compounds, 388 pages
1969 Shubik, P. & Hartwell, J.L., edited by Peters, J.A., Supplement 2, Literature for the years 1954-1960 on 1048 compounds, 655 pages
1971 National Cancer Institute, Literature for the years 1968-1969 on 882 compounds, 653 pages
1973 National Cancer Institute, Literature for the years 1961-1967 on 1632 compounds, 2343 pages
1974 National Cancer Institute, Literature for the years 1970-1971 on 750 compounds, 1667 pages
1976 National Cancer Institute, Literature for the years 1972-1973 on 966 compounds, 1638 pages
1980 National Cancer Institute, Literature for the year 1978 on 664 compounds, 1331 pages
1983 National Cancer Institute, Literature for years 1974-1975 on 575 compounds, 1043 pages

9. Pike, M.C. & Roe, F.J.C. (1963) An actuarial method of analysis of an experiment in two-stage carcinogenesis. *Br. J. Cancer*, *17*, 605-610

10. Miller, E.C. (1978) Some current perspectives on chemical carcinogenesis in humans and experimental animals: Presidential address. *Cancer Res.*, *38*, 1479-1496

11. Miller, E.C. & Miller, J.A. (1981) Searches for ultimate chemical carcinogens and their reactions with cellular macromolecules. *Cancer, 47*, 2327-2345

12. Peto, R. (1974) Guidelines on the analysis of tumour rates and death rates in experimental animals. *Br. J. Cancer, 29*, 101-105

13. Peto, R. (1975) Letter to the editor. *Br. J. Cancer, 31*, 697-699

14. Hoel, D.G. & Walburg, H.E., Jr (1972) Statistical analysis of survival experiments. *J. natl Cancer Inst., 49*, 361-372

15. Peto, R., Pike, M.C., Day, N.E., Gray, R.G., Lee, P.N., Parish, S., Peto, J., Richards, S. & Wahrendorf, J. (1980) *Guidelines for simple sensitive significance tests for carcinogenic effects in long-term animal experiments.* In: IARC Monographs on the Evaluation of the Carcinogenic Risk of Chemicals to Humans, Supplement 2, *Long-term and Short-term Screening Assays for Carcinogens: A Critical Appraisal*, Lyon, pp. 311-426

16. Berenblum, I. (1975) *Sequential aspects of chemical carcinogenesis: Skin.* In: Becker, F.F., ed., *Cancer. A Comprehensive Treatise*, Vol. 1, New York, Plenum Press, pp. 323-344

17. Foulds, L. (1969) *Neoplastic Development*, Vol. 2, London, Academic Press

18. Farber, E. & Cameron, R. (1980) The sequential analysis of cancer development. *Adv. Cancer Res., 31*, 125-126

19. Weinstein, I.B. (1981) The scientific basis for carcinogen detection and primary cancer prevention. *Cancer, 47*, 1133-1141

20. Slaga, T.J., Sivak, A. & Boutwell, R.K., eds (1978) *Mechanisms of Tumor Promotion and Cocarcinogenesis*, Vol. 2, New York, Raven Press

21. IARC Working Group (1980) An evaluation of chemicals and industrial processes associated with cancer in humans based on human and animal data: IARC Monographs Volumes 1 to 20. *Cancer Res., 40*, 1-12

22. IARC (1979) *IARC Monographs on the Evaluation of the Carcinogenic Risk of Chemicals to Humans*, Supplement 1, *Chemicals and Industrial Processes Associated with Cancer in Humans*, Lyon

23. Rall, D.P. (1977) *Species differences in carcinogenesis testing.* In: Hiatt, H.H., Watson, J.D. & Winsten, J.A., eds, *Origins of Human Cancer*, Book C, Cold Spring Harbor, NY, Cold Spring Harbor Laboratory, pp. 1383-1390

24. National Academy of Sciences (NAS) (1975) *Contemporary Pest Control Practices and Prospects: The Report of the Executive Committee*, Washington DC

25. Chemical Abstracts Services (1978) *Chemical Abstracts Ninth Collective Index (9CI), 1972-1976*, Vols 76-85, Columbus, OH

26. International Union of Pure and Applied Chemistry (1965) *Nomenclature of Organic Chemistry*, Section C, London, Butterworths

27. WHO (1958) Second Report of the Joint FAO/WHO Expert Committee on Food Additives. Procedures for the testing of intentional food additives to establish their safety and use. *WHO tech. Rep. Ser. No. 144*

28. WHO (1967) Scientific Group. Procedures for investigating intentional and unintentional food additives. *WHO tech. Rep. Ser. No. 348*

29. Sontag, J.M., Page, N.P. & Saffiotti, U. (1976) Guidelines for carcinogen bioassay in small rodents. *Natl Cancer Inst. Carcinog. tech. Rep. Ser. No.1*

30. IARC (1980) *IARC Monographs on the Evaluation of the Carcinogenic Risk of Chemicals to Humans*, Supplement 2, *Long-term and Short-term Screening Assays for Carcinogens: A Critical Appraisal*, Lyon

31. Montesano, R., Bartsch, H. & Tomatis, L., eds (1980) *Molecular and Cellular Aspects of Carcinogen Screening Tests (IARC Scientific Publications No. 27)*, Lyon

32. de Serres, F.J. & Ashby, J., eds (1981) *Evaluation of Short-Term Tests for Carcinogens. Report of the International Collaborative Program*, Amsterdam, Elsevier/North-Holland Biomedical Press

33. Sugimura, T., Sato, S., Nagao, M., Yahagi, T., Matsushima, T., Seino, Y., Takeuchi, M. & Kawachi, T. (1976) *Overlapping of carcinogens and mutagens*. In: Magee, P.N., Takayama, S., Sugimura, T. & Matsushima, T., eds, *Fundamentals in Cancer Prevention*, Tokyo/Baltimore, University of Tokyo/University Park Press, pp. 191-215

34. Bartsch, H., Tomatis, L. & Malaveille, C. (1982) *Qualitative and quantitative comparison between mutagenic and carcinogenic activities of chemicals*. In: Heddle, J.A., ed., *Mutagenicity: New Horizons in Genetic Toxicology*, New York, Academic Press, pp. 35-72

35. Hollstein, M., McCann, J., Angelosanto, F.A. & Nichols, W.W. (1979) Short-term tests for carcinogens and mutagens. *Mutat. Res.*, 65, 133-226

GENERAL REMARKS ON THE SUBSTANCES CONSIDERED

1. Introduction

This thirty-sixth volume of the *IARC Monographs* comprises 14 monographs on some allyl compounds, aldehydes, epoxides and peroxides. Five compounds - acrolein, diglycidyl resorcinol ether, ethylene oxide, propylene oxide and styrene oxide - had been evaluated by previous Working Groups (IARC, 1976, 1979, 1982a); new data that had become available on these compounds have been included in the present monographs and taken into consideration in the re-evaluations.

Four compounds - allyl alcohol, allyl bromide, crotonaldehyde and glutaraldehyde - were included in a tentative list of substances to be evaluated, but consideration of these compounds was postponed since no study of carcinogenicity to experimental animals or to humans was available. Allyl alcohol has been tested in hamsters by oral administration, but a published report of the study was not yet available; glutaraldehyde is presently being tested by skin application in mice and rats (IARC, 1982b).

Many of the compounds from the four generic classes of agents considered in this volume are produced (and, in many cases, have been for several decades) in large quantities. They possess a broad spectrum of utility and potential for widespread exposure, and are characterized by high chemical and biological activity, particularly in their ability to alkylate macromolecules either directly or after metabolic activation. Several of the compounds occur naturally (e.g., allyl isothiocyanate, eugenol) or are widely present in the environment as products of combustion and in cigarette smoke (e.g., acrolein, acetaldehyde) or as metabolic products and as endogenous agents (e.g., acetaldehyde, malonaldehyde, hydrogen peroxide).

The relative chemical instability of some substances (peroxides and aldehydes) considered in this volume may in some cases lead to uncertainties as to the nature and quantity of the chemical actually tested in in-vitro and in-vivo assays.

An appendix is included in which both the qualitative and quantitative results of the genetic and related short-term assays that are summarized within the monographs are displayed graphically in 'activity profiles'.

Allyl compounds

A number of allyl compounds, such as allyl chloride, are widely employed in the production of a variety of important chemical agents. Interest in the potential carcinogenicity of allyl compounds arises from the natural occurrence of a number of these agents (e.g., allyl isothiocyanate and eugenol) in some edible plants and their wide application, in particular, the use of some of these compounds as food additives and flavouring agents. Alkenylbenzenes, including methyl eugenol (1-allyl-3,4-dimethoxybenzene), have been found in the essential oil and juice of oranges treated with harvesting agents (Moshonas & Shaw, 1978); some alkenylbenzenes have been found to have carcinogenic activity (Miller *et al.*, 1983).

Aldehydes

Two of the aldehydes considered in this volume (acetaldehyde and acrolein) are widely used in many industrial processes. Aldehydes occur in natural vegetative processes, and have been found as gaseous by-products of incomplete combustion of wood and coal, in exhaust effluents from gasoline and diesel engines, industrial waste gases and fumes, tobacco smoke and wood fires (Bailey *et al.*, 1981; National Academy of Sciences, 1981; Lipari *et al.*, 1984). Formaldehyde, which has received the most attention because of its widespread occurrence, use and toxic properties, was evaluated previously (IARC, 1982a,c).

Epoxides

Ethylene oxide and propylene oxide are among the industrial chemicals produced in the largest volumes and have a wide variety of uses. Epoxides are directly-acting alkylating agents, reacting with nucleophiles without the need for metabolic transformation; they react with cellular macromolecules such as nucleic acids and proteins.

Peroxides

The commercial organic peroxides and hydrogen peroxide are produced in considerable volume and have many applications, including use in plastics and elastomers, as bleaching agents for fats, oils, waxes, flour and cheese, and in pharmaceutical and cosmetic preparations. Hydrogen peroxide is a normal cellular constituent.

2. Metabolism

Several of the compounds considered are interrelated *via* their metabolic pathways (Fig. 1): the allyl compounds are presumably metabolized *in vivo* to allyl alcohol, which is in turn metabolized to acrolein.

Fig. 1 Metabolic pathways of some of the allyl compounds and aldehydes considered in this volume (marked with an asterisk); →, reported metabolic pathway; ⇢, possible metabolic pathway

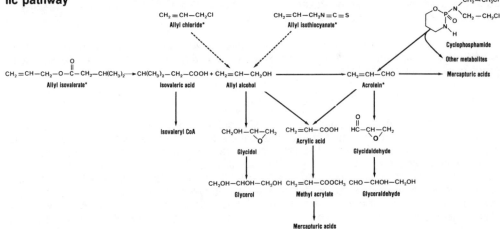

Hydrogen peroxide, acetaldehyde and malonaldehyde either occur endogenously or have been detected in human serum (Zlatkis et al., 1980, 1981), and hydrogen peroxide and acetaldehyde have been detected in expired air (Krotoszynski et al., 1977). Hydrogen peroxide can stimulate lipid peroxidation (Ursini et al., 1981), which results in the production of complex mixtures of aldehydes (Fig. 2).

Fig. 2 Formation of aldehydes from lipid peroxidation (Adapted from Alkino & Ohno, 1981)

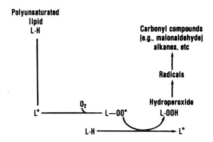

Hydrogen peroxide is evolved by many cellular organelles, for example, by peroxisomes during the oxidation of fatty acids. Analysis of breakdown products from microsomal peroxidation stimulated *in vitro* has shown that an array of carbonyl compounds is produced (Esterbauer, 1982). Malonaldehyde is the most extensively studied of these degradation products.

3. *Epidemiological studies*

The numbers of persons exposed to some of the chemicals considered in this volume are sizeable, and have accrued over several decades of exposure. There are, however, few epidemiological studies that have addressed the potential health risk to humans associated with exposures to these chemicals, some of which have been demonstrated to have carcinogenic and mutagenic effects in experimental systems. The conduct of adequate epidemiological studies is often complicated by several factors, including the relatively small numbers of workers employed at individual production and manufacturing plants, concurrent exposures to other chemicals, and the absence of historical exposure data. Nevertheless, additional epidemiological data are needed and should be pursued to assess whether or not the potential carcinogenic risks identified in experimental systems are present at detectable levels in exposed human populations.

Cytogenetic methods, mainly using human peripheral blood lymphocytes, are being used increasingly to evaluate genetic damage in exposed human populations (e.g., to ethylene oxide). Evaluations of and comparisons among these studies were sometimes difficult because of deficiencies in the study design, and the cytogenetic techniques and statistical methods employed. More standardized methods would facilitate future evaluations.

4. *References*

Akino, T. & Ohno, K. (1981) Phospholipids of the lung in normal, toxic, and diseased states. *Crit. Rev. Toxicol.*, 9, 201-274

Bailey, R.A., Clarke, H.M., Ferris, J.P., Krause, S. & Strong, R.L. (1981) *Chemistry of the Environment*, New York, Academic Press, p. 266

Esterbauer, H. (1982) *Aldehydeic products of lipid peroxidation.* In: McBrien, D.C.H. & Slater, T.F., eds, *Free Radicals, Lipid Peroxidation and Cancer*, London, Academic Press, pp. 101-128

IARC (1976) *IARC Monographs on the Evaluation of Carcinogenic Risk of Chemicals to Man*, Vol. 11, *Cadmium, Nickel, Some Epoxides, Miscellaneous Industrial Chemicals and General Considerations on Volatile Anaesthetics*, pp. 125-129, 157-167, 191-199, 201-208

IARC (1979) *IARC Monographs on the Evaluation of the Carcinogenic Risk of Chemicals to Humans*, Vol. 19, *Some Monomers, Plastics and Synthetic Elastomers, and Acrolein*, pp. 275-283, 479-494

IARC (1982a) *IARC Monographs on the Evaluation of the Carcinogenic Risk of Chemicals to Humans*, Supplement 4, *Chemicals, Industrial Processes and Industries Associated with Cancer in Humans. IARC Monographs, Volumes 1 to 29*, Lyon, pp. 126-128, 131-132, 229-233

IARC (1982b) *Information Bulletin on the Survey of Chemicals Being Tested for Carcinogenicity*, No. 11, Lyon, pp. 163, 200, 237

IARC (1982c) *IARC Monographs on the Evaluation of the Carcinogenic Risk of Chemicals to Humans*, Vol. 29, *Some Industrial Chemicals and Dyestuffs*, Lyon, pp. 345-389

Krotoszynski, B., Gabriel, G. & O'Neill, H. (1977) Characterization of human expired air: A promising investigative and diagnostic technique. *J. chromatogr. Sci.*, *15*, 239-244

Lipari, F., Dasch, J.M. & Scruggs, W.F. (1984) Aldehyde emissions from wood-burning fireplaces. *Environ. Sci. Technol.*, *18*, 326-330

Miller, E.C., Swanson, A.B., Phillips, D.H., Fletcher, T.L., Liem, A. & Miller, J.A. (1983) Structure-activity studies of the carcinogenicities in the mouse and rat of some naturally occurring and synthetic alkenylbenzene derivatives related to safrole and estragole. *Cancer Res.*, *43*, 1124-1134

Moshonas, M.G. & Shaw, P.E. (1978) Compounds new to essential orange oil from fruit treated with abscission chemicals. *J. agric. Food Chem.*, *26*, 1288-1290

National Academy of Sciences (1981) *Formaldehyde and Other Aldehydes*, Washington, DC, National Academy Press, pp. 36-131

Ursini, F., Maiorino, M., Ferri, L., Valente, M. & Gregolin, C. (1981) Hydrogen peroxide and hematin in microsomal lipid peroxidation. *J. inorg. Biochem.*, *15*, 163-169

Zlatkis, A., Poole, C.F., Brazeli, R., Bafus, D.A. & Spencer, P.S. (1980) Volatile metabolites in sera of normal and diabetic patients. *J. Chromatogr.*, *182*, 137-145

Zlatkis, A., Brazell, R.S. & Poole, C.F. (1981) The role of organic volatile profiles in clinical diagnosis. *Clin. Chem.*, *27*, 789-797

THE MONOGRAPHS

ALLYL COMPOUNDS

ALLYL CHLORIDE

1. Chemical and Physical Data

1.1 Synonyms and trade names

Chem. Abstr. Services Reg. No.: 107-05-1

Chem. Abstr. Name: 1-Propene, 3-chloro-

IUPAC Systematic Name: 3-Chloropropene

Synonyms: AC; chlorallylene; chloroallylene; 1-chloropropene-2; 3-chloropropene-1; 1-chloro-2-propene; 3-chloro-1-propene; α-chloropropylene; 3-chloropropylene; 3-chloro-1-propylene; NCI-C04615; 2-propenyl chloride

1.2 Structural and molecular formulae and molecular weight

$$CH_2=CH-CH_2Cl$$

C_3H_5Cl \hfill Mol. wt: 76.5

1.3 Chemical and physical properties of the pure substance

From DeBenedictis (1979), unless otherwise specified

(a) *Description*: Colourless liquid with a pungent, garlic-like odour (Verschueren, 1977)

(b) *Boiling-point*: 44.96°C (Beacham, 1978)

(c) *Freezing-point*: -134.5°C

(d) *Density*: Specific gravity (20°C), 0.9392

(e) *Refractive index*: n_D^{20} 1.4160 (Beacham, 1978)

(f) *Spectroscopy data*: Proton magnetic resonance spectra have been reported (Bothner-By *et al.*, 1966).

(g) *Solubility*: Slightly soluble (0.36 wt %) in water; miscible with chloroform, diethyl ether, ethanol and petroleum ether (Windholz, 1983)

(h) *Viscosity*: 0.336 cP at 20°C

(i) *Volatility*: Vapour pressure, 295.5 mm Hg at 20°C (Beacham, 1978)

(j) *Stability*: Flash-point (closed-cup), -31.7°C

(k) *Reactivity*: Reacts as both an olefin (e.g., additions) and an organic halide (e.g., hydrolysis); undergoes a variety of nucleophilic substitution reactions

(l) *Conversion factor*: 1 ppm = 3.13 mg/m^3 at 760 mm Hg and 25°C (Irish, 1963)

1.4 Technical products and impurities

In 1949, allyl chloride was available in the USA as a single grade with a purity of 97% min. Possible impurities were said to be 2-chloropropene, isopropyl chloride, *n*-propyl chloride and traces of 3,3-dichloropropene (Vesper, 1949).

In 1973, allyl chloride was available in two grades in the USA. A water-washed grade contained 0.01% acidity (as hydrogen chloride) max and an anhydrous grade contained 0.10% acidity max. Both grades met the following specifications: purity, 97% min; apparent specific gravity (20/20°C), 0.935-0.939; and distillation range, 43.0-49.0°C (Bales, 1977).

Allyl chloride is available in western Europe with the following specifications: purity, 97.5% min; water, 200 mg/kg max; specific gravity, 0.932-0.938; and distillation range, 43-50°C.

2. Production, Use, Occurrence and Analysis

2.1 Production and use

(a) *Production*

Allyl chloride was first prepared in 1857 by Cahours and Hofmann by the reaction of allyl alcohol with phosphorus chloride (Vesper, 1949). It was first produced commercially in 1945 from propylene (IARC, 1979a) by means of non-catalytic, high-temperature (500-510°C) chlorination (Beacham, 1978); this method is still used for commercial production (DeBenedictis, 1979).

US production of allyl chloride in 1977 was estimated to have been almost 180 million kg (DeBenedictis, 1979). Only two US companies presently produce it (at three plant locations), and production data are not disclosed (see Preamble, section 8(*b*)(ii)). Separate data on US imports and exports of allyl chloride are not published.

Allyl chloride is produced commercially by one company in France, one company in the German Democratic Republic, three companies in Germany and one company in the Netherlands.

The commercial production of allyl chloride in Japan started in about 1961. Three Japanese companies currently manufacture it by the chlorination of propylene; 1982 production is estimated to have been 30-40 million kg.

(b) Use

Allyl chloride is used almost exclusively as a chemical intermediate. The major use is as an intermediate for epichlorohydrin (see IARC, 1976). It is also used to make sodium allylsulphonate and a series of allyl amines and quaternary ammonium salts as well as the allyl ethers of a variety of alcohols, phenols and polyols, and a number of barbiturate hypnotic agents.

An estimated 150 million kg of allyl chloride were used as a chemical intermediate (probably unisolated) for epichlorohydrin manufacture in the USA in 1982. Epichlorohydrin is used principally for the manufacture of epoxy resins and glycerol (Beacham, 1978).

Sodium allylsulphonate is made by the reaction of allyl chloride with sodium sulphite; it is used as a component in metal plating baths (DeBenedictis, 1979). US production of this chemical was last reported separately in 1973, when it amounted to 843 thousand kg (US International Trade Commission, 1975).

Allyl chloride is used to make mono-, di-, and triallylamine as well as mixed amines containing other alkyl groups (e.g., diallylmethylamine). These amines find use as such, and as intermediates for other chemicals, such as ambuside (a diuretic), diallyl 2-chloroacetamide (a herbicide) (Beacham, 1978), and several quaternary ammonium salts. The most important allyl amine derivative is believed to be diallyl dimethylammonium chloride, which finds use as a comonomer with acrylamide and other monomers in the production of cationic flocculating agents.

Allyl ethers of polyols such as trimethylol propane have reportedly been used commercially in polyester furniture finishes to improve their drying properties. Another ether made from allyl chloride, allyl starch, was formerly made in commercial quantities in the USA for use in surface coatings (Beacham, 1978).

The following six barbiturate hypnotic agents may be made from allyl chloride: aprobarbital, butalbital, methohexital sodium, secobarbital, talbutal and thiamylal sodium, although one source (Swinyard, 1975) has indicated that such products are made from allyl bromide.

Allyl chloride is also used to make allyl isothiocyanate (see p. 55 of this volume), eugenol (see p. 75 of this volume) and 1,2-dibromo-3-chloropropane (see IARC, 1979b). It has reportedly been used in Japan to make allyl esters (e.g., diallyl phthalate).

Other commercial uses which have been reported for allyl chloride include the synthesis of glycerol chlorohydrins, trichloropropane and cyclopropane (Shell Chemical Corp., 1949).

In Japan, allyl chloride is used as a chemical intermediate for epichlorohydrin (the major use) and other chemicals (e.g., pesticides and pharmaceuticals).

Occupational exposure to allyl chloride has been limited by regulation or recommended guidelines in at least 11 countries. The standards are listed in Table 1.

The US Environmental Protection Agency (EPA) (1983) requires that notification be given whenever discharges containing 454 kg or more of allyl chloride are made into waterways,

Table 1. National occupational exposure limits for allyl chloride[a]

Country	Year	mg/m³	ppm	Interpretation[b]	Status
Australia	1978	3	1	TWA	Guideline
Belgium	1978	3	1	TWA	Regulation
Finland	1981	3	1	TWA	Guideline
		9	3	STEL	
German Democratic Republic	1979	3	-	TWA	Regulation
		6	-	Maximum (30 min)	
Germany, Federal Republic of	1984	3	1	TWA[c]	Guideline
Italy	1978	3	1	TWA	Guideline
Netherlands	1978	3	1	TWA	Guideline
Romania	1975	3	-	TWA	Regulation
		6	-	Maximum	
Switzerland	1978	3	1	TWA	Regulation
USA[d]					
OSHA	1978	3	1	TWA	Regulation
		-	300	Maximum (30 min)[e]	
ACGIH	1984/85	3	1	TWA	Guideline
		6	2	STEL	
NIOSH	1976	3	1	TWA	Guideline
		9	3	Ceiling (15 min)	
Yugoslavia	1971	3	1	Ceiling	Regulation

[a]From International Labour Office (1980); National Institute for Occupational Safety and Health (1980); National Finnish Board of Occupational Safety and Health (1981); American Conference of Governmental Industrial Hygienists (1984); Deutsche Forschungsgemeinschaft (1984)

[b]TWA, time-weighted average; STEL, short-term exposure limit

[c]Carcinogenic risk notation added

[d]OSHA, Occupational Safety and Health Administration; ACGIH, American Conference of Governmental Industrial Hygienists; NIOSH, National Institute for Occupational Safety and Health

[e]Skin irritant notation added

but has proposed that this be revised to require notification when discharges containing 2270 kg are made.

As part of the Hazardous Materials Regulations of the US Department of Transportation (1982), shipments of allyl chloride are subject to a variety of labelling, packaging, quantity and shipping restrictions consistent with its designation as a hazardous material.

2.2 Occurrence

(a) Natural occurrence

Allyl chloride has not been reported to occur in nature.

(b) Occupational exposure

It has been estimated that approximately 5000 workers in the USA are potentially exposed to allyl chloride. Occupations involving potential exposure to allyl chloride have been reported to include producers of allyl chloride, epichlorohydrin (crude epichlorohydrin can contain 10-15% allyl chloride), glycerol, diallyldimethylammonium chloride, allyl alcohol (by a process no longer in current use) and medicinal products (National Institute for Occupational Safety and Health, 1976).

Occupational exposure has been reported in a plant producing allyl chloride in the German Democratic Republic. The data are summarized in Table 2.

Table 2. Occupational exposure to allyl chloride at various locations in an allyl chloride plant in the German Democratic Republic[a]

Plant area	Concentration (mg/m^3)
Laboratory	3
Filling	19
Production	53-59
Tank storage	43-310
Pump room	189-350

[a]From Häusler and Lenich (1968)

Results of personnel monitoring at two US allyl chloride plants are summarized in Tables 3 and 4.

Table 3. Levels of allyl chloride at a US allyl chloride manufacturing site[a]

Job classification	No. of samples	Concentration[b] (mg/m^3)		
		High	Low	Average
Control room, operator A	6	2.82	0.59	1.40
Control room, operator C	8	2.91	0.74	1.77
Instrument	4	14.63[c]	0.37	6.70
Laboratory	4	2.20	0.71	1.24
Shift foreman	4	12.49[c]	0.37	4.03
Maintenance	4	18.88[c]	2.42	9.46
Class 2 operator	5	19.00[c]	0.016	5.36
Head packaging operator	2	0.28	0.062	0.16
Chief material-handling technician	2	0.96	0.40	0.58

[a]From National Institute for Occupational Safety and Health (1976)
[b]It is possible that high values are caused by acetone interference.
[c]Potential exposure; protective equipment was worn during sampling operations and process upsets.

Table 4. Results of allyl chloride monitoring at a US manufacturing plant[a]

Job classification	Eight-hour TWA[b]			Peak (up to 15 min)			Comments
	No. of TWAs	Range (mg/m^3)	Mean[c] (mg/m^3)	No. of samples	Range (mg/m^3)	Mean[c] (mg/m^3)	
Loading	8	1.2-9.9	5.9	5	19.2-122.5	60.5	Loading operators wear breathing masks. Drum-loading measurements were taken prior to installation of ventilation system. Tank car and tank truck loading rate is 2-3 h/day for each.
Operators, except for G-300	70	<0.3-11.2	1.46	15	0.3-95.2	35.0	Evaluations are for routine operations and do not include shut-down or start-up periods when full breathing apparatus is worn.
G-300 operators	5	0.3-16.4	-[d]	-	-	-	"
Shift foreman	16	0.3-10.5	1.89	-	-	-	"

[a]From National Institute for Occupational Safety and Health (1976)
[b]TWA, time-weighted average
[c]This represents the arithmetic mean, which is an overestimate of the central tendency of distribution; the data appear to follow a log-normal distribution with a lower geometric mean.
[d]Four of five samples contained <1 mg/m^3.

Occupational exposure to allyl chloride associated with the production of epichlorohydrin has been reported. Data reported on US units in 1976 are summarized in Table 5 (which mainly reflects potential exposures, since protective equipment was generally used) and Table 6. Data on US manufacturing facilities from reports made in 1977 and 1978 are shown in Table 7.

Table 5. Levels of allyl chloride at a US epichlorohydrin unit[a]

Job classification	No. of samples	Concentration[b] (mg/m^3)		
		High	Low	Average
Control room, operator A	5	3.44[c]	0.12	1.52
Instrument	2	3.84[c]	0.96	2.42
Laboratory	6	10.60[c]	0.12	4.96
Shift foreman	3	14.48[c]	1.21	5.86
Epichlorohydrin helper	4	8.40[c]	0.16	2.73
Control finisher	2	4.40[c]	0.84	2.64
Maintenance	13	1.58	0.16	0.62

[a]From National Institute for Occupational Safety and Health (1976)
[b]It is possible that high values are caused by acetone interference.
[c]Potential exposure; protective equipment was worn during sampling operations and process upsets.

Table 6. Occupational exposure to allyl chloride at a US epichlorohydrin-manufacturing site[a]

Job classification	Eight-hour TWA[b]			Peak (up to 15 min)			Comments
	No. of TWAs	Range	Mean[c] (mg/m^3)	No. of samples	Range (mg/m^3)	Mean[c] (mg/m^3)	
Marine cargo inspection	-	-	-	5	<0.3-8.4	2.8	Gauging and inspecting crude epichlorohydrin barges. Exposure is limited to about 13-20 min/barge and 2-3 barges/month. Allyl chloride is a contaminant (10-15%) in crude epichlorohydrin.
Dockman	1	<0.3	-	7	<0.3-18.6	4.7	Connecting and disconnecting barge loading lines on crude epichlorohydrin barges

[a]From National Institute for Occupational Safety and Health (1976)
[b]TWA, time-weighted average
[c]This represents the arithmetic mean, which is an overestimate of the central tendency of distribution; the data appear to follow a log-normal distribution with a lower geometric mean.

Table 7. Occupational exposure to allyl chloride at various locations in epichlorohydrin manufacturing facilities[a]

Job classification	Plant	No. of samples	Time-weighted average (mg/m^3)[b] Range	Median
Chemical operators	A	8	<0.16-2.11	0.56
	B	12	<0.16-27.6	<0.16
	C	6	<0.3-44.2	1.43
	C	5	<0.3-1.59	0.64
	C	5	<0.3-0.95	0.32
	C	5	<0.03-2.06	0.32
Foremen				
Epichlorohydrin production	A	3	<0.16-0.93	0.53
	B	3	<0.16-0.62	0.53
	B	1	0.53	0.53
	C	6	<0.3-0.95	[0.21]
	C	1	--	0.31
Epoxide plant production	C	1	--	<0.3
Tank truck loading	B	1	<0.16	<0.16
	C	1	--	<0.3
Pipe fitters	A	3	<0.16	<0.16
Control room area	B	3	<0.16	<0.16
	C	5	<0.3-0.95	[<0.3]
Glycerol production area	B	1	0.3	<0.16
	C	2	0.32-0.54	[0.44]

[a]From Bales (1978)
[b]Time-weighted average concentration to which workers may be exposed for a normal eight-hour working day of a 40-hour week; figures in square brackets were calculated by the Working Group.

The majority of exposures were within the time-weighted average (TWA) for allyl chloride recommended by the National Institute for Occupational Safety and Health. Although at plants B and C levels for two chemical operators reached 27.6 and 44.2 mg/m^3 (TWA), respectively, these samples were considered abnormal since they were taken during the repairing of processing machinery; the operators were wearing cartridge respirators so that actual exposure was considered to be significantly less.

(c) Air

Diurnal urban air samples, collected in the USA from Denver, CO, Houston, TX, Riverside, CA, and St Louis, MO, were found to contain <16 ng/m^3 of allyl chloride, whereas samples from Pittsburgh, PA, were found to contain a mean of 64 ng/m^3. None was detected in Chicago, IL, or Staten Island, NY (Singh et al., 1982).

2.3 Analysis

Typical methods for the analysis of allyl chloride in various matrices are summarized in Table 8.

Table 8. Methods for the analysis of allyl chloride

Sample matrix	Sample preparation	Assay procedure[a]	Limit of detection	Reference
Air	Trap onto Tenax-GC; desorb thermally	GC/FID	not given	Brown & Purnell (1979)
	Trap onto charcoal; desorb with benzene	GC/FID	1.8-7.19 mg/m^3 (range of validation)	National Institute of Occupational Safety and Health (1977)
	Concentrate on a glass wool trap at liquid oxygen temperature; desorb thermally	GC/EC	<3 ng/m^3 ($\pm$ 15%)	Singh et al. (1982)
	--	GC	0.2 mg/m^3	Yu et al. (1981)
	Trap onto Tenax-GC; desorb thermally	GC/MS	83 ng/m^3	Krost et al. (1982)
Waste water	Sparge with an inert gas; trap volatiles onto Tenax-GC; desorb thermally	GC/MS	2 µg/l	Spingarn et al. (1982)

[a]Abbreviations: GC/FID, gas chromatography/flame ionization detection; GC/EC, gas chromatography/electron capture detection; GC, gas chromatography; GC/MS, gas chromatography/mass spectroscopy

3. Biological Data Relevant to the Evaluation of Carcinogenic Risk to Humans

3.1 Carcinogenicity studies in animals

(a) Oral administration

Mouse: Groups of 50 male and 50 female B6C3F$_1$ mice, five weeks old, were given doses of 172 or 199 and 129 or 258 mg/kg bw (low- or high-dose males and females, respectively) allyl chloride (technical grade; purity, 98%) per day in corn oil by gavage on five days per week for 78 weeks. Groups of 20 animals of each sex received corn oil alone and served as vehicle controls, and a further 20 animals of each sex served as untreated controls. Animals were maintained without further exposure through week 92. There was excessive mortality in male mice; 48% of mice in the high-dose group had died by week 27, and the 10 mice surviving longer than 48 weeks were killed at week 56. Survival rates in other groups at the end of the study were 8/20, 14/20 and 23/50 in untreated controls, vehicle controls and the low-dose group, respectively. Of the females, 70-90% of mice were still alive at the end of the study. Treatment-related lesions were observed in the forestomachs of animals of both sexes. A metastasizing squamous-cell carcinoma was found in 2/46 low-dose male mice, but not in high-dose (0/50), vehicle-control (0/20) or untreated (0/18) males. Acanthosis and hyperkeratosis of the forestomach were found in 9/46 low-dose and 19/50 high-dose, but not in control males. In females, a squamous-cell carcinoma was found in 2/48 low-dose, but not in high-dose (0/45), vehicle-control (0/19) or untreated (0/20) groups. Squamous-cell papillomas were observed in three high-dose and in one low-dose female. Acanthosis and hyperkeratosis of the forestomach occurred in 17 low-dose and 25 high-dose females, but not in controls. The incidence of forestomach tumours in male and female mice was not statistically different from that in controls (National Cancer Institute, 1977; Weisburger, 1977). [The Working Group noted the high mortality in treated males.]

Rat: Groups of 50 male and 50 female Osborne-Mendel rats, six weeks of age, were given initial doses of 70 and 140, and 55 and 110 mg/kg bw (low-dose and high-dose males and females, respectively) allyl chloride (technical grade; purity, 98%) per day in corn oil by gavage on five days per week. Due to toxicity the doses were reduced on two occasions; average time-weighted doses over a 78-week treatment period were 57 and 77 mg/kg bw for low- and high-dose males and 55 and 73 mg/kg bw for low- and high-dose females, respectively. All surviving animals were maintained without further treatment up to a maximum of 110 weeks. A group of 20 rats of each sex was treated with corn oil alone and served as vehicle controls, and a further group of 20 rats of each sex served as untreated controls. In the high-dose group, 50% of males had died by week 14 and 50% of females by week 38; the number of animals at risk for developing tumours was insufficient for analysis of these results. In the low-dose group, 50% of males were still alive at 77 weeks and 50% of females at 99 weeks; no increased incidence of tumours related to treatment was observed (National Cancer Institute, 1977; Weisburger, 1977).

(b) Skin application

Mouse: Groups of 30 female Ha:ICR Swiss mice, six to eight weeks of age, received skin applications of 31 or 94 mg allyl chloride (technical grade) [purity unspecified] in 0.2 ml acetone three times per week for 440-594 days. No skin tumour was observed in treated animals, and the incidence of other tumours did not differ significantly from that in controls. No skin tumour occurred in 30 controls treated with 0.1 ml acetone alone (Van Duuren *et al.*, 1979).

In a two-stage mouse-skin assay, groups of 30 female Ha:ICR Swiss mice, aged six to eight weeks, received a single skin application of 94 mg allyl chloride (technical grade) [purity unspecified] in 0.2 ml acetone, followed 14 days later by thrice-weekly applications of 5 µg phorbol myristyl acetate [12-*O*-tetradecanoylphorbol 13-acetate, TPA] in 0.2 ml acetone for life (median survival, 428-576 days). Seven papillomas were found in 6/90 control animals treated three times weekly with 5 µg TPA alone. A total of 10 papillomas was found in 7/30 treated mice ($p < 0.025$). The first tumour appeared at day 197 in the allyl chloride-treated group and at day 449 in the TPA-treated controls (Van Duuren *et al.*, 1979).

(c) Intraperitoneal administration

Mouse: Groups of 10 male and 10 female A/St mice, six to eight weeks of age, received intraperitoneal injections in tricaprylin of allyl chloride (technical grade, without further purification) three times weekly for eight weeks at total-dose levels of 15.6, 38.4 and 76.8 mmol/kg bw (1.2, 2.9 and 5.9 g/kg bw). Controls received tricaprylin alone. All animals were killed 24 weeks after the first injection, when the numbers of survivors were 16/20, 20/20, 20/20 and 20/20 in the control, low-, medium- and high-dose groups, respectively. The numbers of lung adenomas seen grossly per mouse in animals of both sexes combined were 0.19 ± 0.1, 0.60 ± 0.20, 0.50 ± 0.27 and 0.60 ± 0.15 in the control, low-, medium- and high-dose groups, respectively. The incidence of lung tumours in the high-dose group differed statistically from that in controls by one of two statistical tests ($p < 0.05$) (Theiss *et al.*, 1979).

3.2 Other relevant biological data

(a) Experimental systems

Toxic effects

The oral LD_{50}s of allyl chloride (purity, >99 %) have been reported to be 425 mg/kg bw in mice and 460 mg/kg bw in rats. In mice, the oral LD_{50} of a commercial-grade sample (purity, 90%) was 550 mg/kg bw. Inhalation studies in a static exposure system gave the following LC_{50} values for two-hour exposures: female mice, 11 500 mg/m^3; rats, approximately 11 400 mg/m^3; and guinea-pigs, 5800 mg/m^3 (Lu et al., 1982). Exposure by inhalation to 2000 ppm (6200 mg/m^3) allyl chloride vapour for four hours was lethal for 1/6 rats (Smyth & Carpenter, 1948).

Allyl chloride has strong irritating properties. Inhalation of vapours produces inflammatory and necrotizing effects in the respiratory ducts and lung damage is the usual cause of death in rats and guinea-pigs. The major systemic effects are degenerative changes of kidney and to a lesser extent of the liver (Adams et al., 1940). Long-term inhalation studies, in which rabbits and cats were exposed to 206 mg/m^3 allyl chloride vapour for six hours per day on six days per week for three months, resulted in the development in rabbits of flaccid paralysis with muscular atrophy, which were in part reversible after cessation of exposure; cats were affected to a lesser extent. Exposure to 17.5 mg/m^3 under comparable conditions was tolerated without adverse effects (Lu et al., 1982).

All of six rabbits injected subcutaneously with 50 mg/kg bw allyl chloride three times for one week followed by 100 mg/kg bw three times a week for up to 11 weeks developed peripheral neuropathy (He et al., 1980).

Inhalation exposure of rats, guinea-pigs and rabbits to 8 ppm (24.8 mg/m^3) allyl chloride vapour in a dynamic-flow system for seven hours per day, on five days per week for five weeks caused histological damage to the livers and kidneys, whereas exposure to 3 ppm (9.3 mg/m^3) for six months under the same conditions caused no observable toxic effect in rats, guinea-pigs, rabbits or dogs (Torkelson et al., 1959).

In a bioassay gavage study (described in section 3.1), rats received daily time-weighted average doses of 57 or 77 (males) and 55 or 73 (females) mg/kg bw allyl chloride in corn oil for 78 weeks. A slight depression of body-weight gain was observed in the high-dose groups, but a dose-dependent reduction in mean survival times was seen in both the high- and low-dose groups. Hunched appearance and some respiratory distress were also reported. Mice received daily time-weighted average doses of 172 or 199 (males) and 129 or 258 (females) mg/kg bw. No depression of body weight was observed, but a significant decrease in mean survival times, accompanied by loss of equilibrium and abdominal distention, was reported in high-dose males (National Cancer Institute, 1977).

Effects on reproduction and prenatal toxicity

Groups of 25-39 Sprague-Dawley rats and 20-25 New Zealand white rabbits were exposed via inhalation to 0, 30 or 300 ppm (0, 93 or 930 mg/m^3) allyl chloride (purity, 98.6%) vapour for seven hours per day on gestation days 6-15 (rats) or 6-18 (rabbits). Exposure to 300 ppm resulted in decreased maternal weight gain during the first two or three days of exposure. Only minor alterations of the foetal skeleton were seen, including delayed ossification of the sternebrae and vertebral centra among rats at the highest exposure level (John et al., 1983).

A group of 10-15 Sprague-Dawley rats received intraperitoneal injections of 80 mg/kg bw allyl chloride [purity unspecified] in corn oil on days 1-15 of gestation. Foetuses were examined on day 21 of gestation. Maternal heart, liver, spleen and kidney weights were significantly increased by treatment, but no histopathological change was evident. There were significant increases in the occurrence of oedematous foetuses and foetuses with short snout and protruding tongue in treated litters (Hardin et al., 1981).

[The Working Group noted that these studies differed in the route of administration.]

Absorption, distribution, excretion and metabolism

Male rats dosed subcutaneously with allyl chloride (1 ml of a 10% v/v solution in arachis oil) excreted mercapturic acids in the urine which were identified as 3-hydroxypropylmercapturic acid and allylmercapturic acid and its sulphoxide (Kaye *et al.*, 1972). This suggests that allyl chloride is metabolized through two different pathways (see also Fig. 1, General Remarks on the Substances Considered, p. 32). Allylglutathione and *S*-allyl-L-cysteine have been detected in the bile of a rat given allyl chloride (Kaye *et al.*, 1972).

Allyl chloride in the presence of NADPH and oxygen partially destroyed rat hepatic microsomal cytochrome P-450 with loss of haem (Patel *et al.*, 1981).

Allyl chloride exerts direct alkylating properties *in vitro* (Eder *et al.*, 1980, 1982a,b).

Mutagenicity and other short-term tests (see also 'Appendix: Activity Profiles for Short-Term Tests', p. 329).

Allyl chloride is positive in the bacterial polA$^+$/polA$^-$ DNA-repair assay with *Escherichia coli* (McCoy *et al.*, 1978).

The mutagenicity of allyl chloride for *Salmonella typhimurium* TA1535 and TA100 but not TA1538 can be demonstrated if precautions are taken to prevent its escape due to volatility (McCoy *et al.*, 1978; Bignami *et al.*, 1980; Eder *et al.*, 1980; Norpoth *et al.*, 1980; Simmon, 1981; Eder *et al.*, 1982a,b); its activity is greatly decreased by the presence of an exogenous metabolic system (Eder *et al.*, 1980, 1982a,b).

Allyl chloride is mutagenic to *Streptomyces coelicolor* but not to *Aspergillus nidulans* (Bignami *et al.*, 1980); it induces gene conversions in *Saccharomyces cerevisiae* (McCoy *et al.*, 1978).

(b) Humans

Toxic effects

Workers exposed to concentrations of allyl chloride ranging from 1-113 ppm (3-350 mg/m^3) for 16 months were reported to have developed liver damage, as determined by serum enzyme activities, which was shown to be reversible after cessation or minimization of exposure (Häusler & Lenich, 1968). In another study of exposure to unknown concentrations of allyl chloride, workers were reported to have impaired kidney function (Ali-Zade, 1979). Both motor and sensory neurotoxic damage at the distal parts of the extremities (similar to that seen after exposure to *n*-hexane or tri-*ortho*-cresyl phosphate) was reported in 17 industrial workers, which was reversible after cessation of exposure and treatment; however, this effect recurred after return to work. The same symptoms were induced in rabbits exposed under similar laboratory conditions (He *et al.*, 1980).

Effects on reproduction and prenatal toxicity

Potential adverse effects of allyl chloride (as well as the structurally similar epichlorohydrin and 1,3-dichloropropene) on male fertility were investigated in employees of a glycerol factory in Texas, USA (Venable *et al.*, 1980). In general, no difference in a variety of sperm counts was found between 64 exposed and 63 control workers, although a subgroup of 10 workers exposed to all three chemicals had a significantly lowered sperm count in comparison to the remainder of the study group. The eight-hour time-weighted average for all of these chemicals was estimated to have been <1 ppm (<3.1 mg/m^3) during the five years immediately preceding the study. Two other subgroups, exposed to epichlorohydrin and allyl chloride or 1,3-dichloropropane and allyl chloride, showed no evidence of decreased sperm counts. [The Working Group noted the multiple exposures evaluated, the possible exposure of controls to other chemicals, as well as possible variations in individual sample collections and processing time, making this study inadequate for the evaluation of the effects of allyl chloride on semen quality.]

No data were available to the Working Group on absorption, distribution, excretion and metabolism or on mutagenicity and chromosomal effects.

3.3 Case reports and epidemiological studies of carcinogenicity to humans

No data were available to the Working Group.

4. Summary of Data Reported and Evaluation

4.1 Exposure data

Allyl chloride has been produced commercially since 1945 and is used almost exclusively as a chemical intermediate, principally in the production of epichlorohydrin.

4.2 Experimental data

Allyl chloride has been tested for carcinogenicity by intragastric intubation in mice and rats, by skin application in mice, both by repeated application and in a two-stage assay, and by intraperitoneal injection in mice. Following its oral administration to mice, a nonsignificant increase in the incidence of squamous-cell papillomas and carcinomas of the forestomach was observed; the experiment in rats was inadequate for evaluation. No skin tumour was observed in mice following repeated skin applications; however, a single application followed by treatment with 12-O-tetradecanoylphorbol 13-acetate gave some evidence that allyl chloride acts as an initiator. Following its intraperitoneal injection to strain A mice, a slight increase in the incidence of lung adenomas was observed.

Inhalation exposure to allyl chloride of high purity did not induce teratogenicity in rats or rabbits.

Allyl chloride caused DNA damage in bacteria, and was mutagenic to bacteria and fungi.

Overall assessment of data from short-term tests: allyl chloride[a]

	Genetic activity			Cell transformation
	DNA damage	Mutation	Chromosomal effects	
Prokaryotes	+	+		
Fungi/Green plants		+		
Insects				
Mammalian cells (*in vitro*)				
Mammals (*in vivo*)				
Humans (*in vivo*)				
Degree of evidence in short-term tests for genetic activity: *Limited*				Cell transformation: No data

[a]The groups into which the table is divided and the symbol + are defined on pp. 17-18 of the Preamble; the degrees of evidence are defined on p. 18.

4.3 Human data

No case report or epidemiological study of the carcinogenicity of allyl chloride to humans was available to the Working Group.

4.4 Evaluation[1]

There is *inadequate evidence* for the carcinogenicity of allyl chloride in experimental animals.

In the absence of epidemiological data, no evaluation could be made of the carcinogenicity of allyl chloride to humans.

5. References

Adams, E.M., Spencer, H.C. & Irish, D.D. (1940) The acute vapor toxicity of allyl chloride. *J. ind. Hyg. Toxicol.*, 22, 79-86

Ali-Zade, G.A. (1979) Renal function in workers exposed to some chemical substances (allyl chloride and metallic mercury) (Russ.). *Gig. Tr. prof. Zabol.*, 10, 22-25

American Conference of Governmental Industrial Hygienists (1984) *TLVs Threshold Limit Values for Chemical Substances in the Work Environment Adopted by ACGIH for 1984-85*, Cincinnati, OH, p. 9

---

[1]For definitions of the italicized terms, see the Preamble, pp. 15-16.

Bales, R.E. (1977) *Epichlorohydrin: Industrial Hygiene Survey at Shell Chemical Co. Deer Park Manufacturing Complex, Houston, Texas (IWS-58.10, PB82-100595)*, Springfield, VA, National Technical Information Service, pp. 10, 19, 26, 30

Bales, R.E. (1978) *Epichlorohydrin Manufacture and Use: Industrial Hygiene Survey*, Cincinnati, OH, National Institute for Occupational Safety and Health, pp. 19, 25, 26, 27

Beacham, H.H. (1978) Allyl compounds. In: Grayson, M., ed., *Kirk-Othmer Encyclopedia of Chemical Technology*, 3rd ed., Vol. 2, New York, John Wiley & Sons, pp. 97-108

Bignami, M., Conti, G., Conti, L., Crebelli, R., Misuraca, F., Puglia, A.M., Randazzo, R., Sciandrello, G. & Carere, A. (1980) Mutagenicity of halogenated aliphatic hydrocarbons in *Salmonella typhimurium*, *Streptomyces coelicolor* and *Aspergillus nidulans*. *Chem.-biol. Interactions*, 30, 9-23

Bothner-By, A.A., Castellano, S., Ebersole, S.J. & Günther, H. (1966) The proton magnetic resonance spectra of olefins. V. 3-Chloro- and 3-methoxypropenes. *J. Am. chem. Soc.*, 88, 2466-2468

Brown, R.H. & Purnell, C.J. (1979) Collection and analysis of trace organic vapour pollutants in ambient atmospheres. The performance of a Tenax-GC adsorbent tube. *J. Chromatogr.*, 178, 79-90

DeBenedictis, A. (1979) Chlorocarbons, α-hydrocarbons (allyl chloride). In: Grayson, M., ed., *Kirk-Othmer Encyclopedia of Chemical Technology*, 3rd ed., Vol. 5, New York, John Wiley & Sons, pp. 763-773

Deutsche Forschungsgemeinschaft (1984) *Maximal Work Place Concentrations and Biological Tolerance Values for Compounds in the Work Place* (Ger.), Part XX, Weinheim, Verlag Chemie GmbH, p. 24

Eder, E., Neudecker, T., Lutz, D. & Henschler, D. (1980) Mutagenic potential of allyl and allylic compounds. Structure-activity relationship as determined by alkylating and direct in vitro mutagenic properties. *Biochem. Pharmacol.*, 29, 993-998

Eder, E., Henschler, D. & Neudecker, T. (1982a) Mutagenic properties of allylic and α,β-unsaturated compounds: Consideration of alkylating mechanisms. *Xenobiotica*, 12, 831-848

Eder, E., Neudecker, T., Lutz, D. & Henschler, D. (1982b) Correlation of alkylating and mutagenic activities of allyl and allylic compounds: Standard alkylation test vs. kinetic investigation. *Chem.-biol. Interactions*, 38, 303-315

Hardin, B.D., Bond, G.P., Sikov, M.R., Andrew, F.D., Beliles, R.P. & Niemeier, R.W. (1981) Testing of selected workplace chemicals for teratogenic potential. *Scand. J. Work Environ. Health*, 7, 66-75

Häusler, M. & Lenich, R. (1968) Effect of chronic occupational allyl chloride exposure (Ger.). *Arch. Toxikol.*, 23, 209-214

He, F., Shen, D., Guo, Y. & Lu, B. (1980) Toxic polyneuropathy due to chronic allyl chloride intoxication. A chemical and experimental study. *China med. J.*, 93, 177-182

IARC (1976) *IARC Monographs on the Evaluation of Carcinogenic Risk of Chemicals to Man*, Vol. 11, *Cadmium, Nickel, Some Epoxides, Miscellaneous Industrial Chemicals and General Considerations on Volatile Anaesthetics*, Lyon, pp. 131-139

IARC (1979a) *IARC Monographs on the Evaluation of the Carcinogenic Risk of Chemicals to Humans*, Vol. 19, *Some Monomers, Plastics and Synthetic Elastomers, and Acrolein*, Lyon, pp. 213-230

IARC (1979b) *IARC Monographs on the Evaluation of the Carcinogenic Risk of Chemicals to Humans*, Vol. 20, *Some Halogenated Hydrocarbons*, Lyon, pp. 83-96

International Labour Office (1980) *Occupational Exposure Limits for Airborne Toxic Substances*, 2nd (rev.) ed. (*Occupational Safety and Health Series No. 37*), Geneva, pp. 38-39

Irish, D.D. (1963) Halogenated hydrocarbons: I. Aliphatic. In: Patty, F.A., ed., *Industrial Hygiene and Toxicology*, 2nd rev. ed., New York, Interscience, p. 1317

John, J.A., Gushow, T.S., Ayres, J.A., Hanley, T.R., Jr, Quast, J.F. & Rao, K.S. (1983) Teratologic evaluation of inhaled epichlorohydrin and allyl chloride in rats and rabbits. *Fundam. appl. Toxicol.*, 3, 437-442

Kaye, C.M., Clapp, J.J. & Young, L. (1972) The metabolic formation of mercapturic acids from allyl halides. *Xenobiotica*, 2, 129-139

Krost, K.J., Pellizzari, E.D., Walburn, S.G. & Hubbard, S.A. (1982) Collection and analysis of hazardous organic emissions. *Anal. Chem.*, 54, 810-817

Lu, B., Dong, S., Yu, A., Xian, Y., Geng, T. & Chui, T. (1982) Studies on the toxicity of allyl chloride. *Ecotoxicol. environ. Saf.*, 6, 19-27

McCoy, E.C., Burrows, L. & Rosenkranz, H.S. (1978) Genetic activity of allyl chloride. *Mutat. Res.*, 57, 11-15

National Cancer Institute (1977) *Bioassay of Allyl Chloride for Possible Carcinogenicity (DHEW Publ. No. (NIH) 78-1323)*, Washington DC, US Government Printing Office

National Finnish Board of Occupational Safety and Health (1981) *Airborne Contaminants in the Work Places (Safety Bulletin No. 3)*, Helsinki, p. 7

National Institute for Occupational Safety and Health (1976) *Criteria for a Recommended Standard--Occupational Exposure to Allyl Chloride (DHEW Publication No. (NIOSH)76-204, PB 267071)*, Springfield, VA, National Technical Information Service

National Institute for Occupational Safety and Health (1977) *NIOSH Manual of Analytical Methods*, Part II, 2nd ed., Vol. 2 (*DHEW Publication No. (NIOSH) 77-157-B*), Washington DC, US Government Printing Office, pp. S116-1 to S116-8

National Institute for Occupational Safety and Health (1980) *Summary of NIOSH Recommendations for Occupational Health Standards*, Rockville, MD

Norpoth, K., Reisch, A. & Heinecke, A. (1980) Biostatistics of Ames-test data. In: Norpoth, K.H. & Garner, R.C., eds, *Short-Term Test Systems for Detecting Carcinogens*, New York, Springer-Verlag, pp. 312-322

Patel, J.M., Ortiz, E. & Leibmann, K.C. (1981) Destruction of hepatic cytochrome P-450 by allylic industrial toxicants (Abstract No. 2325) *Fed. Proc.*, 40, 636

Shell Chemical Corp. (1949) *Allyl Chloride and Other Allyl Halides (Technical Publication SC: 49-8)*, San Francisco, CA

Simmon, V.F. (1981) Applications of the Salmonella/microsome assay. In: Stich, H.F. & San, R.H.C., eds, *Short-Term Tests for Chemical Carcinogens*, New York, Springer-Verlag, pp. 120-126

Singh, H.B., Salas, L.J. & Stiles, R.E. (1982) Distribution of selected gaseous organic mutagens and suspect carcinogens in ambient air. *Environ. Sci. Technol.*, 16, 872-880

Smyth, H.F., Jr & Carpenter, C.P. (1948) Further experience with the range finding test in the industrial toxicology laboratory. *J. ind. Hyg. Toxicol.*, *30*, 63-68

Spingarn, N.E., Northington, D.J. & Pressely, T. (1982) Analysis of volatile hazardous substances by GC/MS. *J. Chromatogr.*, *20*, 286-288

Swinyard, E.A. (1975) Sedatives and hypnotics. In: Osol, A., ed., *Remington's Pharmaceutical Sciences*, 15th ed., Easton, PA, Mack Publishing Co., pp. 997-1003

Theiss, J.C., Shimkin, M.B. & Poirier, L.A. (1979) Induction of pulmonary adenomas in strain A mice by substituted organohalides. *Cancer Res.*, *39*, 391-395

Torkelson, T.R., Wolf, M.A., Oyen, F. & Rowe, V.K. (1959) Vapor toxicity of allyl chloride as determined on laboratory animals. *Am. ind. Hyg. Assoc. J.*, *20*, 217-223

US Department of Transportation (1982) Performance-oriented packagings standards. *US Code Fed. Regul., Title 49*, Parts 171, 172, 173, 178; *Fed. Regist.*, *47* (No. 73), pp. 16268, 16273

US Environmental Protection Agency (1983) Notification requirements; Reportable quantity adjustments. *US Code Fed. Regul., Title 40*, Part 302; *Fed. Regist.*, *48* (No. 102), pp. 23552, 23572

US International Trade Commission (1975) *Synthetic Organic Chemicals, US Production and Sales, 1973* (*USITC Publication 728*), Washington DC, US Government Printing Office, p. 202

Van Duuren, B.L., Goldschmidt, B.M., Loewengart, G., Smith, A.C., Melchionne, S., Seldman, I. & Roth, D. (1979) Carcinogenicity of halogenated olefinic and aliphatic hydrocarbons in mice. *J. natl Cancer Inst.*, *63*, 1433-1439

Venable, J.R., McClimans, C.D., Flake, R.E. & Dimick, D.B. (1980) A fertility study of male employees engaged in the manufacture of glycerine. *J. occup. Med.*, *22*, 87-91

Verschueren, K. (1977) *Handbook of Environmental Data on Organic Chemicals*, New York, Van Nostrand Reinhold Co., p. 86

Vesper, H.G. (1949) Chlorine compounds, organic. Allyl chloride. In: Kirk, R.E & Othmer, D.F., eds, *Encyclopedia of Chemical Technology*, Vol. 3, New York, The Interscience Encyclopedia, Inc., pp. 800-806

Weisburger, E.K. (1977) Carcinogenicity studies on halogenated hydrocarbons. *Environ. Health Perspect.*, *21*, 7-16

Windholz, M. (1983) *The Merck Index*, 10th ed., Rahway, NJ, Merck & Co., pp. 44-45

Yu, A., Xian, Y., Zhao, G. & Fu, Y. (1981) Study on gas chromatographic determination of 3-chloropropene and epichlorohydrin in air (Chin.) *Zhongguo Yixue Kexueyuan Xuebao*, *3*, 209-212 [*Chem. Abstr.*, *96*, 90786v]

ALLYL ISOTHIOCYANATE

1. Chemical and Physical Data

1.1 Synonyms and trade names

Chem. Abstr. Services Reg. No.: 57-06-7

Chem. Abstr. Name: 1-Propene, 3-isothiocyanato-

IUPAC Systematic Name: Allyl isothiocyanate

Synonyms: AITC; AITK; allyl isorhodanide; allyl isosulphocyanate; allylisothiocyanate; allyl mustard oil; allylsenevol; allylsenföl; allyl sevenolum; allyl thiocarbonimide; artificial mustard oil; artificial oil of mustard; FEMA No. 2034; 3-isothiocyanatopropene; 3-isothiocyanato-1-propene; isothiocyanic acid allyl ester; NCI-C50464; Oil of Mustard BPC 1949, Synthetic; oleum sinapis; oleum sinapis volatile; 2-propenyl isothiocyanate; propylene-3-isothiocyanate; synthetic mustard oil; synthetic mustard oil volatile; volatile mustard oil; volatile oil of mustard

Trade Names: Carbospol; Redskin

1.2 Structural and molecular formulae and molecular weight

$$CH_2=CH-CH_2N=C=S$$

C_4H_5NS Mol. wt: 99.2

1.3 Chemical and physical properties of the pure substance

From Windholz (1983), unless otherwise specified

(a) *Description*: Colourless or pale-yellow liquid with a pungent, irritating odour and an acrid taste

(b) *Boiling-point*: 148-154°C

(c) *Melting-point*: -102.5°C (Furia & Bellanca, 1975)

(d) *Density*: 1.013-1.020

(e) *Refractive index*: n_D^{20} 1.5268-1.5280

(f) *Spectroscopy data*: Infrared spectral data have been reported (National Research Council, 1981)

(g) *Solubility*: Slightly soluble in water; limited solubility in 70% aqueous ethanol (0.1 l in 1 l); miscible with ethanol and most organic solvents

(h) *Volatility*: Vapour pressure, 40 mm Hg at 67.4°C (Verschueren, 1977)

(i) *Stability*: Flash-point, 46.1°C (Hawley, 1981); tends to darken on ageing (Furia & Bellanca, 1975); converts to methyl allylthiocarbamate in methanol and to unidentified products in water (Ina *et al.*, 1981)

(j) *Conversion factor*: 1 ppm = 4.1 mg/m^3 at 760 mm Hg and 25°C [calculated by the Working Group]

1.4 Technical products and impurities

In 1972, allyl isothiocyanate was available in the USA as a chemical grade with a purity of 93-94%. A food grade was available with the following specifications: purity, 93% min; arsenic, 0.0003% max; lead, 0.001% max. It was also required to pass a test for phenol content (Food and Drug Research Laboratories, 1972).

In the USA, to meet the requirements of the Food Chemicals Codex, allyl isothiocyanate must pass an infrared identification test and meet the following specifications: purity, 93% min; refractive index (20°C), 1.527-1.531; specific gravity (25°C), 1.013-1.020; distillation range, 148-154°C; and pass a test for phenols (National Research Council, 1981).

In western Europe, allyl isothiocyanate is available as a clear, almost colourless or pale-yellow liquid with the following specifications: assay, 93% min; density (20°C), 1.015-1.020; and refractive index, n_D^{20} 1.527-1.530. It has a typical flash-point (closed-cup) of 46°C (Bush Boake Allen Ltd, undated).

2. Production, Use, Occurrence and Analysis

2.1 Production and use

(a) *Production*

Allyl isothiocyanate can be prepared by synthetic methods or by isolation from natural sources. Apparently it was first synthesized by Dulière in 1920 by the reaction of allyl iodide and potassium thiocyanate (Windholz, 1983). It is probably produced commercially by the reaction of allyl chloride with sodium or potassium thiocyanate at elevated temperatures.

Natural allyl isothiocyanate (so-called volatile mustard oil) is made by macerating the dried ripe seeds (free from fixed oil) of the mustard plants *Brassica nigra* or *B. juncea* with water, followed by distillation. The allyl isothiocyanate is freed from the glucoside sinigrin, in which it occurs naturally, by the action of the water and an enzyme, myrosinase (Food and Drug Research Laboratories, 1972). The content of allyl isothiocyanate in the resulting volatile mustard oil is more than 90% (US Food and Drug Administration, 1975).

Synthetic allyl isothiocyanate was first produced commercially in the USA in 1937 (US Tariff Commission, 1938).

Approximately 15 thousand kg of allyl isothiocyanate were used by the US food industry in 1970 (National Toxicology Program, 1982), but separate total US production data have never been published. Two US companies reported total combined production in the range of 9.1-90.8 thousand kg in 1977, and another producing company claimed confidentiality for its production volume (NIH/EPA Chemical Information System, 1984). Only one US company currently manufactures allyl isothiocyanate. Separate data on US imports and exports of this compound are not published.

Allyl isothiocyanate is produced commercially by one company each in Germany and the UK.

One Japanese company currently manufactures allyl isothiocyanate by the reaction of allyl chloride with sodium thiocyanate, and production in 1982 is estimated to have been two thousand kg.

(b) Use

Allyl isothiocyanate is used principally as a flavouring agent. It is also used as a denaturant for ethanol and as a rubefacient (counterirritant) in medicine. Several other uses have been proposed, but their present commercial significance is unknown.

Allyl isothiocyanate is used extensively as a flavouring agent at low concentrations (normally 85 mg/kg max) in pickled products, condiments, meats and spice flavours. A survey in 1977 on the use of food additives in US industry (National Research Council/National Academy of Sciences, 1979) reported the information shown in Table 1.

Allyl isothiocyanate is one of numerous denaturants approved for use (at a level of 10 lb/100 gallons [12 g/l] of alcohol) in the USA in specially denatured alcohol Formula No. 38-B. Although the volume of this formula used in the USA each year is published, no information was available on the amount made with allyl isothiocyanate.

Allyl isothiocyanate has been used in external analgesic products as a counterirritant. From 1962 to 1972, nearly 15 million package units of such products were marketed in the USA (US Food and Drug Administration, 1979).

Table 1. Results of a US survey on the use of food additives[a]

Food additive	Year of first use	Usage (1976) (thousand kg)	No. of users	No. of food categories	Use level (%) Usual	Use level (%) Highest
Allyl isothiocyanate	1950	2.2	13	9[b]	10^{-9}-0.13	10^{-9}-0.5
Allyl isothiocyanate (distillate)	-	4.5	<4	2[c]	0.005-0.05	0.005-0.05
Allyl isothiocyanate (essential oil)	1966	0.15	<4	6[d]	10^{-6}-0.01	5×10^{-6}-0.1

[a]From National Research Council/National Academy of Sciences (1979)
[b]The users provided more than 30 responses, of which 18 indicated use in condiments/relish
[c]Dressings and condiments/relish
[d]The users provided more than 14 responses, of which four indicated use in fish and seafood

Allyl isothiocyanate has also been reported to have been used as a gas in warfare (Windholz, 1983); as a fumigant (Hawley, 1981); as a fungicide and insecticidal fumigant, as a repellent for cats and dogs, and as an ingredient in some model airplane cements to deter glue sniffers (Gosselin et al., 1982). It has also been used as a preservative in animal feed (Říhová, 1982).

The principal use for allyl isothiocyanate in western Europe is as a flavouring agent. In Japan, it is used as a flavouring agent and as a pharmaceutical agent.

Allyl isothiocyanate has been approved for use as a synthetic flavouring substance and adjuvant in foods, provided it is used in the minimum quantity required to produce its intended effect (US Food and Drug Administration, 1980).

Allyl isothiocyanate is not permitted for use as a food additive by the Council of Europe.

The Bureau of Alcohol, Tobacco and Firearms of the US Department of the Treasury (1983) lists allyl isothiocyanate among the approved denaturants for specially denatured alcohol Formula No. 38-B.

In 1979, an advisory review panel on over-the-counter drugs of the US Food and Drug Administration (FDA) concluded that allyl isothiocyanate is safe and effective for use as an external analgesic when used within certain dosage restrictions (US Food and Drug Administration, 1979). In 1983, the FDA published a tentative final monograph in which it proposed a rule indicating that external analgesic drug products for over-the-counter use were generally recognized as safe and effective and not misbranded if they contained 0.5-5% allyl isothiocyanate as the counterirritant active ingredient (US Food and Drug Administration, 1983).

2.2 Occurrence

Natural occurrence

Allyl isothiocyanate has been reported to occur in the plant kingdom in the glucoside sinigrin. During processing to yield essential oils or in cooking, sinigrin is hydrolysed by myrosinase to allyl isothiocyanate (Food and Drug Research Laboratories, 1972; Hassan et al., 1980). The following plants of the *Cruciferae* family have been found to contain allyl isothiocyanate: *Brassica nigra* L. (black mustard), *Sinapis alba* (white mustard) (Food and Drug Research Laboratories, 1972); *Alliaria officinalis* (wild garlic), *Alyssum spp.* (madwort), *Armoracia lapathifolia* (horseradish), *Barbarea vulgaris* (winter cress), *Cakile maritima* (sea rocket), *Capsella bursa-pastoris* (shepherd's purse), *Crambe maritima* (sea kale), *Diplotaxis muralis* (white wall rocket), *Erucastrum gallium*, *Erysimum cheiranthoides* (blister cress), *Hesperis matronalis* (rocket), *Iberis sempervirens* (candytuft), *Sinapis arvensis* (wild mustard), *Sisymbrium spp.* (hedge mustard), *Raphanus sativus* (radish) (Mitchell & Jordan, 1974); *Thlaspi arvense* (penny cress) (Mitchell & Jordan, 1974; Furia & Bellanca, 1975); *Brassica oleracea botrytis* (cauliflower), *Brassica rapa* (turnip) (Cole, 1980); *Brassica oleracea capitata* (cabbage) (Tookey et al., 1980); *Brassica juncea* (black mustard), *Cocloleria aroracia* (horseradish), *Wasabia japonica* (wasabi) (Ina, 1982); *Brassica oleracea italica* (broccoli), *Brassica oleracea* (kale) (National Toxicology Program, 1982); *Brassica oleracea alboglabra* (kairan) at 96 mg/g (Uda et al., 1982a); *Brassica napus* seeds (Uda et al., 1982b). Naturally-derived mustard oil was found to contain 97.8% allyl isothiocyanate (Hassan et al., 1980).

2.3 Analysis

Typical methods for the analysis of allyl isothiocyanate in various matrices are summarized in Table 2.

Table 2. Methods for the analysis of allyl isothiocyanate

Sample matrix	Sample preparation	Assay procedure[a]	Limit of detection	Reference
Mustard and rapeseed oils	Add to a solution of allylthiourea and potassium ferricyanide in dilute acetic acid	S	2.5 μg/ml	Mukhopadhyay & Bhattacharyya (1983)
Cauliflower (*Brassica oleracea botrytis*) and turnip (*Brassica rapa*)	Two alternative procedures: (1) adsorb volatiles on Porapak Q and Tenax-GC; desorb either thermally or by extraction with diethyl ether and concentrate under nitrogen; or (2) hydrolyse chopped plant material in water and extract with dichloromethane	GC/FID	Not given	Cole (1980)
Radish (*Raphanus sativus*)	Mix juice with ethanol and ammonium hydroxide; neutralize with acetic acid; incubate with a modified Grote reagent	S (600 nm)	20 μg/ml	Esaki & Onozaki (1980)
Mustard seeds and mustard preparations	Extract with 70% methanol	HPLC/UV	Not given (abstract)	Henning (1981)
Cruciferous (Chinese and Japanese) vegetables	Blanch with boiling water and extract with methanol; extract with diethyl ether and filter through an Amberlite-IR 4B column; treat eluate with aqueous potassium hydroxide and with myrosinase and ascorbic acid; steam distil and extract distillate with diethyl ether and concentrate	GC/MS	Not given (abstract)	Uda *et al.* (1982a)
Mayonnaise (oil component)	Freeze-break the emulsion; evaporate oil onto a Tenax-GC column; desorb thermally	GC/FID	Not given	Min (1981); Min & Tickner (1982)
Mustard oil	Dissolve in carbon tetrachloride containing N,N dimethylaniline as the internal standard	PMR	Not given	Hassan *et al.* (1980)

[a]Abbreviations: S, spectrophotometry; GC/FID, gas chromatography/flame ionization detection; HPLC/UV, high-performance liquid chromatography/ultraviolet detection; GC/MS, gas chromatography/mass spectroscopy; PMR, proton magnetic resonance

3. Biological Data Relevant to the Evaluation of Carcinogenic Risk to Humans

3.1 Carcinogenicity studies in animals

(a) Oral administration

Mouse: Groups of 50 male and 50 female B6C3F$_1$ mice, 57 days old, were administered 0, 12 or 25 mg/kg bw allyl isothiocyanate (commercial grade; purity, >93%, found by chromatography and nuclear magnetic resonance spectroscopy, with six unidentified impurities) in corn oil by gavage five times per week for 103 weeks. Survival was comparable among the groups: at the end of the study period, 26, 24 and 27 males and 16, 25 and 18 females in the control, low-dose and high-dose groups, respectively, were still alive. Allyl isothiocyanate did not increase the incidence of tumours in treated mice of either sex (Dunnick et al., 1982; National Toxicology Program, 1982).

Rat: Groups of 50 male and 50 female Fischer 344/N rats, 39 days old, were administered 0, 12 or 25 mg/kg bw allyl isothiocyanate (commercial grade; purity, >93%, found by chromatography and nuclear magnetic resonance spectroscopy, with six unidentified impurities) in corn oil by gavage five times a week for 103 weeks. Survival was comparable in all groups, with 29-33 animals per treated group and 35-37 controls surviving at the end of the study. In males, transitional-cell papillomas occurred in the urinary bladder in 0/49 control, 2/49 low-dose and 4/49 high-dose animals ($p < 0.05$, trend test). Epithelial hyperplasia in the urinary bladder also occurred in 0/49 control, 1/49 low-dose and 6/49 high-dose males (not the same animals that had papillomas). One transitional-cell papilloma and one epithelial hyperplasia were found in high-dose female rats. Subcutaneous-tissue fibrosarcomas occurred in 0/50 control, 0/50 low-dose and 3/50 high-dose females ($p < 0.05$, trend test) (Dunnick et al., 1982; National Toxicology Program, 1982).

(b) *Skin application*

Mouse: Mustard oil, which contains >90% allyl isothiocyanate (US Food and Drug Administration, 1975), was tested in a two-stage mouse-skin assay. Three control groups of 12-16 'S' strain mice [actual strain, age and sex unspecified] received 0.3 ml of 0.1-0.15% 7,12-dimethylbenz[a]anthracene (DMBA) on the skin of the back followed by no treatment (two groups) or 21 days later by twice weekly applications for 12 weeks and weekly applications for 15 weeks of acetone. A fourth group of 16 mice received an initial skin application of 0.3 ml of 0.1% DMBA followed 39 days later by applications of a 3-4.5% solution of mustard oil in acetone weekly for 20 weeks. The incidence of skin papillomas among animals surviving to the end of the experiment was 4/21, 1/12 and 1/16 in the combined DMBA control groups with no secondary treatment, in controls receiving DMBA followed by acetone and in treated mice receiving DMBA and mustard oil, respectively (Gwynn & Salaman, 1953). [The Working Group noted the extremely short duration of the study and the limited number of animals used.]

3.2 Other relevant biological data

(a) *Experimental systems*

Toxic effects

The oral LD_{50} of allyl isothiocyanate dissolved in corn oil was reported to be 339 mg/kg bw in rats (Jenner et al., 1964); in mice, the subcutaneous LD_{50} of a 10% solution of allyl isothiocyanate in corn oil was found to be 80 mg/kg bw (Klesse & Lukoschek, 1955).

Allyl isothiocyanate has been described as a strong irritant to skin and mucous membranes (Gosselin et al., 1982).

Acute and chronic effects from short- and long-term oral administration have been reported in the context of a bioassay study (described in section 3.1): a single administration of the compound in corn oil by gavage caused growth retardation and dose-related, non-specific signs of toxicity at dose levels of 200 and 400 mg/kg bw in rats and 100-800 mg/kg bw in mice. A 14-day experiment, in which mice and rats received doses of 3-50 and 25-400 mg/kg bw, respectively, resulted in dose-dependent thickening of the stomach mucosa and, in rats, in adhesion of the stomach wall to the peritoneum; mice given the highest dose developed a thickening of the urinary-bladder wall. Lethality began with doses of 200 mg/kg bw in rats and of 50 mg/kg bw in mice. In a chronic study lasting 103 weeks, daily doses of 12 or 25 mg/kg bw caused a slight, dose-related decrease in body-weight gain and mean survival time; an increased rate of cytoplasmic vacuolization was noted in the livers of male mice (National Toxicology Program, 1982).

A reduction in blood clotting and prothrombin times, an increase in total plasma and liver triglycerides and cholesterol, and a decrease in D-amino acid oxidase (Muztar et al., 1979a,b) and xanthine oxidase were reported after administration to rats of 0.1% allyl isothiocyanate in the diet for 30 days (Huque & Ahmad, 1975).

Allyl isothiocyanate exerts a slight goitrogenic activity. Studies in rats (bw, 150-320 g) given 2-4 mg as a single dose by gavage in water showed inhibition of iodine uptake into the thyroid gland (Langer & Štolc, 1963). This effect may be due, according to in-vitro studies, to an inhibition of inorganic iodide storage, as well as organic binding of iodine (Langer & Greer, 1968).

Effects on reproduction and prenatal toxicity

The teratogenic potential of allyl isothiocyanate was evaluated in mice, rats, hamsters and rabbits (Food and Drug Research Laboratories, 1973). Groups of 23-25 CD-1 mice were treated with 0, 0.3, 1.3, 6.0 or 28.0 mg/kg bw allyl isothiocyanate [purity unspecified] in corn oil by oral gavage on gestation days 6-15. Foetuses were examined on day 17 for malformations. Groups of 25 Wistar rats received 0, 0.2, 0.85, 4.0 or 18.5 mg/kg bw in corn oil by oral intubation on gestation days 6-15. Foetuses were examined for malformations on day 20. Groups of 25-27 golden hamsters received doses of 0, 0.2, 1.1, 5.1 or 23.8 mg/kg bw in corn oil by oral intubation on days 6-10 of gestation. Foetuses were examined on day 14 for malformations. Groups of 11-14 Dutch-belted rabbits received doses of 0, 0.123, 0.6, 2.8 or 12.3 mg/kg bw in corn oil by oral intubation on days 6-18 of gestation. Foetuses were delivered by caesarean section on day 29. No evidence of maternal toxicity or treatment-related malformation was observed in any species. In mice, there appeared to be an increase in dead and resorbed foetuses in the high-dose group, although no statistical analysis of the data was presented (at the highest dose level, 38/276 implantation sites were dead or resorbed compared to 15/264 in the control group, and the average number of live pups per litter was 9.92 compared to 11.3).

Groups of pregnant Wistar rats (29 animals in the control group, five in the low-dose and unstated for the high-dose groups) were given 0, 60 or 120 mg/kg bw of allyl isothiocyanate [purity unspecified] in corn oil by oral intubation on days 12 or 13 of gestation as part of an effort to determine structure-activity relationships for chemicals similar to the teratogen ethylenethiourea. Despite the occurrence of maternal toxicity at the high dose, no adverse effect on the foetuses was found (Ruddick et al., 1976).

Two groups of six and eight pregnant Holtzman rats received 50 or 100 mg/kg bw allyl isothiocyanate [purity unspecified], respectively, by subcutaneous administration (the vehicle was either propylene glycol or distilled water) on days 8 and 9 of gestation. A group of 54 pregnant rats served as controls. Maternal toxicity was evident with the high dose. Foetuses were examined on day 20; those in the low-dose group weighed significantly less than controls, while an increased incidence of resorptions was seen in the high-dose group. No treatment-related malformation was observed (Nishie & Daxenbichler, 1980).

Absorption, distribution, excretion and metabolism

Fischer 344 rats and B6C3F$_1$ mice were used to study the tissue distribution and metabolism of allyl isothiocyanate (purity, >98%). When measured 15 min after intravenous injection of 25 mg/kg bw, allyl isothiocyanate-derived radioactivity was found at the highest concentration in the urinary bladder of male rats and mice and in the kidneys of male mice; the bladders of males contained five to ten times more radioactivity than the bladders of females. After oral and intravenous administration, most of the radioactivity was cleared through urine (70-80%), while exhaled air (13-15%) and faeces (3-5%) contained less. The major metabolite detected in urine was *N*-acetyl-*S*-(*N*-allylthiocarbamoyl)-L-cysteine (Ioannou *et al.*, 1984). (See also Fig. 1, General Remarks on the Substances Considered, p. 32.)

Mutagenicity and other short-term tests (see also 'Appendix: Activity Profiles for Short-Term Tests', p. 330).

Allyl isothiocyanate was negative in the *Bacillus subtilis* rec$^+$/rec$^-$ DNA-repair assay (Oda *et al.*, 1978).

Allyl isothiocyanate [purity unspecified] was mutagenic to *Salmonella typhimurium* TA98 and TA100 using the preincubation procedure; the presence of an exogenous metabolic system (S9) from the livers of polychlorinated biphenyl-induced rats had no effect on this activity (Yamaguchi, 1980). However, addition of an exogenous metabolic system was reported to abolish a weak response in strain TA100 following a modified treatment procedure with allyl isothiocyanate (purity, 99.8%) (Eder *et al.*, 1980, 1982a,b). Using the standard plate incorporation test, Kasamaki *et al.* (1982) reported [data not given] that allyl isothiocyanate (purity, 90-95%) was not mutagenic in strains TA98 and TA100 in the presence or absence of a rat-liver S9. Negative results were obtained in strains TA1535, TA1537 and TA1538 (Yamaguchi, 1980). Allyl isothiocyanate (purity, 95%) induces mutations in *Escherichia coli* WP67; this activity requires the presence of an exogenous metabolic system; rodent, goat and monkey liver were tested (Říhová, 1982).

Allyl isothiocyanate induces chromosomal aberrations in root-tip cells of *Allium cepa* (Sharma & Sharma, 1962).

Allyl isothiocyanate was reported to induce sex-linked recessive lethal mutations in *Drosophila melanogaster* (Auerbach & Robson, 1944). However, in an abstract, Schalet and Herskowitz (1954) reported no induction of sex-linked recessive lethal mutations.

Allyl isothiocyanate has been reported to induce chromosomal aberrations in Chinese hamster B241 cells. [The Working Group noted the extremely low effective dose (5 nM) reported in this study.] An exogenous metabolic system derived from Aroclor-induced rats had no effect on the activity (Kasamaki *et al.*, 1982).

Allyl isothiocyanate does not induce dominant lethal mutations in mice when doses of up to 19 mg/kg bw are given intraperitoneally (Epstein *et al.*, 1972).

Mustard oil, which is reported to contain >90% allyl isothiocyanate, was reported not to induce 'genetic effects' in *Saccharomyces cerevisiae* in a host-mediated assay with mice treated with up to 130 mg/kg bw (US Food and Drug Administration, 1975). [Data were not available to the Working Group.]

Mustard oil did not induce chromosomal aberrations in cultured human embryonic lung cells nor in the bone marrow of rats given up to 100 mg/kg bw (US Food and Drug Administration, 1975 [no data given]). It has been reported to induce chromosomal aberrations in the root tips of wheat (Swaminathan & Natarajan, 1956, 1959).

The induction by mustard oil of sex-linked lethal mutations has been reported in *Drosophila melanogaster* (Auerbach & Robson, 1947). It did not induce dominant lethal mutations in rats at levels of up to 100 mg/kg bw (US Food and Drug Administration, 1975 [no data given]).

(b) Humans

Toxic effects

Allyl isothiocyanate has been found to produce irritation of mucous membranes and eczematous or vesicular skin reactions (Gaul, 1964). Contact dermatitis was reported in a waitress who handled salad plants; patch tests in this woman with radishes and with allyl isothiocyanate produced positive reactions (Mitchell & Jordan, 1974).

No data were available to the Working Group on effects on reproduction and prenatal toxicity, on absorption, distribution, excretion and metabolism, or on mutagenicity and chromosomal effects.

3.3 Case reports and epidemiological studies of carcinogenicity to humans

No data were available to the Working Group.

4. Summary of Data Reported and Evaluation

4.1 Exposure data

Allyl isothiocyanate occurs widely in natural products in the glucoside sinigrin. Synthetic allyl isothiocyanate has been produced commercially since 1937. Allyl isothiocyanate is also prepared from the seeds of mustard plants, *Brassica nigra* and *B. juncea*. It is used principally as a flavouring agent in a variety of foods. Exposures can also occur from its use as an alcohol denaturant and in external analgesic products.

4.2 Experimental data

Allyl isothiocyanate was tested for carcinogenicity by gastric intubation in mice of one strain and in rats of one strain. In mice, no increase in the incidence of tumours was observed. An increased incidence of epithelial hyperplasia and transitional-cell papillomas of the urinary bladder was observed in male rats only, and some subcutaneous fibrosarcomas occurred in female rats given the high dose.

Allyl isothiocyanate was not teratogenic to mice, rats, hamsters or rabbits, but resorptions were seen in mice and rats.

Allyl isothiocyanate did not induce DNA damage in bacteria. It induced mutations in bacteria and insects and chromosomal aberrations in plants. It did not induce dominant lethal mutations in mice.

Overall assessment of data from short-term tests: allyl isothiocyanate[a]

	Genetic activity			Cell transformation
	DNA damage	Mutation	Chromosomal effects	
Prokaryotes	–	+		
Fungi/Green plants			+	
Insects		+		
Mammalian cells (in vitro)				
Mammals (in vivo)			–	
Humans (in vivo)				
Degree of evidence in short-term tests for genetic activity: Limited				Cell transformation: No data

[a]The groups into which the table is divided and the symbols are defined on pp. 17-18 of the Preamble; the degrees of evidence are defined on p. 18.

4.3 Human data

No case report or epidemiological study of the carcinogenicity of allyl isothiocyanate to humans was available to the Working Group.

4.4 Evaluation[1]

There is *limited evidence* for the carcinogenicity of allyl isothiocyanate to experimental animals.

In the absence of epidemiological data, no evaluation could be made of the carcinogenicity of allyl isothiocyanate to humans.

5. References

Auerbach, C. & Robson, J.M. (1944) Production of mutations by allyl isothiocyanate. *Nature*, *154*, 81

Auerbach, C. & Robson, J.M. (1947) Tests of chemical substances for mutagenic action. *Proc. roy. Soc. Edinburgh*, *62B*, 284-291

Cole, R.A. (1980) The use of porous polymers for the collection of plant volatiles. *J. Sci. Food Agric.*, *31*, 1242-1249

---

[1]For definitions of the italicized terms, see the Preamble, pp. 15-16.

Dunnick, J.K., Prejean, J.D., Haseman, J., Thompson, R.B., Giles, H.D. & McConnell, E. (1982) Carcinogenesis bioassay of allyl isothiocyanate. *Fundam. appl. Toxicol.*, *2*, 114-120

Eder, E., Neudecker, T., Lutz, D. & Henschler, D. (1980) Mutagenic potential of allyl and allylic compounds. *Biochem. Pharmacol.*, *29*, 993-998

Eder, E., Henschler, D. & Neudecker, T. (1982a) Mutagenic properties of allylic and α,β-unsaturated compounds: Consideration of alkylating mechanisms. *Xenobiotica*, *12*, 831-848

Eder, E., Neudecker, T., Lutz, D. & Henschler, D. (1982b) Correlation of alkylating and mutagenic activities of allyl and allylic compounds: Standard alkylation test vs. kinetic investigation. *Chem.-biol. Interactions*, *38*, 303-315

Epstein, S.S., Arnold, E., Andrea, J., Bass, W. & Bishop, Y. (1972) Detection of chemical mutagens by the dominant lethal assay in the mouse. *Toxicol. appl. Pharmacol.*, *23*, 288-325

Esaki, H. & Onozaki, H. (1980) Colorimetric determination of pungent taste substances in radish root (Jpn.). *Eiyo to Shokuryo*, *33*, 161-167 [*Chem. Abstr.*, *94*, 63857w]

Food and Drug Research Laboratories (1972) *GRAS (Generally Recognized as Safe) Food Ingredients - Oil of Mustard and Allyl Isothiocyanate, Final Report 1920-1972* (*Report No. FDABF-GRAS-015; PB 221 215*), Prepared for the Food and Drug Administration, Springfield, VA, National Technical Information Service

Food and Drug Research Laboratories (1973) *Teratologic Evaluation of FDA 71-26 (Oil of Mustard)* (*PB-223 812*), Washington DC, National Technical Information Service

Furia, T.E. & Bellanca, N., eds (1975) *Fenaroli's Handbook of Flavor Ingredients*, 2nd ed., Vol. 2, Cleveland, OH, CRC Press Inc., p. 19

Gaul, L.E. (1964) Contact dermatitis from synthetic oil of mustard. *Arch. Dermatol.*, *90*, 158-159

Gosselin, R.E., Hodge, H.C., Smith, R.P. & Gleason, M.N., eds (1982) *Clinical Toxicology of Commercial Products. Acute Poisoning*, 4th ed., Baltimore, MD, The Williams & Wilkins Co., p. II-216

Gwynn, R.H. & Salaman, M.H. (1953) Studies on co-carcinogenesis. SH-reactors and other substances tested for co-carcinogenic action in mouse skin. *Br. J. Cancer*, *7*, 482-489

Hassan, M.M.A., Aboutabl, E.A. & El-Obeid, H.A. (1980) PMR assay of essential oils. III. Assay of allylisothiocyanate in mustard oil. *Spectrosc. Lett.*, *13*, 199-204

Hawley, G.G., ed. (1981) *The Condensed Chemical Dictionary*, 10th ed., New York, Van Nostrand Reinhold Co., p. 36

Henning, W. (1981) Determination of mustard glucosides and their decomposition products in mustard seeds and nutritional mustard preparations by ion-pair high performance liquid chromatography (Ger.). *Dtsch Lebensm.-Rundsch.*, *77*, 313-318 [*Chem. Abstr.*, *95*, 202l65e]

Huque, T. & Ahmad, P. (1975) Effect of allyl isothiocyanate on blood and urine levels of uric acid and glucose in rats. *Bangladesh J. biol. agric. Sci.*, *4*, 12-13

Ina, K. (1982) Volatile components of wasabi, horseradish and mustard (Jpn.). *Koryo*, *136*, 44-52 [*Chem. Abstr.*, *97*, 143282g]

Ina, K., Nobukuni, M., Sano, A. & Kishima, I. (1981) Studies on the volatile components of wasabi and horse radish. III. Stability of allyl isothiocyanate (Jpn.). *Nippon Shokuhin Kogyo Gakkaishi*, *28*, 627-637 [*Chem. Abstr.*, *96*, 141379z]

Ioannou, Y.M., Burka, L.T. & Matthews, H.B. (1984) Allyl isothiocyanate: Comparative disposition in rats and mice. *Toxicol. appl. Pharmacol.*, *75*, 173-181

Jenner, P.M., Hagan, E.C., Taylor, J.M., Cook, E.L. & Fitzhugh, O.G. (1964) Food flavourings and compounds of related structure. I. Acute oral toxicity. *Food Cosmet. Toxicol.*, *2*, 327-343

Kasamaki, A., Takahashi, H., Tsumura, N., Niwa, J., Fujita, T. & Urasawa, S. (1982) Genotoxicity of flavoring agents. *Mutat. Res.*, *105*, 387-392

Klesse, P. & Lukoschek, P. (1955) Investigations of the bacteriostatic action of some mustard oils (Ger.). *Arzneimittelforschung*, *5*, 505-507

Langer, P. & Greer, M.A. (1968) Antithyroid activity of some naturally occurring isothiocyanates *in vitro*. *Metabolism*, *17*, 596-605

Langer, P. & Štolc, V. (1963) Goitrogenic activity of allylisothiocyanate - A widespread natural mustard oil. *Endocrinology*, *76*, 151-155

Min, D.B. (1981) Correlation of sensory evaluation and instrumental gas chromatographic analysis of edible oils. *J. Food Sci.*, *46*, 1453-1456

Min, D.B. & Tickner, D.B. (1982) Preliminary gas chromatographic analysis of flavor compounds in mayonnaise. *J. Am. Oil Chem. Soc.*, *59*, 226-228

Mitchell, J.C. & Jordan, W.P. (1974) Allergic contact dermatitis from the radish, *Raphanus sativus*. *Br. J. Dermatol.*, *91*, 183-189

Mukhopadhyay, S. & Bhattacharyya, D.K. (1983) Colorimetric estimation of allyl isothiocyanate content in mustard and rapeseed oils. *Fette Seifen Anstrichm.*, *85*, 309-311 [*Chem. Abstr.*, *99*, 138236g]

Mutzar, A.J., Ahmad, P., Huque, T. & Slinger, S.J. (1979a) A study of the chemical binding of allyl isothiocyanate with thyroxine and of the effect of allyl isothiocyanate on lipid metabolism in the rat. *Can. J. Physiol. Pharmacol.*, *57*, 385-389

Mutzar, A.J., Huque, T., Ahmad, P. & Slinger, S.J. (1979b) Effect of allyl isothiocyanate on plasma and urinary concentrations of some biochemical entities in the rat. *Can. J. Physiol. Pharmacol.*, *57*, 504-509

National Research Council (1981) *Food Chemicals Codex*, 3rd ed., Washington DC, National Academy Press, pp. 353, 356-357, 583, 678

National Research Council/National Academy of Sciences (1979) *The 1977 Survey of Industry on the Use of Food Additives*, Part 1 of 3 (*PB80-113418*), Springfield, VA, National Technical Information Service, pp. 624-627

National Toxicology Program (1982) *Carcinogenesis Bioassay of Allyl Isothiocyanate (CAS No. 57-06-7) in F334/N Rats and B6C3F$_1$ Mice (Gavage Study)* (NTP No. 81-36; NIH Publication No. 83-1790), Washington DC, US Department of Health and Human Services

NIH/EPA Chemical Information System (1984) *Toxic Substance Control Act (TSCA) Plant and Production*, Washington DC, Information Sciences Corporation

Nishie, K. & Daxenbichler, M.E. (1980) Toxicology of glucosinolates, related compounds (nitriles, R-goitrin, isothiocyanates) and vitamin U in cruciferae. *Food Cosmet. Toxicol.*, 18, 159-172

Oda, Y., Hamano, Y., Inoue, K., Yamamoto, H., Niihara, T. & Kunita, N. (1978) Mutagenicity of food flavours in bacteria (Jpn.). *Osaka-Furitsa Koshu Eisei Kenkyu Hokoku, Shokuhim Eisei Hen* (Osaka Pref. Inst. publ. Health, Food Microbiol.), 9, 177-181

Říhová, E. (1982) Mutagenic effects of allyl isothiocyanate in *Escherichia coli* WP 67. *Folia microbiol.*, 27, 25-31

Ruddick, J.A., Newsome, W.H. & Nash, L. (1976) Correlation of teratogenicity and molecular structure: Ethylenethiourea and related compounds. *Teratology*, 13, 263-266

Schalet, A. & Herskowitz, I.H. (1954) Chemical mutagenesis of mature *Drosophila* sperm treated in sperm baths and postcopulatory vaginal douches (Abstract). *Genetics*, 39, 992

Sharma, A.K. & Sharma, A. (1962) A study of the importance of nucleic acids in controlling chromosome breaks induced by different compounds. *Nucleus*, 5, 127-136

Swaminathan, M.S. & Natarajan, A.T. (1956) Chromosome breakage induced by vegetable oils and edible fats. *Curr. Sci.*, 12, 382-384

Swaminathan, M.S. & Natarajan, A.T. (1959) Cytological and genetic changes induced by vegetable oils in *Triticum*. *J. Hered.*, 50, 177-187

Tookey, H.L., Daxenbichler, M.E., VanEtten, C.H., Kwolek, W.F. & Williams, P.H. (1980) Cabbage glucosinolates: Correspondence of patterns in seeds and leafy heads. *J. Am. Soc. hortic. Sci.*, 105, 714-717 [*Chem. Abstr.*, 93, 164397f]

Uda, Y., Ozawa, Y., & Maeda, Y. (1982a) Volatile isothiocyanates from glucosinolates of cruciferous vegetables introduced from China (Jpn.). *Nippon Nogei Kagaku Kaishi*, 56, 1057-1060 [*Chem. Abstr.*, 98, 87864e]

Uda, Y., Ozawa, Y. & Maeda, Y. (1982b) Volatile hydrolysis products of glucosinolates occurring in leaves seeds from two varieties of artificial *Brassica napus*. *Agric. biol. Chem.*, 46, 3097-3099 [*Chcm. Abstr.*, 98, 50446u]

US Department of the Treasury (1983) Formulas for denatured alcohol and rum. *US Code Fed. Regul., Title 27*, Parts 21, 212; *Fed. Regist., 48* (No. 107), pp. 24672-24675, 24680-24681, 24687

US Food and Drug Administration (1975) *Evaluation of Health Aspects of Mustard and Oil of Mustard and Food Ingredients (PB-254 528)*, Federation of American Societies for Experimental Biology, Springfield, VA, National Technical Information Service

US Food and Drug Administration (1979) External analgesic drug products for over-the-counter human use; Establishment of a monograph and notice of proposed rulemaking. *US Code Fed. Regul., Title 21*, Part 348; *Fed. Regist., 44* (No. 234), pp. 69768-69771, 69791-69792

US Food and Drug Administration (1980) Food and drugs. *US Code Fed. Regul., Title 21*, Part 172.515

US Food and Drug Administration (1983) External analgesic drug products for over-the-counter use; Tentative final monograph. *US Code Fed. Regul., Title 21*, Part 348; *Fed. Regist., 48* (No. 27), pp. 5852-5853, 5865-5869

US Tariff Commission (1938) *Dyes and Other Synthetic Organic Chemicals in the United States, 1937 (Report No. 132, Second Series).*, Washington DC, US Government Printing Office, p. 49

Verschueren, K. (1977) *Handbook of Environmental Data on Organic Chemicals*, New York, Van Nostrand Reinhold Co., p. 88

Windholz, M., ed. (1983) *The Merck Index*, 10th ed., Rahway, NJ, Merck & Co., p. 45

Yamaguchi, T. (1980) Mutagenicity of isothiocyanates, isocyanates and thioureas on *Salmonella typhimurium. Agric. biol. Chem., 44*, 3017-3018

ALLYL ISOVALERATE

1. Chemical and Physical Data

1.1 Synonyms and trade names

Chem. Abstr. Services Reg. No.: 2835-39-4

Chem. Abstr. Name: Butanoic acid, 3-methyl-, 2-propenyl ester

IUPAC Systematic Name: Allyl isovalerate

Synonyms: Allyl isovalerianate; allyl 3-methylbutyrate; FEMA No. 2045; isovaleric acid, allyl ester; 3-methylbutanoic acid, 2-propenyl ester; 3-methylbutyric acid, allyl ester; 2-propenyl isovalerate; 2-propenyl 3-methylbutanoate

1.2 Structural and molecular formulae and molecular weight

$$CH_2=CH-CH_2-O-\overset{\overset{O}{\|}}{C}-CH_2-CH(CH_3)_2$$

$C_8H_{14}O_2$
Mol. wt: 142.2

1.3 Chemical and physical properties of the pure substance

From Furia and Bellanca (1975), unless otherwise specified

(a) *Description*: Colourless liquid (Opdyke, 1979)

(b) *Boiling-point*: 89-90°C

(c) *Refractive index*: 1.4162 at 21°C

(d) *Spectroscopy data*: Infrared spectral data have been reported (Opdyke, 1979).

1.4 Technical products and impurities

No information was available to the Working Group.

2. Production, Use, Occurrence and Analysis

2.1 Production and use

(a) Production

Allyl isovalerate can be prepared by the reaction of allyl alcohol with isovaleric acid (Opdyke, 1979).

Although allyl isovalerate has reportedly been used since the 1950s (Opdyke, 1979), commercial production has been reported in the USA only since 1973 (US International Trade Commission, 1975). Since that time only one US company has produced it, and production data are not disclosed [see Preamble, section 8(b)(ii)]. Separate data on US imports and exports of allyl isovalerate are not published.

Allyl isovalerate is not produced commercially in western Europe or Japan.

(b) Use

Allyl isovalerate has been used as a raw material to impart a fruit-like (apple, cherry) aroma (Furia & Bellanca, 1975). It has been reported to be used in soaps, detergents, creams, lotions and perfumes (Opdyke, 1979). However, a survey of US industry on the use of food additives in 1977 did not indicate that it is used in foods, and a recent compilation of chemicals used in cosmetics did not include it.

Allyl isovalerate has been approved for use as a synthetic flavouring substance and adjuvant in foods provided it is used in the minimum quantity required to produce its intended effect (US Food and Drug Administration, 1980).

Allyl isovalerate is not permitted for use as a food additive by the Council of Europe.

2.2 Occurrence

Natural occurrence

Allyl isovalerate has not been reported to occur as such in nature.

2.3 Analysis

No information on methods for the analysis of allyl isovalerate were available to the Working Group.

3. Biological Data Relevant to the Evaluation of Carcinogenic Risk to Humans

3.1 Carcinogenicity studies in animals

Oral administration

Mouse: Groups of 50 male and 50 female B6C3F$_1$ mice, 50 days of age, received 0, 31 or 62 mg/kg bw allyl isovalerate (purity, 95.6%) in corn oil by gavage five times per week for 103 weeks. Survivors were killed at 112-114 weeks of age. The numbers of males surviving to the end of the experiment were 29 controls, 31 in the low-dose group and 31 in the high-dose group; the numbers of females surviving were 32 controls, 17 in the low-dose group and 24 in the high-dose group; the main cause of death in females was a genital-tract infection. Squamous-cell papillomas were observed in the forestomach of male mice in 0/50 controls, and in 1/50 low-dose and 3/48 high-dose animals ($p < 0.05$, incidental tumour test); and epithelial hyperplasia was seen in the forestomach of 1/50 controls, 1/50 low-dose and 7/48 high-dose male mice. Forestomach lesions observed in control, low-dose and high-dose female mice, respectively, were: squamous-cell papillomas, 1/50, 0/50 and 2/50; and epithelial hyperplasia, 0/50, 2/50 and 3/50. The incidence of lymphomas was increased in males: 4/50 controls, 6/50 low-dose and 8/50 high-dose, but was not significant by the trend test or by the incidental tumour test; in females, the corresponding incidences were 11/50, 11/50 and 18/50, which, using life-table analysis, gave a dose-response trend, the high dose giving significantly different ($p < 0.05$) tumour incidences from those in controls (National Toxicology Program, 1983).

Rat: Groups of 50 male and 50 female Fischer 344/N rats, 46 days of age, were administered 0, 31 or 62 mg/kg bw allyl isovalerate (purity, 95.6%) in corn oil by gavage five times per week for 103 weeks. Of the males, 34 controls, 30 low-dose and 28 high-dose rats lived to the end of the study and were killed at 112-114 weeks of age; of the females, 38 controls, 36 low-dose and 29 high-dose animals survived to the end of the experiment. The incidence of mononuclear-cell leukaemia in males was 1/50 in controls, 4/50 in low-dose and 7/50 in high-dose animals; in females, it was 4/50, 6/50, and 9/49. Using life-table analysis, both sexes showed a significant ($p < 0.05$) dose-response trend, and the incidence in high-dose males was significantly increased ($p < 0.05$) when compared to controls. Two high-dose male and two high-dose female rats had gliomas (astrocytoma, medulloblastoma) of the central nervous system, while no such tumour was found in low-dose or control animals (National Toxicology Program, 1983).

3.2 Other relevant biological data

(a) *Experimental systems*

Toxic effects

The acute oral LD$_{50}$ of allyl isovalerate in rats is reported to be 230 mg/kg bw, and the dermal LD$_{50}$ in rabbits, 560 mg/kg bw (Opdyke, 1979).

When administered undiluted, allyl isovalerate has low irritating potency on rabbit skin (Opdyke, 1979).

Allyl isovalerate caused cell necrosis in the liver of rats given 60-150 mg/kg bw per day for 10 days by intragastric intubation (Drake, 1975).

In the course of a National Toxicological Program bioassay (described in section 3.1), daily oral doses of 31 or 62 mg/kg bw given on five days per week for 103 weeks did not significantly depress body-weight gain in rats or mice of both sexes, nor was median survival time reduced significantly (National Toxicological Program, 1983).

Effects on reproduction and prenatal toxicity

No data were available to the Working Group.

Absorption, distribution, excretion and metabolism

Allyl isovalerate is hydrolysed *in vivo* in rats to allyl alcohol and isovaleric acid (Drake, 1975). Allyl alcohol is then oxidized to acrolein (see p. 133, this volume). (See also Fig. 1 of General Remarks on the Substances Considered, p. 32).

Mutagenicity and other short-term tests (see also 'Appendix: Activity Profiles for Short-Term Tests', p. 331)

Allyl isovalerate was not mutagenic to *Salmonella typhimurium* TA1535, TA1537, TA98 or TA100 in a preincubation assay, in the presence or absence of metabolic activation systems (S9) from livers of Aroclor-induced rats and hamsters (National Toxicology Program, 1983).

Allyl alcohol, a metabolite of allyl isovalerate, has been tested in several studies for mutagenicity in *S. typhimurium*; activity appears to be dependent upon the experimental conditions used. Negative results were obtained in strains TA1535, TA1537, TA1538, TA98 and TA100 with the standard plate incorporation assay in the presence or absence of an exogenous metabolic system (S9) (Lijinsky & Andrews, 1980; Rosen *et al.*, 1980; Carere & Morpurgo, 1981; Principe *et al.*, 1981). With a 90-minute pulsed exposure, allyl alcohol (purity, 99.9%) was mutagenic to strain TA100; activity was decreased by addition of Aroclor-induced rat liver S9 (Eder *et al.*, 1982; Lutz *et al.*, 1982). The preincubation procedure, with and without S9, yielded negative results with strain TA100 (Lijinsky & Andrews, 1980; Yamaguchi, 1980). Mutagenicity was demonstrated in strain TA1535 in the presence of hamster S9 (Lijinsky & Andrews, 1980).

Allyl alcohol was not mutagenic to *Aspergillus nidulans* or *Streptomyces coelicolor*. The tests were carried out in the absence of an exogenous metabolic system (Carere & Morpurgo, 1981; Principe *et al.*, 1981).

(b) Humans

Toxic effects

The irritating potency of liquid allyl isovalerate on human skin has been described as low. A 'maximization skin sensitization test' in volunteers was negative after administration of 1% allyl isovalerate (Opdyke, 1979).

No data were available to the Working Group on effects on reproduction and prenatal toxicity, on absorption, distribution, excretion and metabolism, or on mutagenicity and chromosomal effects.

ALLYL ISOVALERATE

3.3 Case reports and epidemiological studies of carcinogenicity to humans

No data were available to the Working Group.

4. Summary of Data Reported and Evaluation

4.1 Exposure data

Allyl isovalerate has been used since the 1950s as a fragrance raw material in cosmetics, lotions and perfumes and in certain food products, although it is not known whether it is used in these applications currently.

4.2 Experimental data

Allyl isovalerate was tested for carcinogenicity by intragastric intubation in mice of one strain and in rats of one strain. In mice, it induced papillomas of the forestomach in males and increased the incidence of lymphomas in females. In rats, an increased incidence of mononuclear-cell leukaemia was observed in animals of both sexes.

Allyl isovalerate was not mutagenic to bacteria.

Overall assessment of data from short-term tests: allyl isovalerate[a]

	Genetic activity			Cell transformation
	DNA damage	Mutation	Chromosomal effects	
Prokaryotes		–		
Fungi/Green plants				
Insects				
Mammalian cells (*in vitro*)				
Mammals (*in vivo*)				
Humans (*in vivo*)				
Degree of evidence in short-term tests for genetic activity: *Inadequate*				Cell transformation: No data

[a]The groups into which the table is divided and the symbol - are defined on pp. 17-18 of the Preamble; the degrees of evidence are defined on p. 18.

4.3 Human data

No case report or epidemiological study of the carcinogenicity of allyl isovalerate to humans was available to the Working Group.

4.4 Evaluation[1]

There is *limited evidence* for the carcinogenicity of allyl isovalerate to experimental animals.

In the absence of epidemiological data, no evaluation could be made of the carcinogenicity of allyl isovalerate to humans.

5. References

Carere, A. & Morpurgo, G. (1981) Comparison of the mutagenic activity of pesticides *in vitro* in various short-term assays. *Progr. Mutat. Res.*, 2, 87-104

Drake, J.J.-P. (1975) Safety evaluation of allyl esters. *Int. J. Flavours Food Add.*, 6, 352-356

Eder, E., Henschler, D. & Neudecker, T. (1982) Mutagenic properties of allylic and α,β-unsaturated compounds: Consideration of alkylating mechanisms. *Xenobiotica*, 12, 831-848

Furia, T.E. & Bellanca, N., eds (1975) *Fenaroli's Handbook of Flavor Ingredients*, 2nd ed., Vol. 2, Cleveland, OH, CRC Press Inc., p. 20

Lijinsky, W. & Andrews, A.W. (1980) Mutagenicity of vinyl compounds in *Salmonella typhimurium. Teratog. Carcinog. Mutagenesis*, 1, 259-267

Lutz, D., Eder, E., Neudecker, T. & Henschler, D. (1982) Structure-mutagenicity relationship in α,β-unsaturated carbonylic compounds and their corresponding allylic alcohols. *Mutat. Res.*, 93, 305-315

National Toxicology Program (1983) *Carcinogenesis Studies of Allyl Isovalerate (CAS No. 2835-39-4) in F344/N rats and B6C3F1 mice (Gavage studies)* (NTP TR No. 253), Washington DC, US Department of Health & Human Services

Opdyke, D.L.J. (1979) Monographs on fragrance raw materials. Allyl isovalerate. *Food Cosmet. Toxicol.*, 17 Suppl., 703

Principe, P., Dogliotti, E., Bignami, M., Crebelli, R., Falcone, E., Fabrizi, M., Conti, G. & Comba, P. (1981) Mutagenicity of chemicals of industrial and agricultural relevance in *Salmonella, Streptomyces* and *Aspergillus. J. Sci. Food Agric.*, 32, 826-832

Rosen, J.D., Segall, Y. & Casida, J.E. (1980) Mutagenic potency of haloacroleins and related compounds. *Mutat. Res.*, 78, 113-119

US Food and Drug Administration (1980) Food and drugs. *US Code Fed. Reg., Title 21*, Part 172.515

US International Trade Commission (1975) *Synthetic Organic Chemicals, US Production and Sales, 1973* (ITC Publication 728), Washington DC, US Government Printing Office, p. 122

Yamaguchi, T. (1980) Mutagenicity of isothiocyanates, isocyanates and thioureas on *Salmonella typhimurium. Agric. biol. Chem.*, 44, 3017-3018

---

[1]For definitions of the italicized terms, see the Preamble, pp. 15-16.

EUGENOL

1. Chemical and Physical Data

1.1 Synonyms and trade names

Chem. Abstr. Services Reg. No.: 97-53-0

Chem. Abstr. Name: Phenol, 2-methoxy-4-(2-propenyl)-

IUPAC Systematic Name: 4-Allyl-2-methoxyphenol

Synonyms: 4-Allylcatechol-2-methyl ether; allylguaiacol; 4-allylguaiacol; *p*-allylguaiacol; 1-allyl-4-hydroxy-3-methoxybenzene; 4-allyl-1-hydroxy-2-methoxybenzene; caryophyllic acid; eugenic acid; 1,3,4-eugenol; *p*-eugenol; FEMA No. 2467; 1-hydroxy-4-allyl-2-methoxybenzene; 4-hydroxy-3-methoxyallylbenzene; 1-hydroxy-2-methoxy-4-allylbenzene; 1-hydroxy-2-methoxy-4-propenylbenzene; 1-hydroxy-2-methoxy-4-prop-2-enylbenzene; 2-methoxy-4-allylphenol; 2-methoxy-1-hydroxy-4-allylbenzene; 2-methoxy-4-prop-2-enylphenol; 2-methoxy-4-(2-propenyl)phenol; 2-methoxy-4-(2-propen-1-yl)phenol; NCI-C50453

Trade Name: FA 100

1.2 Structural and molecular formulae and molecular weight

$C_{10}H_{12}O_2$ Mol. wt: 164.2

1.3 Chemical and physical properties of the pure substance

From Windholz (1983), unless otherwise specified

(a) *Description*: Colourless or pale-yellow liquid with an odour of cloves and a spicy, pungent taste

(b) *Boiling-point*: 255°C

(c) *Melting-point*: -9.2 to -9.1°C

(d) *Density*: d_4^{20} 1.0664

(e) *Refractive index*: n_D^{20} 1.5410

(f) *Optical rotation*: -1°30' (Furia & Bellanca, 1975)

(g) *Spectroscopy data*: Infrared spectral data have been reported (National Research Council, 1981)

(h) *Solubility*: Practically insoluble in water; limited solubility in 70% aqueous ethanol (1 ml in 2 ml); soluble in glacial acetic acid and aqueous alkalis; miscible with chloroform, diethyl ether and oils

(i) *Volatility*: Vapour pressure, 10 mm Hg at 123°C (Verschueren, 1977)

(j) *Stability*: Flash-point, approx. 104°C (Furia & Bellanca, 1975); darkens and thickens on exposure to air

(k) *Reactivity*: Rearranges to isoeugenol when treated with strong alkali (Van Ness, 1983)

(l) *Conversion factor*: 1 ppm = 6.71 mg/m^3 at 760 mm Hg and 25°C (Hake & Rowe, 1963)

1.4 Technical products and impurities

In the USA, to meet the requirements of the Food Chemicals Codex, eugenol must pass an infrared identification test, meet the following specifications: phenols by volume, 100%; refractive index, n_D^{20}, 1.540-1.542; solubility in 70% ethanol, 1 ml in 2 ml; specific gravity (25°C), 1.064-1.070; distillation range, 95% min in 250-255°C range; and pass a test for hydrocarbons (National Research Council, 1981).

The specifications for USP grade eugenol are the same as those for the Food Chemicals Codex except that the heavy metals content must not exceed 0.004% max (US Pharmacopeial Convention, Inc., 1980).

Eugenol is available in western Europe with the following specifications: density (20°C), 1.064-1.068; refractive index, n_D^{20}, 1.540-1.542. It has a typical flash-point (closed-cup) of 113°C, and one volume dissolves in two volumes of 70% ethanol at 20°C (Bush Boake Allen Ltd, undated).

Eugenol is available in the UK with a purity of 97% min and as eugenol DQ (dental quality) with a purity of 98% min.

EUGENOL

2. Production, Use, Occurrence and Analysis

2.1 Production and use

(a) *Production*

Eugenol was first prepared in 1919 by Claisen and was first isolated from oil of cloves (*Eugenia caryophyllata*) in 1929 (Windholz, 1983). It can be extracted from clove oil with aqueous potassium hydroxide, followed by liberation with an acid, and distillation in a stream of carbon dioxide (Hawley, 1981). It can be synthesized by the reaction of allyl chloride (see p. 41 of this volume) with guaiacol (Varagnat, 1981).

Commercial production of eugenol was first reported in the USA in 1941-1943 (US Tariff Commission, 1945). Annual US production reached a peak in 1972 when eight companies reported combined production of 215 thousand kg (US Tariff Commission, 1974). In 1982, seven US companies reported combined production of 145 thousand kg (US International Trade Commission, 1983). Separate data on US imports and exports of eugenol are not published.

Eugenol is produced by six companies in France, three companies each in Italy and the UK, and one company each in Germany and the Netherlands. One company in the UK produced 20.4 tonnes of eugenol and 2.7 tonnes of eugenol DQ in 1983.

Eugenol is produced by seven companies in Japan by extraction from clove oil. Japanese imports of clove oil for all purposes amounted to 172 thousand kg in 1982.

(b) *Use*

Eugenol is used principally as a fragrance and flavouring agent, as an analgesic in dental materials and nonprescription drug products, as an insect attractant, and as a chemical intermediate. Several other applications have been reported, the commercial status of which is unknown.

Oil of cloves (which is principally eugenol) has reportedly been in use as a fragrance raw material since the nineteenth century. In 1975, the quantity of eugenol used in fragrances in the USA was estimated to have been less than 45.4 thousand kg per year (Opdyke, 1975).

Clove-bud oil has been reported to be used as a flavouring agent in pharmaceuticals, baked goods, sweets, mouthwashes and chewing gum (Rogers, 1981). A survey of US industry on the use of food additives (National Research Council/National Academy of Sciences, 1979) reported that 1.8 thousand kg of clove-bud oil were used in 1976 as an essential component of many flavours added to foods such as condiments, sweets and chewing gum at usual levels of up to 0.0099%.

Eugenol is used as a component of several dental materials (e.g., dental cements, impression pastes and surgical pastes). Such products are principally combinations of zinc oxide and eugenol in varying ratios (Paffenbarger & Rupp, 1979). They are reported to be widely used in dentistry as temporary filling materials, cavity liners for pulp protection, capping materials, temporary cementation of fixed protheses, impression materials and major ingredients of endodontic sealers (Miller *et al.*, 1978). In addition, eugenol has been used in dentistry for disinfecting root canals (US Food and Drug Administration, 1979). Nonprescription medicines for toothache commonly contain eugenol, and some products for canker-sore treatment may do so (American Pharmaceutical Association, 1982).

A mixture of eugenol and 2-phenylethyl propionate is used as an attractant to trap Japanese beetles (Metcalf, 1981).

Eugenol esters (acetate, benzoate and formate) and its methyl ether, all of which find use as flavouring agents, are made commercially from eugenol. It can also be used in the synthesis of vanillin, and the use of oil of cloves as a source of eugenol was the most popular commercial route to synthetic vanillin during the last quarter of the nineteenth century and the first quarter of this century (Van Ness, 1983), although it is not believed to be in use today.

Eugenol is one of numerous denaturants approved for use (at a level of 10 lbs/100 gallons [12 g/l] of alcohol) in the USA in specially denatured alcohol Formula No. 38-B. Although the volume of this formula used in the USA each year is published, no information is available on the amount made with eugenol.

Eugenol can be used as an antioxidant in inks (Burachinsky et al., 1981), and it has been reported to be useful as a fungicide in pharmaceuticals and cosmetics (Turner, 1966); no indication was found that eugenol is being used commercially for such purposes at present. Eugenol was formerly used internally in human medicine as an antiputrescent, but is no longer employed for this purpose. It has also been used in the treatment of flatulent colic (US Food and Drug Administration, 1979).

In western Europe, eugenol is used mainly in dental and oral hygiene products, and to a lesser degree in perfumes. In Japan, eugenol is used as a fragrance raw material and, of much lesser importance, in pharmaceuticals.

Clove and its derivatives (oil, buds, leaves, etc, including eugenol) are approved in the USA for use in foods as flavouring agents and adjuvants (US Food and Drug Administration, 1980a). Eugenol is also approved for use in the manufacture of textiles and textile fibres which are intended for use in contact with food surfaces (US Food and Drug Administration, 1980b).

The Council of Europe in 1984 permitted the use of eugenol as a food additive.

The Joint FAO/WHO Expert Committee on Food Additives established a conditional acceptable daily intake of eugenol for humans of 0-5 mg/kg bw (WHO, 1967).

An advisory review panel on over-the-counter drugs of the US Food and Drug Administration (FDA) (1979) concluded that eugenol was safe but that there was insufficient evidence to decide whether it was effective for use as an external analgesic. In 1982, another FDA advisory review panel (US Food and Drug Administration, 1982) concluded that eugenol and clove oil were safe and effective for use at specific dose levels as a dental analgesic for the relief of toothache provided that the eugenol content of the preparation was 85-87% and that the product was labelled 'Do not use if you are allergic to eugenol'. In 1983, the US Food and Drug Administration (1983) proposed a rule requiring that any external analgesic products containing eugenol not be marketed unless they were the subject of an approved new drug appplication.

The Bureau of Alcohol, Tobacco and Firearms of the US Department of the Treasury (1983) lists eugenol USP among the approved denaturants for specially denatured alcohol Formula No. 38-B.

2.2 Occurrence

(a) Natural occurrence

Eugenol is widely distributed in the plant kingdom, where it is mainly found as a component of essential oils. Typical concentrations of eugenol in various plants and derived oils are summarized in Table 1.

The Flavor and Extract Manufacturers' Association of the United States (1978) has reported the occurrence of eugenol, without specific concentrations, in the following food sources: cocoa, Japanese ginger oil, loganberries, mace essential oil, sweet marjoram, dried

Table 1. Concentration of eugenol in various plants and derived oils

Plant or derived oil	Concentration (g/kg, except where noted)	Reference
Alfalfa (*Medicago sativa*) oil	7	Buttery & Kamm (1980)
Almond hull (*Prunus amygdalus*) oil		Buttery et al. (1980)
var. Mission	9	
var. Nonpareil	80	
Artemisia scoparia oil	200	Sarin et al. (1982)
Bilberry (*Vaccinium myrtillus*)	0.05 mg/kg (estimated)	FEMA[a] (1978)
Cinnamon (*Cinnamomum cassia*) bark oil	120	Lockwood (1979)
Cinnamon (*Cinnamomum zeylanicum*)		Rabha et al. (1979)
leaf oil	945	
bark oil	420	
Clove (*Eugenia caryophyllata*) oil	896-908	Chiang et al. (1983)
	749	Gracza (1980)
buds	918-921	Yu & Fang (1981)
leaves	785-824	
Common woodworm (*Artemisia absinthium* L.) essential oil	4	Chialva et al. (1982)
Cranberry (*Vaccinium macrocarpon*), press residue	0.002 mg/kg (estimated)	FEMA[a] (1978)
Dry elder (*Sambucus nigra* L.) flowers		Toulemonde & Richard (1983)
essential oil	6	
isopentane extract	43	
Fennel (*Foeniculum vulgare* Miller)		Fujita et al. (1980)
root essential oil	2	
seedling essential oil	3	
Glossocardia bosvallia DC essential oil	68	Sharma & Garg (1980)
Fleabane (*Erigeron philadelphicus*)		Miyazawa et al. (1981)
essential oils from:		
flowers	9	
leaves	10	
stalks	8	
roots	1	
Jasmine (*Jasminium grandiflorum* L.)		Verzele et al. (1981)
oil originating from:		
France	11	
Italy	8	
Algeria	35	
Juniper (*Juniperus virginiana* L.) leaf oil (grown in Washington DC)	<5	Adams et al. (1981)
Lingonberry (*Vaccinium vitus-idaea minus*), press residue	0.008 mg/kg (estimated)	FEMA[a] (1978)
Mikania micrantha essential oil	4.8	Nicollier & Thompson (1981)
Basil (*Ocimum gratissimum*)		Nizharadze & Bagaturiya (1980)
light oils	542-639	
heavy oils	828-940	
Basil (*Ocimum suave* Wild) oils	715	Chogo & Crank (1981)
Oak (*Quercus dentata* Thunb.) leaf essential oil	41	Kameoka et al. (1983)
Roman woodworm (*Artemisia pontica* L.) essential oil	1	Chialva et al. (1982)
Savory (*Satureja hortensis* L.) essential oil	>1	Chialva et al. (1982)
Sweet basil (*Ocimum basilicum* L.) essential oil	40-279	Fleisher (1981)

[a]FEMA, Flavor and Extract Manufacturers' Association of the United States

mushrooms, nutmeg, yellow passion fruit, black pepper, peppermint, pimento berry oil and tomatoes. Additional reported occurrences of eugenol are as follows: allspice tincture (Gracza, 1980); *Alpinia galanga* oil (Rui *et al.*, 1982); *Apium graveolens* seed essential oil (Gupta & Baslas, 1978); *Artemisia glacialis* (glacier woodworm) essential oil (Shirokov *et al.*, 1980); *Bupleurum chinense* D.C. essential oil (Pu *et al.*, 1983); *Capsicum spp.* (red pepper) (Keller *et al.*, 1981); *Castanea creata* Sieb et Zucc (chestnut) flower (Yamaguchi & Shibamoto, 1980); *Cinnamomum pauciflorum* Nees) leaf essential oil (Lin & Hua, 1980); corn silage (Sakata *et al.*, 1982); *Curcuma longa* (Chen *et al.*, 1983); *Cytisus scoparius* Link flower essential oil (Kurihara & Kikuchi, 1980a); fermented plum juice (Ismail *et al.*, 1980); *Homalomena occulta* oil (Rui *et al.*, 1982); *Jasminium odoratissimum* oil (Cheng & Chao, 1979); *Juglans regia* leaf oil (Nahrstedt *et al.*, 1981); *Laurus nobilis* L. leaf (Yoshida, 1979); *Ligustrum japonicum* Thunb. flower essential oil (Kikuchi *et al.*, 1981a); *Ligustrum obtusifolium* Sieb et Zucc flower essential oil (Kikuchi, 1981); *Ligustrum ovalifolium* Hassk flower essential oil (Kurihara & Kikuchi, 1980b); *Lonicera japonica* flower essential oil (Wu & Fang, 1980); *Magnolia salicifolia* Maxim. bud (Kikuchi *et al.*, 1981b); *Melaxis accuminata* bulb essential oil (Gupta *et al.*, 1978); *Menyantes triforiata* essential oils (Sakai *et al.*, 1979); Natu tobacco (Nagaraj & Chakraborty, 1979); *Ocimum micranthum* Wild oil (Khosla *et al.*, 1980); *Trachycarpus excelsa* and *T. fortune:* (Oh *et al.*, 1979); and *Vetiveria zizamioides* G. root essential oil (Shibamoto & Nishimura, 1982).

The potential for the simultaneous occurrence of structurally-related natural agents should be noted. For example, eugenol and safrole (IARC, 1976) are both found in essential oils of nutmeg (Bejnarowicz & Kirch, 1963) and in *Magnolia salicifolia* Maxim. (Fujita & Fujita, 1972).

(b) Occupational exposure

On the basis of the 1974 National Occupational Hazard Survey, the National Institute for Occupational Safety and Health (1980, 1981) estimated that 33 000 US workers in 17 industries were exposed to eugenol.

(c) Water and sediment

Eugenol has been identified in the final effluent of a US municipal wastewater treatment plant (Ellis *et al.*, 1982). It has also been found at a level of 47 µg/kg in the untreated waste stream from a paper mill, but it was not detected in the influent stream or the treated effluent stream from this mill or in any of the streams at four other US/Canadian paper mills (Turoski *et al.*, 1983).

(d) Food, beverages and animal feeds

Eugenol has been identified in commercially available alcoholic beverages within the following ranges (mg/l): whiskies, <0.01-0.58; cognacs and brandies, <0.01-0.27; dark rums, <0.01-1.36; and white rums, <0.01-0.18. Since eugenol has been reported to be the main component of the phenolic fraction of an ethanolic extract of oak, the eugenol content of various whiskies is dependent on the type of barrel used for ageing and the number of times it has been used (Lehtonen, 1983).

In 1970, 51 companies reported use of eugenol in the food categories shown in Table 2. Daily per-capita consumption of eugenol was estimated to have been 0.605 mg. Eugenol has also been reported in smoked pork bellies at a level of 1.0 mg/kg and at an estimated level of 0.075 mg/kg in rye crispbread (Flavor and Extract Manufacturers' Association of the United States, 1978).

Table 2. Levels of eugenol in various food categories[a]

Food category	Number of firms reporting usage	Level (mg/kg) Usual	Level (mg/kg) Maximum
Baked goods	23	15	21.3
Beverages (non-alcoholic)	23	1.33	2.19
Soft sweets	18	10.3	14.9
Frozen dairy products	15	1.82	3.79
Gelatin	13	1.55	2.75
Hard sweets	11	29.5	43.6
Chewing gum	9	94.6	221
Meat products	4	50.8	102
Beverages (alcoholic)	4	0.5	1.0
Condiment relishes	<3	49	100
Confectionary frosting	<3	625	750
Miscellaneous	<3	520	814

[a]From Flavor and Extract Manufacturers' Association of the United States (1978)

Eugenol has also been reported in buckwheat flour, which is used to make buckwheat noodles (Yajima *et al.*, 1983), and has been identified as a component of the steam volatile oils obtained from blueberries (Horvat *et al.*, 1983).

2.3 Analysis

Typical methods for the analysis of eugenol in various matrices are summarized in Table 3.

Table 3. Methods for the analysis of eugenol

Sample matrix	Sample preparation	Assay procedure[a]	Reference
Almond hulls	Steam distil/extract with hexane; freeze out water and concentrate	GC/MS; GC/IR	Buttery *et al.* (1980)
Japanese chestnut (*Castanea creata* Sieb et Zucc) flowers	Steam distil/extract with dichloromethane; dry over sodium sulphate and concentrate	GC/FID	Yamaguchi & Shibamoto (1980)
Cinnamon bark and leaf oil	Dissolve in butan-1-ol	GC/FID	Analytical Methods Committee (1981)
Ocimum suave Wild (tropical shrub)	Two alternative procedures to isolate oils: (1) steam distil; or (2) extract with 95% ethanol, filter and concentrate, re-extract with diethyl ether, separate into acidic, basic and neutral components	GC/MS	Chogo & Crank (1981)
Flowers, leaves, stalks and roots of *Erigeron philadelphicus*	Steam distil; extract with diethyl ether; concentrate under nitrogen; separate into neutral, acidic and phenolic fractions	GC/MS	Miyazawa *et al.* (1981)
Leaves of juniper trees (*Juniperus virginiana*)	Steam distil; dry over sodium sulphate	GC/MS	Adams *et al.* (1981)
Essential oils of jasmine (*Jasminium grandiflorum* L.)	Two alternative procedures: (1) vacuum evaporation and cold trapping; or (2) head-space analysis	GC/MS	Verzele *et al.* (1981)

Sample matrix	Sample preparation	Assay procedure[a]	Reference
Alfalfa (*Medicago sativa*)	Steam distil/extract with hexane; separate from water and concentrate	GC/MS	Buttery & Kamm (1980)
Dried roots of *Vetiveria zizamioides*	Steam distil, dissolve in diethyl ether and wash (sodium bicarbonate, then sodium hydroxide); extract aqueous layer at pH 5-6 with diethyl ether; dry over sodium sulphate and concentrate	GC/MS; NMR	Shibamoto & Nishimura (1982)
Clove (*Eugenia caryophyllata*) oil	Aspirate oil droplets into a microcapillary tube and dissolve in chloroform	HPLC/UV	Hashimoto *et al.* (1981)
	React with *m*-aminophenol and sodium metaperiodate	S	Sastry *et al.* (1982)
	Direct analysis	NMR	Chiang *et al.* (1983)
Sweet basil (*Ocimum basilicum* L.) oil	Steam distil; extract aqueous layer with chloroform; dry over sodium sulphate and concentrate	GC	Fleisher (1981)
Mikania micranthia	Steam distil; fractionate on aluminium oxide column with hexane	GC/FID	Nicollier & Thompson (1981)
Alcoholic beverages (whisky, dark rum, white rum, cognac, armagnac and other brandies	React with 1-fluoro-2,4-dinitrobenzene to form a 2,4-dinitrophenyl ether derivative	GC/EC[b]	Lehtonen (1980, 1983)
Elder (*Sambucus nigra* L.) flowers	Three alternative procedures: (1) steam distil and concentrate; fractionate on Florisil column with petroleum ether/ petroleum ether-diethyl ether/methanol; (2) soak with ethanol, extract with isopentane; or (3) reflux with petroleum ether, concentrate and dissolve in ethanol	GC/MS; IR	Toulemonde & Richard (1983)
Buckwheat flour	Steam distil/extract with diethyl ether, fractionate into basic, acidic, weakly acidic and neutral components	GC/FID; GC/MS	Yajima *et al.* (1983)
Tomatoes	Two alternative procedures: (1) steam distil, saturate with sodium chloride and extract with diethyl ether, concentrate, separate into neutral, basic, phenolic and acidic fractions; or (2) heat, trap headspace volatiles on Tenax-GC, desorb thermally	GC/FID; GC/MS	Chung *et al.* (1983)
Foeniculum vulgare Miller	Steam distil and fractionate using an alumina column	GC/MS	Fujita *et al.* (1980)
Ligustrum obtusifolium Sieb et Zucc	Extract with diethyl ether, steam distil, separate into neutral and phenolic fractions	GC	Kikuchi (1981)

Sample matrix	Sample preparation	Assay procedure[a]	Reference
Water	Two alternative procedures: (1) extract at pH 11 with dichloromethane, extract at pH 6 with dichloromethane, combine extracts, dry over magnesium sulphate and concentrate; or (2) adsorb on XAD-2 and XAD-8 resin, desorb with diethyl ether, dry over magnesium sulphate and concentrate	GC/MS	Ellis et al. (1982)
Paper mill influent and effluent streams	Extract with dichloromethane at pH 3; evaporate extract		Turoski et al. (1983)

[a]Abbreviations: GC/MS, gas chromatography/mass spectrometry; GC/IR, gas chromatography/infrared detection; GC/FID, gas chromatography/flame ionization detection; NMR, nuclear magnetic resonance; HPLC/UV, high-performance liquid chromatography/ultraviolet detection; S, spectrophotometry; GC, gas chromatography; GC/EC, gas chromatography/electron capture detection; IR, infrared detection
[b]Limit of detection, 0.001 mg/l

3. Biological Data Relevant to the Evaluation of Carcinogenic Risk to Humans

3.1 Carcinogenicity studies in animals

(a) Oral administration

Mouse: Groups of 50 male and 50 female B6C3F$_1$ mice, six to seven weeks old, were fed diets containing USP-extra-grade eugenol (purity, >99%, with up to four trace impurities) at levels of 0, 3000 or 6000 mg/kg of diet for 103 weeks. Survival at 106 weeks was 41/50, 35/50 and 35/50 among control, low-dose and high-dose males, respectively; survival in females varied between 80-90%. The incidences of hepatocellular adenomas in males were 4/50 in control, 13/50 in low-dose and 10/49 in high-dose animals; and for females: 0/50 in control, 4/49 in low-dose and 3/49 in high-dose animals. The incidences of hepatocellular carcinomas in males were 10/50 in control, 20/50 in low-dose and 9/49 in high-dose animals; and for females: 2/50 in control, 3/49 in low-dose and 6/49 in high-dose animals. The total numbers of male mice with hepatocellular tumours were 14/50 in control, 28/50 in low-dose and 18/49 in high-dose animals; and those of females: 2/50 in control, 7/49 in low-dose and 9/49 in high-dose animals. For hepatocellular tumours in female mice, a trend test was significant ($p = 0.02$), as was a pair-wise comparison test between the high-dose and control groups ($p = 0.02$). For male mice, the trend test was not significant, but the pair-wise comparison test between the low-dose group and the control group was significant ($p = 0.004$). Tumours at other sites were not increased in treated animals compared with controls (National Toxicology Program, 1983). [The Working Group noted that the incidence of hepatocellular tumours in historical controls reported from the same laboratory showed little variation between experiments and that the incidences in male and female controls in the present experiment fell within this range, thus supporting the observed increase in incidences in treated animals.]

Groups of 30 female Charles River CD-1 mice, approximately eight weeks old, were fed a diet containing 0.5% eugenol (purity, >98%) for 12 months and maintained for an additional eight months before killing. In addition, some of the groups were given 0 or 0.05% phenobarbital in the drinking-water throughout the experiment. Control groups of 30 mice were also set up, some of which received 0.05% phenobarbital in the drinking-water. Survival at 18 months was 30/30 in untreated controls, 29/30 in phenobarbital controls, 29/30 in the eugenol group and 24/30 in the eugenol and phenobarbital group. There was no increase in the incidence of tumours in treated animals (Miller et al., 1983). [The Working Group noted the short duration of the experiments.]

A group of 59 male and 55 female Charles River CD-1 mice, four days old, received 2.5 µmol/g bw (0.4 mg/g bw) eugenol (purity, >98%) in trioctanoin by stomach tube twice weekly for five weeks; the experiment was terminated at 14 months. A group of 65 males and 52 females served as controls. No increased tumour incidence was found (Miller et al., 1983). [The Working Group noted the short treatment period, the short duration of the experiment and that survival data were not reported, although, under the same conditions, safrole and estragole gave positive results.]

Rat: Groups of 50 male Fischer 344 rats, six to seven weeks old, were fed diets containing 'USP-extra-grade' eugenol (purity, >99%, with up to four trace impurities) at concentrations of 3000 or 6000 mg/kg of diet for 103 weeks, while groups of 50 female Fischer 344 rats, six to seven weeks old, were fed diets containing eugenol at 6000 or 12 500 mg/kg of diet for 103 weeks. Groups of 40 rats of each sex served as controls. In both male and female rats, survival was better among high-dose animals than controls, and 52-88% of the high-dose animals survived to the end of the experiment. An increased incidence of endometrial stromal polyps was observed in treated females (controls, 6/40; low-dose, 6/50; high-dose, 16/50; $p = 0.02$ by a trend test), but the incidence in the high-dose group was not significant by a pair-wise comparison test (National Toxicology Program, 1983).

(b) *Skin application*

Mouse: In a two-stage mouse-skin assay, groups of 20 female ICR/Ha Swiss mice, eight weeks old, received either a single initiating skin application of 150 µg 7,12-dimethylbenz[a]anthracene (DMBA) in 0.1 ml acetone followed two to three weeks later by applications of 5 mg eugenol (purified by vacuum distillation [final purity not given]) in 0.1 ml acetone three times weekly, or treatment with eugenol alone three times per week. The experiments were terminated after 63 weeks. No skin tumour was found in 13/20 mice receiving eugenol alone and surviving to the end of the experiment. Mice receiving DMBA followed by eugenol had three papillomas (first appearing at 270 days); 14/20 mice survived to the end of the experiment. In 20 mice receiving DMBA followed by acetone, no skin papilloma was found, and nine mice survived to the end of the experiment. In the positive controls receiving DMBA followed by 25 µg croton resin, 19 mice had papillomas and three had carcinomas; 6/20 mice survived to the end of the experiment (Van Duuren et al., 1966). [The Working Group noted the short duration of the experiment and the small number of animals used.]

(c) *Intraperitoneal administration*

Mouse: A group of 52 male CD-1 mice received four intraperitoneal injections of eugenol (purity, >98%) in trioctanoin on days 1, 8, 15 and 22 after birth (total dose, 9.45 µmol [1.6 mg]). A group of 59 male mice that received 0.01 ml/g bw trioctanoin only served as vehicle controls and another group of 50 male mice served as untreated controls. Mortality data were not given. The study was terminated at 12 months; no increase in the incidence of tumours was observed (Miller et al., 1983). [The Working Group noted the short duration of the experiment, although under the same conditions, safrole and estragole gave positive results.]

(d) Carcinogenicity studies on possible metabolites

Mouse: A group of 53 male CD-1 mice received four *intraperitoneal* injections of 2',3'-epoxyeugenol in trioctanoin on days 1, 8, 15 and 22 after birth (total dose, 9.45 µmol [1.5 mg]). A group of 59 male mice that received 0.01 ml/g bw trioctanoin only served as vehicle controls, and another group of 50 male mice served as untreated controls. No mortality data were given. The study was terminated at 12 months; no increase in the incidence of tumours was observed (Miller *et al.*, 1983). [The Working Group noted the short duration of the experiment, although under the same conditions, safrole and estragole gave positive results.]

Groups of 40 eight-week old female CD-1 mice were treated *topically* on four days a week for six weeks with applications of 0.15 ml redistilled acetone containing 11.2 µmol [2 mg] 2',3'-epoxyeugenol or 0.15 ml acetone alone. Starting one week after the last dose of epoxide, all of the mice were treated topically twice weekly with 0.15 ml of 0.6% croton oil in acetone until termination of the experiment at 40 weeks, by which time 38-40 mice were alive in each group. 2',3'-Epoxyeugenol induced benign skin tumours (average number of tumours per mouse, 0.9) in 40% of mice, whereas benign skin tumours (average number of tumours per mouse, 0.1) were found in only 7% of acetone- and croton oil-treated controls (Miller *et al.*, 1983).

Rat: Groups of 20 male Fischer rats, five weeks of age, were given a total dose of 2 mmol [360 mg] per rat of 2',3'-epoxyeugenol in trioctanoin or trioctanoin alone as 20 *subcutaneous* injections of 0.2 ml in the right-hind leg. At 20 months, 16-19 rats per group were still alive; the surviving rats were killed at 24 months. Treated rats had two sarcomas at the injection site, while none were found in the controls. The incidences of tumours at other sites were not increased. The results were not significant (Miller *et al.*, 1983). [The Working Group noted that the degree of exposure was inadequate.]

3.2 Other relevant biological data

(a) Experimental systems

Toxic effects

Acute toxicity studies have shown oral LD_{50}s for eugenol of 2680 mg/kg bw (Jenner *et al.*, 1964) and 1930 mg/kg bw (Sober *et al.*, 1950) in rats, 3000 mg/kg bw in mice, and 2130 mg/kg bw in guinea-pigs (Jenner *et al.*, 1964).

In guinea-pigs, the possible skin sensitization potency of eugenol has been predicted to be 20% (Itoh, 1982) and 30 or 70% (according to the concentration) (Maurer *et al.*, 1979) in a maximization test. Eugenol exerts a low irritating effect on mucous membranes. A 5% emulsion of eugenol administered on a dog's tongue for five minutes produced erythema, ulcers and inflammatory infiltration (Lilly *et al.*, 1972).

Intravenous administration of varying doses (0.05-0.15 ml of a 1:20 or 1:60 dilution) in dogs led to a transient fall in blood pressure and a reduction of myocardial contractile force (Sticht & Smith, 1971). After single oral doses of 500 mg/kg bw eugenol, 2/4 dogs with predominant symptoms of vomiting died; all animals receiving doses of 250 mg/kg bw survived (Lauber & Hollander, 1950). Single and repeated oral administration of a 5% aqueous eugenol emulsion to dogs caused degeneration of the gastric mucosal cells (Hollander & Goldfischer, 1949).

Oral doses increasing from 1400-4000 mg/kg bw administered to rats over 34 days resulted in slight liver enlargement with yellow discolouration. Moderately severe hyperplasia and hyperkeratosis associated with focal ulceration were seen in the forestomach (Hagan et al., 1965).

During prescreening for a National Toxicology Program bioassay (described in section 3.1) in Fischer 344 rats and B6C3F1 mice, administration of increasing doses of eugenol in the diet over 14 days resulted in loss in body-weight gain at 25 000 mg/kg in rats and at 12 500 mg/kg in mice, and death at 100 000 mg/kg in male and female mice, female rats and 1/5 male rats. In the chronic feeding study, lasting 103 weeks, in which dietary levels of eugenol of 3000 or 6000 and 6000 or 12 500 mg/kg were fed to male and female rats, respectively, and 6000 or 3000 mg/kg to mice, no change in body-weight gain or in mean survival time was observed (National Toxicology Program, 1983).

Effects on reproduction and prenatal toxicity

No data were available to the Working Group.

Absorption, distribution, excretion and metabolism

No penetration of mouse skin was demonstrated after dermal application of eugenol (Meyer & Meyer, 1959).

Eugenol has been reported to be metabolized to 2',3'-epoxyeugenol and 2'3'-dihydroxy-2'3'-dihydroeugenol by cultured epithelial cells derived from rat livers (Delaforge et al., 1976; Dorange et al., 1977). The 2',3'-epoxide has also been detected when eugenol is incubated in the presence of microsomes derived from the livers of female mice (Miller et al., 1979).

Two metabolites of eugenol, 3-piperidyl-1-(3'-methoxy-4'-hydroxyphenyl)-1-propanone and 3-pyrrolidinyl-1-(3'-methoxy-4'-hydroxyphenyl)-1-propanone, have been isolated from rat urine (Green & Savage, 1978).

Inhibition of glucuronidation reactions was observed in liver slices incubated with varying concentrations of eugenol (Hartiala et al., 1966).

A slight increase in pentobarbital and ethanol sleeping-time was observed in rats that received intraperitoneal doses of 50 and 100 mg/kg bw eugenol, respectively (Seto & Keup, 1969).

Mutagenicity and other short-term tests (see also 'Appendix: Activity Profiles for Short-Term Tests', p. 332).

Reports concerning the activity of eugenol in the *Bacillus subtilis* rec$^+$/rec$^-$ DNA-repair assay are contradictory: positive results were reported by Sekizawa and Shibamoto (1982) and no activity was found by Yoshimura et al. (1981), both in the absence of a metabolic system (S9).

Eugenol was not mutagenic to *Escherichia coli* WP2 *uvr*A when tested in the presence or absence of S9 derived from the livers of Aroclor-induced rats (Sekizawa & Shibamoto, 1982).

Eugenol was not mutagenic to *Salmonella typhimurium* TA1530, TA1535, TA1537, TA1538, TA98 or TA100 in the presence of S9 from the livers of polychlorinated biphenyl-induced rats, or Aroclor-induced Syrian hamsters or mice. The procedures used included the standard plate incorporation assay, disc diffusion, pre-incubation and pulsed exposure (Delaforge et al., 1977; Dorange et al., 1977; Green & Savage, 1978; Miller et al., 1979; Rockwell & Raw, 1979; Swanson et al., 1979; Eder et al., 1980; Florin et al., 1980; Nestmann *et al.*, 1980; Rapson *et al.*, 1980; Yoshimura *et al.*, 1981; Eder *et al.*, 1982a,b; Pool & Lin, 1982; Sekizawa & Shibamoto, 1982; To *et al.*, 1982; Haworth *et al.*, 1983; National Toxicology Program, 1983). However, supplementation with 3'-phosphoadenosine-5'-phosphosulphate and S9 from the livers of Aroclor-induced rats was reported to result in significant, although not dose-dependent, mutagenicity to *S. typhimurium* TA1535 (To *et al.*, 1982).

β-Glucuronidase-treated urine (300 μl) of Sprague-Dawley rats given 0.5 ml eugenol by intubation was not mutagenic to *S. typhimurium* TA100 or TA98 in the presence of S9 derived from the liver of the same strain of rat (Rockwell & Raw, 1979). Similarly, eugenol was not mutagenic to *S. typhimurium* TA1950, TA1951, TA1952 or TA1964 in the host-mediated assay in which male C3H/HeJ mice were given 200 mg/kg bw intramuscularly (Green & Savage, 1978).

Eugenol [concentration not indicated] induced neither mutation nor gene conversion in *Saccharomyces cerevisiae* (Nestmann & Lee, 1983).

Eugenol induced chromosomal aberrations in Chinese hamster ovary cells in the absence of an exogenous metabolic system (Stich *et al.*, 1981). In a second study, chromosomal aberrations were induced by eugenol in Chinese hamster ovary cells only in the presence of S9 from Aroclor-induced rats; a small increase in the incidence of sister chromatid exchanges was also observed in the presence or absence of S9 (National Toxicology Program, 1983).

2',3'-Epoxyeugenol was mutagenic to *S. typhimurium* TA1535 and TA100 in the absence of an exogenous metabolic system but not to strains TA1537, TA1538 or TA98 in the presence or absence of a metabolic system (Delaforge *et al.*, 1977; Dorange *et al.*, 1977; Miller *et al.*, 1979; Swanson *et al.*, 1979).

Two metabolites of eugenol present in rat urine, 3-piperidyl-1-(3'-methoxy-4'-hydroxyphenyl)-1-propanone and 3-pyrrolidinyl-1-(3'-methoxy-4'-hydroxyphenyl)-1-propanone, were not mutagenic to *S. typhimurium* in the presence or absence of S9 from the livers of male C3H/HeJ mice. Similarly these metabolites were not mutagenic to *S. typhimurium* in the host-mediated assay in male C3H/HeJ mice at doses of 200 mg/kg bw (Green & Savage, 1978).

(b) Humans

Toxic effects

Many reports have been published on the high potential of eugenol and of clove-leaf oil (approximately 85% eugenol) for skin sensitization (for a review, see Rothenstein *et al.*, 1983). Patch tests for eugenol in patients suffering from 'cosmetic dermatitis' were positive in 2.6% (4/155) of cases (Itoh, 1982).

No data were available to the Working Group on effects on reproduction and prenatal toxicity, on absorption, distribution, excretion and metabolism, or on mutagenicity and chromosomal effects.

3.3 Case reports and epidemiological studies of carcinogenicity to humans

No data were available to the Working Group.

4. Summary of Data Reported and Evaluation

4.1 Exposure data

Eugenol occurs widely as a component of essential oils and is a major constituent of clove oil. It has been used since at least the nineteenth century, primarily as a flavouring agent, in a variety of foods and pharmaceutical products, and as an analgesic in dental materials.

4.2 Experimental data

Eugenol was tested in mice of one strain and in rats of one strain by oral administration of a diet containing a high dose of eugenol. In mice, there was a significant increase in the incidence of liver tumours in females; in males, the increase was significant only for those receiving the lower dose. In rats, no increased incidence of tumours was observed. Other studies in mice by oral administration, skin application and intraperitoneal injection were inadequate for an evaluation of carcinogenicity, mainly due to the short duration of treatment.

Eugenol gave both positive and negative results in tests for DNA damage in bacteria. It was not mutagenic in several studies in bacteria. The compound was not active in a host-mediated assay in mice, nor was the urine of rats treated with eugenol mutagenic. Eugenol induced chromosomal aberrations and a small increase in sister chromatid exchanges in mammalian cells *in vitro*.

In one two-stage mouse-skin assay, 2',3'-epoxyeugenol, an in-vitro metabolite of eugenol, showed initiating activity.

Overall assessment of data from short-term tests: eugenol[a]

	Genetic activity			Cell transformation
	DNA damage	Mutation	Chromosomal effects	
Prokaryotes	?	—		
Fungi/Green plants				
Insects				
Mammalian cells (*in vitro*)			+	
Mammals (*in vivo*)				
Humans (*in vivo*)				
Degree of evidence in short-term tests for genetic activity: *Inadequate*				Cell transformation: No data

[a]The groups into which the table is divided and the symbols are defined on pp. 17-18 of the Preamble; the degrees of evidence are defined on p. 18.

2',3'-Epoxyeugenol was mutagenic to bacteria. Two urinary metabolites of eugenol, 3-piperidyl-1-(3'-methoxy-4'-hydroxyphenyl)-1-propanone and 3-pyrrolidinyl-1-(3'-methoxy-4'-hydroxyphenyl)-1-propanone were not mutagenic to bacteria or in the host-mediated assay.

4.3 Human data

No case report or epidemiological study of the carcinogenicity of eugenol to humans was available to the Working Group.

4.4 Evaluation[1]

There is *limited evidence* for the carcinogenicity of eugenol to experimental animals.

In the absence of epidemiological data, no evaluation could be made of the carcinogenicity of eugenol to humans.

5. References

Adams, R.P., von Rudloff, E. & Hogge, L. (1981) The volatile terpenoids of *Juniperus blancoi* and its affinities with other entire leaf margin junipers of North America. *J. nat. Prod.*, 44, 21-26

American Pharmaceutical Association (1982) *Handbook of Nonprescription Drugs*, 7th ed., Washington DC, pp. 463, 465, 468-469

Analytical Methods Committee (1981) Application of gas-liquid chromatography to the analysis of essential oils. Part IX. Determination of eugenol in oil of cinnamon bark. *Analyst*, 106, 456-460

Bejnarowicz, E.A. & Kiroh, E.П. (1960) Gas chromatographic analysis of oil of nutmeg. *J. pharm. Sci.*, 52, 988-993

Burachinsky, B.V., Dunn, H. & Ely, J.K. (1981) Inks. In: Grayson, M., ed., *Kirk-Othmer Encyclopedia of Chemical Technology*, 3rd ed., Vol. 13, New York, John Wiley & Sons, p. 379

Buttery, R.G. & Kamm, J.A. (1980) Volatile components of alfalfa: Possible insect host plant attractants. *J. agric. Food Chem.*, 28, 978-981

Buttery, R.G., Soderstrom, E.L., Seifert, R.M., Ling, L.C. & Haddon, W.F. (1980) Components of almond hulls: Possible navel orangeworm attractants and growth inhibitors. *J. agric. Food Chem.*, 28, 353-356

Chen, Y., Yu, J. & Fang, H. (1983) Studies on Chinese curcuma. III. Comparison of the volatile oil and phenolic constituents from the rhizome and the tuber of *Curcuma longa* (Chin.). *Zhongyao Tongbao*, 8, 27-29 [*Chem. Abstr.*, 99, 19664k]

---

[1]For definitions of the italicized terms, see Preamble, pp. 15-16.

Cheng, Y.-S. & Chao, Y.-L. (1979) *Absolute of Siu-eng flower* (Jasminium odoratissimum). In: *7th International Congress on Essential Oils*, Vol. 7, Tokyo, Japan Flavor Fragrance Manufacturers' Association, pp. 467-469 [*Chem. Abstr.*, *92*, 37772d]

Chialva, F., Gabri, G., Liddle, P.A.P. & Ulian, F. (1982) Qualitative evaluation of aromatic herbs by direct headspace GC analysis. Applications of the method and comparison with the traditional analysis of essential oils. *J. high Resolut. Chromatogr.*, *5*, 182-188

Chiang, H.C., Wang, P.L. & Huang, K.F. (1983) Quantitative analysis of clove oil by NMR spectrometry. *J. Chin. chem. Soc.*, *30*, 117-120 [*Chem. Abstr.*, *99*, 110521x]

Chogo, J.B. & Crank, G. (1981) Chemical composition and biological activity of the Tanzanian plant *Ocimum suave*. *J. nat. Prod.*, *44*, 308-311

Chung, T.-Y., Hayase, F. & Kato, H. (1983) Volatile components of ripe tomatoes and their juices, purées and pastes. *Agric. biol. Chem.*, *47*, 343-351

Delaforge, M., Janiaud, P., Dorange, J.L., Morizot, J.P. & Padieu, P. (1977) Metabolic activation of a natural promutagen, eugenol, in replicative cultures of adult rat liver epithelial cells (Fr.). *C.R. Soc. Biol. (Paris)*, *171*, 100-107

Dorange, J.-L., Delaforge, M., Janiaud, P. & Padieu, P. (1977) Mutagenic action of metabolites of the epoxide-diol pathway of safrole and analogues. Study with *Salmonella typhimurium* (Fr.). *C.R. Soc. Biol.*, *171*, 1041-1048

Eder, E., Neudecker, T., Lutz, D. & Henschler, D. (1980) Mutagenic potential of allyl and allylic compounds. *Biochem. Pharmacol.*, *29*, 993-998

Eder, E., Henschler, D. & Neudecker, T. (1982a) Mutagenic properties of allylic and α,β-unsaturated compounds: Consideration of alkylating mechanisms. *Xenobiotica*, *12*, 831-848

Eder, E., Neudecker, T., Lutz, D. & Henschler, D. (1982b) Correlation of alkylating and mutagenic activities of allyl and allylic compounds: Standard alkylation test vs. kinetic investigation. *Chem.-biol. Interactions*, *38*, 303-315

Ellis, D.D., Jone, C.M., Larson, R.A. & Schaeffer, D.J. (1982) Organic constituents of mutagenic secondary effluents from wastewater treatment plants. *Arch. environ. Contam. Toxicol.*, *11*, 373-382

Flavor and Extract Manufacturers' Association of the United States (1978) *Scientific Literature Review of Eugenol and Related Substances in Flavor Usage*, Vol. 1 (*PB-283 501*), Washington DC, US Food and Drug Administration, pp. 57-61, 65-67

Fleisher, A. (1981) Essential oils from two varieties of *Ocimum basilicum* L. grown in Israel. *J. Sci. Food Agric.*, *32*, 1119-1122

Florin, I., Rutberg, L., Curvall, M. & Enzell, C.R. (1980) Screening of tobacco smoke constituents for mutagenicity using the Ames' test. *Toxicology*, *18*, 219-232

Fujita, S., Asami, Y. & Nozaki, K. (1980) The constituents of the essential oils from *Foeniculum vulgare* Miller (miscellaneous contributions to the essential oil of the plants from various territories). Part XLVI (Jpn.). *Nippon Nogei Kogaku Kaishi*, 54, 765-767 [*Chem. Abstr.*, 94, 90020z]

Fujita, S.-I. & Fujita, Y. (1972) Comparative biochemical and chemo-taxonomical studies of the essential oils of *Magnolia salicifolia* Maxim. *Chem. pharm. Bull.*, 20, 2251-2255

Furia, T.E. & Bellanca, N., eds (1975) *Fenaroli's Handbook of Flavor Ingredients*, 2nd ed., Vol. 2, Cleveland, OH, CRC Press Inc., p. 198

Gracza, L. (1980) HPLC determination of phenylpropane derivatives in drugs and drug preparations. Part 3. Analysis of phenylpropane derivatives (Ger.). *Dtsch. Apoth.-Ztg.*, 120, 1859-1863 [*Chem. Abstr.*, 94, 20478f]

Green, N.R. & Savage, J.R. (1978) Screening of safrole, eugenol, their ninhydrin positive metabolites and selected secondary amines for potential mutagenicity. *Mutat. Res.*, 57, 115-121

Gupta, R. & Baslas, R.K. (1978) Chromatographic separation and identification of some constituents of essential oil of the seed of *Apium graveolens*. *Indian Perfum.*, 22, 175-178 [*Chem. Abstr.*, 92, 82218p]

Gupta, R., Agarwal, M. & Baslas, R.K. (1978) Chromatographic separation and identification of various constituents of essential oil from the bulb of *Melaxis accuminata*. *Indian Perfum.*, 22, 287-288 [*Chem. Abstr.*, 92, 82214j]

Hagan, E.C., Jenner, P.M., Jones, W.I., Fitzhugh, O.G., Long, E.L., Brouwer, J.G. & Webb, W.K. (1965) Toxic properties of compounds related to safrole. *Toxicol. appl. Pharmacol.*, 7, 18-24

Hake, C.L. & Rowe, V.K. (1963) Ethers. In: Patty, F.A., ed., *Industrial Hygiene and Toxicology*, 2nd rev. ed., Vol. 2, New York, Interscience, pp. 1690-1692

Hartiala, K.J.W., Pulkkinen, M. & Ball, P. (1966) Inhibition of β-D-glucosiduronic acid conjugation by eugenol. *Nature*, 210, 739-740

Hashimoto, Y., Kawanishi, K., Tomita, H., Uhara, Y. & Moriyasu, M. (1981) Histochemical chromatography. A new technique for identifying crystal and oil components in cell tissue by combination of a micromanipulator and HPLC. *Anal. Lett.*, 14, 1525-1529

Hawley, G.G., ed. (1981) *The Condensed Chemical Dictionary*, 10th ed., New York, Van Nostrand Reinhold Co., p. 445

Haworth, S., Lawlor, T., Mortelmans, K., Speck, W. & Zeiger, E. (1983) *Salmonella* mutagenicity test results for 250 chemicals. *Environ. Mutagenesis*, 5 (Suppl. 1), 3-142

Hollander, F. & Goldfischer, R.L. (1949) Histological study of the destruction and regeneration of the gastric mucous barrier following application of eugenol. Preliminary report. *J. natl Cancer Inst.*, 10, 339-354

Horvat, R.J., Senter, S.D. & Dekazos, E.D. (1983) GLC-MS analysis of volatile constituents in Rabbiteye blueberries. *J. Food Sci.*, 48, 278-279

IARC (1976) *IARC Monographs on the Evaluation of Carcinogenic Risk of Chemicals to Man*, Vol. 10, *Some Naturally Occurring Substances*, Lyon, pp. 231-244

Ismail, H., Williams, A.A. & Tucknott, O.G. (1980) The flavor components of plum. An examination of the aroma components present in a distillate obtained from fermented plum juice. *Z. Lebensm.-Unters. Forsch.*, *171*, 24-27 [*Chem. Abstr.*, *93*, 166042k]

Itoh, M. (1982) Sensitization potency of some phenolic compounds with special emphasis on the relationship between chemical structure and allergenicity. *J. Dermatol.*, *9*, 223-233

Jenner, P.M., Hagan, E.C., Taylor, J.M., Cook, E.L. & Fitzhugh, O.G. (1964) Food flavourings and compounds of related structure. I. Acute oral toxicity. *Food Cosmet. Toxicol.*, *2*, 327-343

Kameoka, H., Tsujino, H. & Aso, Y. (1983) Constituents of essential oil of leaves of *Quercus dentata* Thunb. (Jpn.). *Nippon Nogei Kagaku Kaishi*, *57*, 135-138 [*Chem. Abstr.*, *98*, 176173w]

Keller, U., Flath, R.A., Mon, T.R. & Teranishi, R. (1981) *Volatiles from red pepper* (Capsicum spp.). In: *Quality of Selected Fruits and Vegetables* (ACS Symp. Ser. *170*), Washington DC, American Chemical Society, pp. 137-146

Khosla, M.K., Pushpangadan, P., Thappa, R.K. & Sobti, S.N. (1980) Search for new aroma chemicals from *Ocimum* species. III. Studies on the genetic variability for essential oil and other allied characters of South American species *O. micranthum* Wild. *Indian Perfum.*, *24*, 148-152 [*Chem. Abstr.*, *94*, 99827a]

Kikuchi, M. (1981) The constituents of the flower of *Ligustrum obtusifolium* Sieb et Zucc (Jpn.). *Nippon Nogei Kagaku Kaishi*, *55*, 821-823 [*Chem. Abstr.*, *95*, 217663q]

Kikuchi, M., Kataoka, H. & Kurihara, T. (1981a) Studies on the constituents of flowers. XV. On the components of the flower of *Ligustrum japonicum* Thunb. (Jpn.). *Yakugaku Zasshi*, *101*, 575-578 [*Chem. Abstr.*, *95*, 111727y]

Kikuchi, T., Kadota, S., Yanada, K., Watanabe, K., Yosizaki, M. & Kimura, M. (1981b) Studies on the constituents of crude drug 'Shin-i' (dried buds of *Magnolia salicifolia* Maxim.) (Jpn.). *Wakanyaku Shinpojumu* (*Kiroku*), *14*, 101-104 [*Chem. Abstr.*, *96*, 223081k]

Kurihara, T. & Kikuchi, M. (1980a) Studies on the constituents of flowers. XIII. On the components of the flower of *Cytisus scoparius* Link (Jpn.). *Yakugaku Zasshi*, *100*, 1054-1057 [*Chem. Abstr.*, *93*, 235183w]

Kurihara, T. & Kikuchi, M. (1980b) Studies on the constituents of flowers. XIV. On the components of the flower of *Ligustrum ovalifolium* Hassk (Jpn.). *Yakugaku Zasshi*, *100*, 1161-1163 [*Chem. Abstr.*, *94*, 27415s]

Lauber, F.U. & Hollander, F. (1950) Toxicity of the mucigogue, eugenol, administered by stomach tube to dogs. *Gastroenterology*, *15*, 481-486

Lehtonen, M. (1980) Gas chromatographic determination of phenols as 2,4-dinitrophenyl ethers using glass capillary columns and an electron-capture detector. *J. Chromatogr.*, *202*, 413-421

Lehtonen, M. (1983) Gas-liquid chromatographic determination of volatile phenols in matured distilled alcoholic beverages. *J. Assoc. off. anal. Chem.*, 66, 62-70

Lilly, G.E., Cutcher, J.L. & Jendresen, M.D. (1972) Reaction of oral mucous membranes to selected dental materials. *J. biomed. Mater. Res.*, 6, 545-551

Lin, Z.-K. & Hua, Y.-F. (1980) Studies on the essential oils in the leaves of *Cinnamonum pauciflorum* Nees (Chin.). *Chih Wu Hsueh Pao*, 22, 252-256 [*Chem. Abstr.*, 94, 52680g]

Lockwood, G.B. (1979) Phenylpropanoids from a Nigerian sample of *Cinnamonum cassia*. *J. Pharm. Pharmacol.*, 31 (Suppl.), 1-8 [*Chem. Abstr.*, 92, 143237t]

Maurer, T., Thomann, P., Weirich, E.G. & Hess, R. (1979) Predictive evaluation in animals of the contact allergenic potential of medically important substances. II. Comparison of different methods of cutaneous sensitization with 'weak' allergens. *Contact Dermatitis*, 5, 1-10

Metcalf, R.L. (1981) *Insect control technology*. In: Grayson, M., ed., *Kirk-Othmer Encyclopedia of Chemical Technology*, 3rd ed., Vol. 13, New York, John Wiley & Sons, p. 482

Meyer, F. & Meyer, E. (1959) Percutaneous absorption of ethereal oils and their ingredients (Ger.). *Arzneimittelforschung*, 9, 516-519

Miller, E.C., Swanson, A.B., Phillips, D.H., Fletcher, T.L., Liem, A. & Miller, J.A. (1983) Structure-activity studies of the carcinogenicities in the mouse and rat of some naturally occurring and synthetic alkenylbenzene derivatives related to safrole and estragole. *Cancer Res.*, 43, 1124-1134

Miller, J.A., Swanson, A.B. & Miller, E.C. (1979) *The metabolic activation of safrole and related naturally occurring alkenylbenzenes in relation to carcinogenesis by these agents*. In: Miller, E.C., Miller, J.A., Hirono, I., Sugimura, T. & Takayama, S., eds, *Naturally Occurring Carcinogens-Mutagens and Modulators of Carcinogenesis*, Tokyo, Japan Scientific Societies Press, pp. 111-125

Miller, R.A., Bussell, N.E., Ricketts, C.K. & Jordi, H. (1978) *Analysis and Purification of Eugenol*, Washington DC, Army Institute of Dental Research, p. 1

Miyazawa, M., Tokugawa, M. & Kameoka, H. (1981) The constituents of the essential oil from *Erigeron philadelphicus*. *Agric. biol. Chem.*, 45, 507-510

Nagaraj, G. & Chakraborty, M.K. (1979) Lipophilic constituents of Natu tobacco. II. Phenols. *Tob. Res.*, 5, 147-148 [*Chem. Abstr.*, 92, 143522g]

Nahrstedt, A., Vetter, U. & Hammerschmidt, F.J. (1981) Knowledge of the steam distillate from the leaves of *Juglans regia* (Ger.). *Planta med.*, 42, 313-332 [*Chem. Abstr.*, 95, 156438w]

National Institute for Occupational Safety and Health (1980) *Projected Number of Occupational Exposures to Chemical and Physical Hazards*, Cincinnati, OH, p. 69

National Institute for Occupational Safety and Health (1981) *National Occupational Hazard Survey* (microfiche), Cincinnati, OH, p. 304 (generics excluded)

National Research Council (1981) *Food Chemicals Codex*, 3rd ed., Washington DC, National Academy Press, pp. 353, 380-381, 583, 687

National Research Council/National Academy of Sciences (1979) *The 1977 Survey of Industry on the Use of Food Additives*, Part 1 (*PB80-113418*), Springfield, VA, National Technical Information Service, p. 455

National Toxicology Program (1983) *Carcinogenesis Studies of Eugenol (CAS no. 97-53-0) in F344/N Rats and B6C3F$_1$ Mice (Feed Studies)* (NTP/TR No. 223; NIH No. 84-1779), Washington DC, US Department of Health and Human Services

Nestmann, E.R. & Lee, E.G.-H. (1983) Mutagenicity of constituents of pulp and paper mill effluent in growing cells of *Saccharomyces cerevisiae*. *Mutat. Res.*, *119*, 273-280

Nestmann, E.R., Lee, E.G.-H., Matula, T.I., Douglas, G.R. & Mueller, J.C. (1980) Mutagenicity of constituents identified in pulp and paper mill effluents using the *Salmonella*/mammalian-microsome assay. *Mutat. Res.*, *79*, 203-212

Nicollier, G. & Thompson, A.C. (1981) Essential oil and terpenoids of *Mikania micrantha*. *Phytochemistry*, *20*, 2587-2588

Nizharadze, A.N. & Bagaturiya, N.S. (1980) Eugenol content in essential oils from *Ocimum gratissimum* (Russ.). *Maslo-Zhir. Prom-st.*, *4*, 33 [*Chem. Abstr.*, *93*, 66048y]

Oh, K., Nozawa, T., Ishii, S. & Kameoka, H. (1979) *Essential oils of* Trachycarpus excelsa *and* T. fortunei (Jpn.). In: *Koen Yoshishu-Koryo, Terupen oyobi Seiyu Kagaku ni kansuru Toronkai*, 23rd, Tokyo, Chemical Society of Japan, 58-59 [*Chem. Abstr.*, *92*, 169032x]

Opdyke, D.L.J. (1975) Monographs on fragrance raw materials. Eugenol. *Food Cosmet. Toxicol.*, *13*, 545-546

Paffenbarger, G.C. & Rupp, N.W. (1979) Dental materials. In: Grayson, M., ed., *Kirk-Othmer Encyclopedia of Chemical Technology*, 3rd ed., Vol. 7, New York, John Wiley & Sons, pp. 463-464, 498-499

Pool, B.L. & Lin, P.Z. (1982) Mutagenicity testing in the *Salmonella typhimurium* assay of phenolic compounds and phenolic fractions obtained from smokehouse smoke condensates. *Food Chem. Toxicol.*, *20*, 383-391

Pu, Q., Ji, X., Xu, P., Huang, L., Guo, Q. & Chen, X. (1983) Studies on the constituents of the essential oil of *Bupleurum chinense* DC. (Chin.). *Huaxue Xuebao*, *41*, 559-561 [*Chem. Abstr.*, *99*, 119332u]

Rabha, L.C., Baruah, A.K.S. & Bordoloi, D.N. (1979) Search for aroma chemicals of commercial value from resources of North East India. *Indian Perfum.*, *23*, 178-183 [*Chem. Abstr.*, *93*, 79858n]

Rapson, W.H., Nazar, M.A. & Butsky, V.V. (1980) Mutagenicity produced by aqueous chlorination of organic compounds. *Bull. environ. Contam. Toxicol.*, *24*, 590-596

Rockwell, P. & Raw, I. (1979) A mutagenic screening of various herbs, spices, and food additives. *Nutr. Cancer*, *1*, 10-15

Rogers, J.A., Jr (1981) Oils, essential. In: Grayson, M., ed., *Kirk-Othmer Encyclopedia of Chemical Technology*, 3rd ed., Vol. 16, New York, John Wiley & Sons, p. 322

Rothenstein, A.S., Booman, K.A., Dorsky, J., Kohrman, K.A., Schwoeppe, E.A., Sedlak, R.I., Steltenkamp, R.J. & Thompson, G.R. (1983) Eugenol and clove leaf oil: A survey of consumer patch-test sensitization. *Food Chem. Toxicol.*, 21, 727-733

Rui, H., Yu, Q., He, Q., Yuan, M., Jiang, Y. & Zhou, Y. (1982) A preliminary study on the chemical components of the essential oils of Qian Nian Jian (*Homalomena occulta*), Hong Mu Xiang (*Kadsura longipedunculota*) and Hong Do Kou (*Alpinia galanga*) (Chin.). *Zhongcaoyao*, 13, 43 [*Chem. Abstr.*, 97, 150558e]

Sakai, T., Takagi, K. & Kameoka, H. (1979) *Essential oils of* Menyantes triforiata (Jpn.). In: *Koen Yoshishu-Koryo, Terupen oyobi Seiyu Kagaku ni kansuru Toronkai*, 23rd, Tokyo, Chemical Society of Japan, pp. 206-208 [*Chem. Abstr.*, 92, 152874a]

Sakata, K., Yamamoto, H., Tanaka, H. & Shinozuka, M. (1982) Studies of components of raw corn (*Zea mays* L.) and corn silage. 7. Components of the nonvolatile acidic fraction of raw corn and the volatile phenolic fraction of corn silage (Jpn.). *Nippon Nogei Kagaku Kaishi*, 56, 451-453 [*Chem. Abstr.*, 97, 108762s]

Sarin, Y.K., Puri, S.C., Kapahi, B.K. & Singh, J.P. (1982) Seasonal variability in phenolic constituents of *Artemisia scoparia* oil. *Indian Perfum.*, 26, 145-148 [*Chem. Abstr.*, 99, 67600h]

Sastry, C.S.P., Rao, K.E. & Prasad, U.V. (1982) Spectrophotometric determination of some phenols with sodium metaperiodate and aminophenols. *Talanta*, 29, 917-920 [*Chem. Abstr.*, 98, 100542v]

Sekizawa, J. & Shibamoto, T. (1982) Genotoxicity of safrole-related chemicals in microbial test systems. *Mutat. Res.*, 101, 127-140

Seto, T.A. & Keup, W. (1969) Effects of alkylmethoxybenzene and alkylmethylenedioxybenzene essential oils on pentobarbital and ethanol sleeping time. *Arch. int. Pharmacodyn.*, 180, 232-240

Sharma, G.P. & Garg, B.D. (1980) Chemical examination of the essential oil from *Glossocardia bosvallia* DC. *J. Sci. Res.*, 2, 37-41 [*Chem. Abstr.*, 94, 36103r]

Shibamoto, T. & Nishimura, O. (1982) Isolation and identification of phenols in oil of vetiver. *Phytochemistry*, 21, 793

Shirokov, E.P., Badgaa, D. & Kobozev, I.V. (1980) Essential oil content in plants used in the production of tonics (Russ.). *Izv. Timiryazevsk. S-kh. Akad.*, 3, 187-191 [*Chem. Abstr.*, 93, 41560w]

Sober, H.A., Hollander, F. & Sober, E.K. (1950) Toxicity of eugenol: Determination of LD50 on rats. *Proc. Soc. exp. Biol. Med.*, 73, 148-151

Stich, H.F., Stich, W. & Lam, P.P.S. (1981) Potentiation of genotoxicity by concurrent application of compounds found in betel quid: Arecoline, eugenol, quercetin, chlorogenic acid and Mn^{2+}. *Mutat. Res.*, 90, 355-363

Sticht, F.D. & Smith, R.M. (1971) Eugenol: Some pharmacologic observations. *J. dent. Res.*, *50*, 1531-1535

Swanson, A.B., Chambliss, D.D., Blomquist, J.C., Miller, E.C. & Miller, J.A. (1979) The mutagenicities of safrole, estragole, eugenol, *trans*-anethole, and some of their known or possible metabolites for *Salmonella typhimurium* mutants. *Mutat. Res.*, *60*, 143-153

To, L.P., Hunt, T.P. & Andersen, M.E. (1982) Mutagenicity of *trans*-anethole, estragole, eugenol, and safrole in the Ames *Salmonella typhimurium* assay. *Bull. environ. Contam. Toxicol.*, *28*, 647-654

Toulemonde, B. & Richard, H.M.J. (1983) Volatile constituents of dry elder (*Sambucus nigra* L.) flowers. *J. agric. Food Chem.*, *31*, 365-370

Turner, N.J. (1966) *Fungicides*. In: Kirk, R.E. & Othmer, D.F., eds, *Kirk-Othmer Encyclopedia of Chemical Technology*, 2nd ed., Vol. 10, John Wiley & Sons, pp. 234-235

Turoski, V.E., Woltman, D.L. & Vincent, B.F. (1983) Determination of organic priority pollutants in the paper industry by GC/MS. *Tappi J.*, *66*, 89-90

US Department of the Treasury (1983) Formulas for denatured alcohol and rum. *US Code Fed. Regul., Title 27*, Parts 21, 212; *Fed. Regist.*, *48* (No. 107), pp. 24672-24675, 24680-24681, 24687

US Food and Drug Administration (1979) External analgesic drug products for over-the-counter human use; Establishment of a monograph and notice of proposed rulemaking. *US Code Fed. Regul., Title 21*, Part 348; *Fed. Regist.*, *44* (No. 234), pp. 69768-69771, 69851-69852

US Food and Drug Administration (1980a) Food and drugs. *US Code Fed. Regul., Title 21*, Part 184.1257

US Food and Drug Administration (1980b) Food and drugs. *US Code Fed. Regul., Title 21*, Part 177.2800

US Food and Drug Administration (1982) Drug products for the relief of oral discomfort for over-the-counter human use; Establishment of a monograph. *US Code Fed. Regul., Title 21*, Part 354; *Fed. Regist.*, *47* (No. 101), pp. 22712-22715, 22718-22719, 22724-22728

US Food and Drug Administration (1983) External analgesic drug products for over-the-counter human use; Tentative final monograph. *US Code Fed. Regul., Title 21*, Part 348; *Fed. Regist.*, *48* (No. 27), pp. 5852-5853, 5865

US International Trade Commission (1983) *Synthetic Organic Chemicals, US Production and Sales, 1982* (*USITC Publication 1422*), Washington DC, US Government Printing Office, pp. 125, 127

US Pharmacopeial Convention, Inc. (1980) *The United States Pharmacopeia*, 20th rev., Rockville, MD, p. 315

US Tariff Commission (1945) *Synthetic Organic Chemicals, US Production and Sales, 1941-43 (Report No. 153, Second Ser.)*, Washington DC, US Government Printing Office, p. 109

US Tariff Commission (1974) *Synthetic Organic Chemicals, US Production and Sales, 1972 (TC Publication 681)*, Washington DC, US Government Printing Office, pp. 122, 124

Van Duuren, B.L., Sivak, A., Segal, A., Orris., L. & Langseth, L. (1966) The tumor-promoting agents of tobacco leaf and tobacco smoke condensate. *J. natl Cancer Inst.*, 37, 519-526

Van Ness, J.H. (1983) *Vanillin.* In: Grayson, M., ed., *Kirk-Othmer Encyclopedia of Chemical Technology*, 3rd ed., Vol. 23, New York, John Wiley & Sons, pp. 710-711

Varagnat, J. (1981) *Hydroquinone, resorcinol, and catechol.* In: Grayson, M., ed., *Kirk-Othmer Encyclopedia of Chemical Technology*, 3rd ed., Vol. 13, New York, John Wiley & Sons, pp. 43-44

Verschueren, K. (1977) *Handbook of Environmental Data on Organic Chemicals*, New York, Van Nostrand Reinhold Co., p. 340

Verzele, M., Maes, G., Vuye, A., Godefroot, M., Van Alboom, M., Vervisch, J. & Sandra, P. (1981) Chromatographic investigation of jasmin absolutes. *J. Chromatogr.*, 205, 367-386

WHO (1967) *Toxicological Evaluation of Some Flavouring Substances and Non-Nutritive Sweetening Agents (FAO Nutrition Meetings Report Series No. 44A; WHO/Food Add./68.33)*, Geneva, pp. 41-43

Windholz, M., ed. (1983) *The Merck Index*, l0th ed., Rahway, NJ, Merck & Co., p. 563

Wu, Y.-L. & Fang, H.-J. (1980) Constituents of the essential oil from the flowers of *Lonicera japonica* Thunb. (Chin.). *Hua Hsueh Hsueh Pao*, 38, 573-580 [*Chem. Abstr.*, 94, 99774f]

Yajima, I., Yanai, T., Nakamura, M., Sakakibara, H., Uchida, H. & Hayashi, K. (1983) Volatile flavor compounds of boiled buckwheat flour. *Agric. biol. Chem.*, 47, 729-738

Yamaguchi, K. & Shibamoto, T. (1980) Volatile constituents of the chestnut flower. *J. agric. Food Chem.*, 28, 82-84

Yoshida, T. (1979) *Essential oil of* Laurus nobilis L. (Jpn.). In: *Koen Yoshishu-Koryo, Terupen oyobi Seiyu Kagaku ni kansuru Toronkai*, 23rd, Tokyo, Chemical Society of Japan, pp. 184-185 [*Chem. Abstr.*, 92, 143282d]

Yoshimura, H., Nakamura, M. & Koeda, T. (1981) Mutagenicity screening of anesthetics for fishes. *Mutat. Res.*, 90, 119-124

Yu, J. & Fang, H. (1981) Studies on the essential oils of clove buds and clove leaves (Chin.). *Zhongcaoyao*, 12, 340-342 [*Chem. Abstr.*, 96, 139645q]

ALDEHYDES

ACETALDEHYDE

1. Chemical and Physical Data

1.1 Synonyms and trade names

Chem. Abstr. Services Reg. No.: 75-07-0

Chem. Abstr. Name: Acetaldehyde

IUPAC Systematic Name: Acetaldehyde

Synonyms: Acetic aldehyde; acetylaldehyde; 'aldehyde'; ethanal; ethylaldehyde; ethyl aldehyde; NCI-C56326

1.2 Structural and molecular formulae and molecular weight

$$CH_3-CHO$$

C_2H_4O Mol. wt: 44.1

1.3 Chemical and physical properties of the pure substance

From Hagemeyer (1978), unless otherwise specified

(a) *Description*: Colourless, mobile liquid with a pungent, suffocating odour

(b) *Boiling-point*: 20.16°C

(c) *Melting-point*: -123.5°C

(d) *Density*: d_4^{20} 0.7780

(e) *Refractive index*: n_D^{20} 1.33113

(f) *Spectroscopy data:* Infrared spectral data have been reported (National Research Council, 1981)

(g) *Solubility*: Miscible with water and most common organic solvents

(h) *Volatility*: Vapour pressure, 755 mm Hg at 20°C

(i) *Stability*: Flash-point (closed-cup), -38°C; decomposes above 400°C to form principally methane and carbon monoxide

(j) *Reactivity*: A highly reactive compound which undergoes numerous condensation, addition and polymerization reactions

(k) *Conversion factor*: 1 ppm = 1.8 mg/m^3 at 760 mm Hg and 25°C (Fassett, 1963)

1.4 Technical products and impurities

Acetaldehyde is available in the USA with the following typical specifications: purity, 99.5% min; acidity (as acetic acid), 0.1% max; and specific gravity, 0.780-0.790 at 15.6°C (Eastman Kodak Company, 1982). Typical properties are: boiling-point, 21°C; freezing-point, -123°C; autoignition temperature, 193°C; specific gravity, 0.79; refractive index, n_D^{20} 1.33; flash- and fire-points, -39°C; and miscible with water, diethyl ether and ethanol (Eastman Kodak Company, 1983).

In the USA, to meet the requirements of the Food Chemicals Codex, acetaldehyde must pass an infrared identification test and meet the following specifications: purity, 99.0% min; acidity (as acetic acid), 0.1% max; nonvolatile residue, 0.006% max; and specific gravity (0°/20°C), 0.804-0.811 (National Research Council, 1981).

Acetaldehyde is available as a technically pure grade (purity, 99.5%) in western Europe.

2. Production, Use, Occurrence and Analysis

2.1 Production and use

(a) *Production*

Acetaldehyde was first prepared in 1774 by Scheele by oxidation of ethanol using manganese dioxide and sulphuric acid. Although it has been produced commercially in the USA by the controlled hydration of acetylene and by the air oxidation of butane as well as being recovered as a by-product or co-product of certain processes to produce acrylonitrile, neoprene and glycerin, these sources of acetaldehyde have either been abandoned or are used for the production of minor quantities. Until 1968, most acetaldehyde produced in the USA was made by the partial oxidation of ethanol over a silver catalyst; however, currently less than 5% of US capacity is based on this process. The liquid-phase oxidation of ethylene using a catalytic solution of palladium and copper chlorides was first used commercially in the USA in 1960 (Hagemeyer, 1978), and an estimated 96% of the acetaldehyde is now made by this process.

Acetaldehyde was first produced commercially (by the controlled hydration of acetylene) in 1916. This process was subsequently replaced by one involving the oxidation of ethanol, and US production by this method grew from 63.5 thousand tonnes in 1940 to 408 thousand tonnes by 1960. US production of acetaldehyde peaked in 1969 when 748 thousand tonnes were made (Hagemeyer, 1978) and has steadily fallen since. The total production of the two US companies producing acetaldehyde in 1982 amounted to 281 thousand tonnes. US imports of acetaldehyde have been negligible for many years and amounted to only 1.4

thousand kg in 1982 (US Department of Commerce, 1983). Separate data on US exports are not published.

Acetaldehyde is produced by three companies in Germany, two companies in Spain and one company each in France, Italy and Switzerland. Total acetaldehyde production in western Europe on 1 January 1983 was more than 0.5 million tonnes, and production capacity is estimated to have been nearly 1 million tonnes. Most of this was based on the catalytic oxidation of ethylene; less than 10% was based on partial oxidation of ethanol, and a very small percentage was based on the hydration of acetylene.

Acetaldehyde is produced (by oxidation of ethylene) by seven companies in Japan. Their combined production is estimated to have been 278 thousand tonnes in 1982, down from an estimated 323 thousand tonnes in 1981. Japanese imports and exports of acetaldehyde are negligible.

(b) Use

Acetaldehyde is used principally as a chemical intermediate, predominantly for the manufacture of acetic acid. The use pattern for the estimated 281 thousand tonnes of acetaldehyde used in the USA in 1982 was as follows: acetic acid, 61%; pyridine and pyridine bases, 9%; peracetic acid, 8%; pentaerythritol, 7%; 1,3-butylene glycol, 2%; chloral, 1%; and other applications (including use as a food additive and exports), 12%.

An estimated 25-30% of all US capacity for the production of acetic acid is currently based on synthesis from acetaldehyde. This percentage has been constantly decreasing from 1970 when it was 38%. Acetic acid is an important chemical intermediate used principally to make vinyl acetate (see IARC, 1979), cellulose acetate and other acetic esters as well as being used as a solvent in the manufacture of terephthalic acid and dimethyl terephthalate.

Reaction of acetaldehyde (alone or in combination with formaldehyde) with ammonia is used to produce pyridine and pyridine bases that have alkyl substituents on the pyridine ring. These chemicals are used as intermediates in the synthesis of herbicides, insecticides, fungicides, tyre-cord adhesives, pharmaceuticals and cosmetics ingredients.

Oxidation of acetaldehyde with oxygen using tree-radical type catalysts is used to make peracetic acid, which finds use in the production of epoxy compounds and a number of other applications.

Alkaline condensation of acetaldehyde and formaldehyde is used to produce pentaerythritol. This polyol is used principally to make alkyd resins. Smaller quantities are used for the production of fatty acid esters used as synthetic lubricants and in other applications.

Small amounts of acetaldehyde are used as a food additive. A survey of US industry on the use of food additives reported that 8.6 thousand kg of acetaldehyde were used in 1976 as an important component of many flavours added to foods, such as milk products, baked goods, fruit juices, candies, desserts and soft drinks, at usual levels of up to 0.047% (National Research Council/National Academy of Sciences, 1979).

Among the other uses of acetaldehyde in the USA are the synthesis of crotonaldehyde, various acetals used as flavour and fragrance chemicals, acetaldehyde 1,1-dimethylhydrazone, acetaldehyde cyanohydrin, acetaldehyde oxime, various acetic esters (by the Tishchenko reaction of acetaldehyde alone or in combination with other aldehydes such as n- and iso-butyraldehyde), its cyclic trimer (paraldehyde) and tetramer (metaldehyde, a molluscicide widely used to kill slugs and snails), various polymers (including the homopolymer), halogenated derivatives other than chloral, acetaldol and the sodium sulphite addition product.

Acetaldehyde is one of the denaturants approved for use (at a level of 10 lbs/100 gallons [12 g/l] of alcohol) in the USA in specially denatured alcohol Formula No. 29. Although the volume of this formula used in the USA each year is published, no information is available on the amount made with acetaldehyde.

Acetaldehyde was formerly used in the USA as a chemical intermediate for 2-ethyl-1-butanol, glyoxal, acrolein (see p. 133 of this volume) and acetaldehyde-aniline condensate (a rubber-processing chemical; see IARC, 1982).

Acetaldehyde has been used to make aniline dyes, plastics, synthetic rubber and for silvering mirrors and hardening gelatin fibres (Windholz, 1983). It has also been used in the production of polyvinyl acetal resins, as a preservative for fruit and fish, in fuel compositions and for the prevention of mould growth on leather (Hayes, 1963).

The use pattern for the estimated 706 thousand tonnes of acetaldehyde used in western Europe in 1980 was as follows: acetic acid, 62%; ethyl acetate, 19%; pentaerythritol, 5%; synthetic pyridines, 3%; and all other uses, 11%.

The consumption pattern for acetaldehyde in Japan in 1982 is estimated to have been as follows: acetic acid, 40%; ethyl acetate, 30%; pentaerythritol, 3%; and all other applications (including the synthesis of n-butanol and peracetic acid), 27%.

Acetaldehyde is used as a chemical intermediate in a process for producing butadiene from ethanol which is currently used by plants in India and China.

Occupational exposure to acetaldehyde has been limited by regulation or recommended guidelines in at least 15 countries. The standards are listed in Table 1.

The Bureau of Alcohol, Tobacco and Firearms of the US Department of the Treasury (1983) lists acetaldehyde among the approved denaturants for specially denatured alcohol Formula No. 29.

Acetaldehyde is generally recognized as safe by the US Food and Drug Administration (1980) for its intended use as a synthetic flavouring substance and adjuvant. It is also approved for use as a component of phenolic resins in moulded articles intended for repeated use in contact with nonacid food.

The US Environmental Protection Agency (EPA) (1982) has exempted acetaldehyde from a tolerance on its residues on apples and strawberries when used after harvest as a storage fumigant. The EPA has also identified acetaldehyde as a toxic waste and requires that persons who generate, transport, treat, store or dispose of it comply with the regulations of a federal hazardous waste management programme. Distillation bottoms and side cuts from the production of acetaldehyde from ethylene are included in a list of hazardous wastes in which the trimer of acetaldehyde, paraldehyde, was identified as one of the hazardous constituents (US Environmental Protection Agency, 1980). The EPA also requires that notification be given whenever discharges containing 454 kg or more of acetaldehyde are made into waterways (US Environmental Protection Agency, 1983).

Table 1. National occupational exposure limits for acetaldehyde[a]

Country	Year	Concentration mg/m³	ppm	Interpretation[b]	Status
Australia	1978	180	100	TWA	Guideline
Belgium	1978	180	100	TWA	Regulation
Czechoslovakia	1976	200	-	TWA	Regulation
		400	-	Ceiling (10 min)	
Finland	1981	90	50	TWA	Guideline
		135	75	STEL	
German Democratic Republic	1977	100	-	TWA	Regulation
		100	-	Maximum (30 min)	
Germany, Federal Republic of	1984	90	50	TWA	Guideline
Italy	1978	100	55	TWA	Guideline
Netherlands	1978	180	100	TWA	Guideline
Poland	1976	100	-	Ceiling	Regulation
Romania	1975	100	-	TWA	Guideline
		200	-	Maximum	
Sweden	1981	45	25	TWA	Guideline
		90	50	STEL	
Switzerland	1978	180	100	TWA	Regulation
USA[c]					
OSHA	1978	360	200	TWA	Regulation
		-	10 000	Maximum (30 min)	
ACGIH	1984/85	180	100	TWA	Guideline
		270	150	STEL	
USSR	1977	5	-	Maximum[d]	Regulation
Yugoslavia	1971	360	200	Ceiling	Regulation

[a]From International Labour Office (1980); National Finnish Board of Occupational Safety and Health (1981); National Swedish Board of Occupational Safety and Health (1981); American Conference of Governmental Industrial Hygienists (1984); Deutsche Forschungsgemeinschaft (1984)
[b]TWA, time-weighted average; STEL, short-term exposure limit
[c]OSHA, Occupational Safety and Health Administration; ACGIH, American Conference of Governmental Industrial Hygienists
[d]The USSR has also established a maximum of 0.2 mg/m³ for acetaldehyde tetramer

As part of the Hazardous Materials Regulations of the US Department of Transportation (1982), shipments of acetaldehyde are subject to a variety of labelling, packaging, quantity and shipping restrictions consistent with its designation as a hazardous material.

2.2 Occurrence

(a) Natural occurrence

Acetaldehyde is reported to be a metabolic intermediate in higher plants and a product of alcohol fermentation. It has been identified as a volatile component of mature cotton leaves and cotton blossoms (Berni & Stanley, 1982) and was detected by Kami (1983) in the essential oil of alfalfa at a concentration of about 0.2%. It was also detected at trace levels in two kinds of mushrooms (*Armillaria mellea* and *Boletus luteus*) (Stepanova & Tsapalova, 1982).

Acetaldehyde has been identified as a component of the steam volatile oils obtained from blueberries (Horvat et al., 1983). It occurs in oak leaves and tobacco leaves. It is also a natural constituent of apples, broccoli, coffee, grapefruit, grapes, lemons, onions, oranges, peaches, pears, pineapple, raspberries and strawberries. It is found in the essential oils of rosemary, balm, clary sage, daffodil, bitter orange, camphor, angelica, fennel, mustard and peppermint (National Academy of Sciences/National Research Council, 1965; Furia & Bellanca, 1975).

(b) *Occupational exposure*

On the basis of the 1974 National Occupational Hazard Survey, the National Institute for Occupational Safety and Health (1980, 1981) estimated that 1700 US workers in four non-agricultural industries were exposed to acetaldehyde. The principal industry in which exposure was found was the industrial organic chemicals industry, and some exposure was noted in the fabricated rubber products and biological products industries.

Aldehydes of low molecular weight, such as acetaldehyde, occur as decomposition products of some polymers; however, acetaldehyde concentrations above the detection limits were not found in the workroom air of four US plants in which polymers were heated for various purposes, when surveyed by the National Institute for Occupational Safety and Health. The plants were in the following industries: textile finishing (limits of detection ranged from 1.2-3.4 mg/m^3) (Rosensteel & Tanaka, 1976); polypropylene bottle production [limit of detection of 1 mg/m^3] (Ahrenholz & Gorman, 1980); fabrication of prosthetics (Chrostek & Shoemaker, 1981); and production of scientific glass gauges (Chrostek, 1981). Acetaldehyde concentrations were below the limit of detection (2.3 mg/m^3) in an industrial hygiene survey at a urea-formaldehyde foam-insulation manufacturing plant in the USA, where acetaldehyde was being used at low concentrations as a reactant in the production of resins (Herrick, 1980).

Acetaldehyde was reported at levels of 1-7 mg/m^3 in the workroom air of an aldehyde factory in the German Democratic Republic after equipment leakages (Bittersohl, 1975).

(c) *Air*

Eimutis *et al.* (1978) estimated that annual US atmospheric emissions of acetaldehyde amounted to 12.2 million kg. The various sources of these emissions are given in Table 2.

Subsequently, Anderson *et al.* (1980) estimated atmospheric emissions of acetaldehyde from 1978 to be 2.2 million kg/year. Total US acetaldehyde emissions to the air in 1978 from all sources have been estimated at 52 million kg, 86% of which was due to wood burning in residences (Lipfert & Dungan, 1982).

Table 2. Sources of emissions of acetaldehyde in the air in the USA[a]

Source	Emissions (1 000 kg/year)
Residential external combustion of wood	5056.4
Coffee roasting	4411.5
Acetic acid manufacture	1460.9
Vinyl acetate manufacture from ethylene	1094.6
Ethanol manufacture	57.8
Acrylonitrile manufacture	51.6
Acetic acid manufacture from butane	20.8
Crotonaldehyde manufacture	4.5
Acetone and phenol manufacture from cumene	1.9
Acetaldehyde manufacture by hydration of ethylene	0.5
Polyvinyl chloride manufacture	0.2
Acetaldehyde manufacture by oxidation of ethanol	0.1

[a]From Eimutis *et al.* (1978)

Pellizzari (1979) has reported the detection of acetaldehyde at unidentified levels in the atmospheres of several cities in the USA. It was detected in the air at several locations in the USA by Arnts and Meeks (1981), as follows: none detected-15.0 µg/m^3 in Tulsa, OK; none detected-16.9 µg/m^3 in Rio Blanco County, CO; and none detected-24.0 µg/m^3 in the Smokie Mountains, TN. Singh et al. (1982) measured acetaldehyde concentrations in air of 0.36-4.68 µg/m^3 in Pittsburgh, PA, and 1.62-6.12 µg/m^3 in Chicago, IL, during April 1981. Air-monitoring data accumulated between 1975 and 1978 show that this compound occurred at mean ambient concentrations of 5-124 µg/m^3 at seven other US locations (Brodzinsky & Singh, 1982). It was also detected in the ambient air in southern California under conditions of moderate to severe photochemical pollution at levels of 5.4-63 µg/m^3 (Grosjean, 1982).

It has been reported that photolysis and reaction with hydroxyl radicals cause a daily loss rate of about 80% of atmospheric acetaldehyde emissions (Singh et al., 1982).

Acetaldehyde was detected at unspecified levels in a study of indoor and outdoor air in and around homes in the areas of Washington, DC and Chicago, IL (Jarke et al., 1981).

Acetaldehyde emissions have been reported from coffee-roasting operations (14-22 mg/m^3), from a lithographic plate coater (0.5-4.1 mg/m^3), from an automobile-spray booth (2.5-3.4 mg/m^3) (Levaggi & Feldstein, 1970), from plants manufacturing acrylic acid (Serth et al., 1978), and from a fat-rendering plant at levels of 3.4-6.8 mg/m^3 (Van Langenhove et al., 1983).

Acetaldehyde was detected in air from several Japanese sources as summarized in Table 3 (Hoshika et al., 1981). It was also detected in ambient air and air near and in a refuse-reclamation area in Japan at levels of 5.4-13.7 µg/m^3 (Aoyama & Yashiro, 1983).

Table 3. Acetaldehyde detected in air from several Japanese sources[a]

Odour source	Concentration (mg/m^3)
Exhaust gas from a corn starch manufacturing works	9.6
Exhaust gas from poultry manure driers	7.5-8.1
Exhaust gas from a metal paint dryer	38.2
Exhaust gas from a shell mould processing plant	0.9-2.0
Exhaust gas from kraft pulp recovery boilers	0.3-0.5
Ambient air near a fishmeal factory	0.06
Sewage treatment plant	0.01-0.02
Night-soil treatment plant	0.005-0.02

[a]From Hoshika et al. (1981)

(d) Water and sediments

Acetaldehyde has been detected in finished drinking-water in several US cities at levels of up to 0.1 µg/l. It has also been detected in US well- and river-water, and effluents from sewage-treatment plants and chemical plants (US Environmental Protection Agency, 1975; Shackelford & Keith, 1976).

Acetaldehyde has been identified as a constituent in the wastes from petroleum refining, coal processing, the oxidation of alcohols, saturated hydrocarbons or ethylene, and the hydration of acetylene. Degradation of hydrocarbons, sewage and solid biological wastes also produces acetaldehyde (Versar, Inc., 1975).

(e) *Food, beverages and animal feeds*

Acetaldehyde has been found in vegetables and fruit (see p. 105).

The acetaldehyde concentration in 18 European beers was reported to range from 2.6-13.5 mg/l (Delcour *et al.*, 1982). Okamoto *et al.* (1981) detected acetaldehyde in commercial wine samples in Japan at levels of 0.2-1.2 mg/l. It has also been identified in Cuba in the aqueous condensate obtained from the concentration of sweet orange juice (Pino, 1982).

Acetaldehyde has been detected in cheese, heated skim milk, cooked beef, cooked chicken and rum (National Academy of Sciences/National Research Council, 1965; Furia & Bellanca, 1975).

Trace quantities of acetaldehyde are present in a flavouring used to impart a butter-like flavour to processed foods, especially margarine (US Food and Drug Administration, 1982).

(f) *Tobacco and marijuana smoke*

Hoffmann *et al.* (1975) detected acetaldehyde in the smoke of tobacco cigarettes (980 µg/cigarette) and marijuana cigarettes (1200 µg/cigarette).

Vapour-phase deliveries of acetaldehyde in smoke from several types of cigarettes ranged from 1.14-1.37 mg/cigarette when detected by high-performance liquid chromatography and 0.87-1.22 mg/cigarette when detected by gas chromatography. Three types of low-tar cigarettes delivered 0.09-0.27 mg/cigarette (Manning *et al.*, 1983).

(g) *Pyrolysis products*

Typical emissions of acetaldehyde from burning wood have been estimated at about 0.7 g/kg; total annual acetaldehyde emissions in the USA from the residential burning of wood have been estimated at 45 million kg (Lipfert & Dungan, 1982). Fireplace emissions ranged from 0.083-0.200 g/kg of wood burned (Lipari *et al.*, 1984).

According to the results of a Swedish study (Rudling *et al.*, 1982), the acetaldehyde emissions from combustion of wood and wood-chips in small furnaces and stoves were as follows (mg/kg of fuel): prechamber oven (wood chips), 1-72; central heating furnace (wood), 62-620; and fireplace stove (wood), 9-710.

Ramdahl *et al.* (1982) measured acetaldehyde emissions from wood-burning stoves in Norway and reported emissions of 14.4 mg/kg of dry wood under normal burning conditions, and as much as 992 mg/kg of dry wood under low efficiency (air-starved) combustion. Burning of charcoal emits significantly less acetaldehyde.

Acetaldehyde has been reported to be a combustion product of plastics (e.g., polyphenylene oxide) (Boettner *et al.*, 1973).

Acetaldehyde was detected as a combustion product of polycarbonate and of hard and soft polyurethane foams of western European origin (Hagen, 1967).

It has been reported to occur in heavy-duty diesel exhaust at levels of 0.05-6.4 mg/m^3 (Hare & Bradow, 1979) and in gasoline exhaust (1.4-8.8 mg/m^3) (Verschueren, 1977). Concentrations of acetaldehyde in automobile exhaust are reported in Table 4.

Table 4. Concentrations of acetaldehyde in automobile exhaust[a]

Vehicle type	Acetaldehyde emissions (mg/m^3)	
	Cold start	Hot start
Prototype ethanol-fuelled vehicle	36.5	18.2
1974 (no catalyst) gasoline-fuelled vehicle	1.2	1.1
1981 (catalyst) gasoline-fuelled engine	0.1	<0.02
1978 (5.7-1, V-8) diesel vehicle without exhaust gas recirculation	0.2	0.2
1980 (5.7-1, V-8) diesel vehicle with exhaust gas recirculation	0.3	0.2

[a]From Lipari & Swarin (1982)

The results of one US study of emissions from gasoline- and diesel-powered vehicles are summarized in Table 5.

Table 5. Acetaldehyde emissions from gasoline- and diesel-powered vehicles[a]

Vehicle type	Acetaldehyde emissions (mg/km)
Light-duty gasoline	
Catalyst-equipped	0.0-0.63
Non-catalyst	1.42-6.74
Light-duty diesel	1.00-10.50
Heavy-duty gasoline	10.21-16.03
Heavy-duty diesel	1.76-45.69

[a]From Carey (1981)

(*h*) *Human tissues and secretions*

Acetaldehyde is formed during the intracellular oxidation of ethanol (Eriksson, 1983).

The concentration of acetaldehyde in the whole blood of four fasting, normal human subjects was reported by Lynch *et al.* (1983) to be 1.30 μmol/l (57 μg/l).

Acetaldehyde has been detected in mother's milk in the USA (Pellizzari *et al.*, 1982).

(*i*) *Other*

In the USSR, acetaldehyde was reported to be formed as a gaseous product during the pulping of aspenwood-chips by oxidative ammonolysis (Kondakova *et al.*, 1981).

2.3 Analysis

Analytical methods for acetaldehyde were summarized briefly by Hagemeyer (1978).

Typical methods for the analysis of acetaldehyde in various matrices are given in Table 6.

Table 6. Methods for the analysis of acetaldehyde

Sample matrix	Sample preparation	Assay procedure[a]	Limits of detection	Reference
Air	Collect in impinger containing aqueous acidic 2,4-dinitrophenylhydrazine and cyclohexane:isooctane (9:1); extract (hexane and dichloromethane); evaporate; dissolve (methanol)	HPLC/UV	Approx. 0.2-4 µg/m^3 for 60-litre samples	Grosjean (1982)
	Collect in cartridge containing Tenax GC or other resin; desorb thermally and collect in a trap cooled by liquid nitrogen	GC/MS	Not given	Krost et al. (1982)
	Trap using liquid argon	GC	0.09-4 µg/m^3	Hoshika et al. (1981)
	Collect on silica gel treated with 2,4-dinitrophenylhydrazine; desorb with carbon tetrachloride	GC/FTD	0.09-0.45 µg/m^3 for samples of 50-100 litres	Aoyama & Yashiro (1983)
Indoor and outdoor air	Collect in stainless steel tube containing a porous polymer based on 2,6-diphenyl-p-phenylene oxide	GC/MS	0.9 µg/m^3 for 2-litre sample	Jarke et al. (1981)
Industrial emissions (air)	Collect in impingers containing sodium bisulphite solution	GC/FID	1.2 mg/m^3 [200-litre samples]	Levaggi & Feldstein (1970); Rosensteel & Tanaka (1976)
Occupational air	Adsorb on activated charcoal; elute (carbon disulphide)	GC	18 mg/m^3 for 5-litre sample	National Institute for Occupational Safety and Health (1974)
	Collect in bubbler containing aqueous Girard T reagent at controlled pH	HPLC	170-670 mg/m^3 (range of detection)	Gunderson & Anderson (1980)
Automobile exhaust	Dilute gases with room air, bubble through midget impingers containing aqueous 2,4-dinitrophenylhydrazine/acetonitrile solution; inject directly	HPLC/UV	18 µg/m^3 for 20-litre samples	Lipari & Swarin (1982)
Flue gases (from wood furnaces & stoves)	Collect with impingers containing 2,4-dinitrophenylhydrazine in hydrochloric acid; extract (dichloromethane); concentrate	GC/FID	Not given	Rudling et al. (1982)
Volatile emissions from wastewater	Collect in stainless-steel tube with liquid oxygen	GC/FID	Not given	Thibodeaux et al. (1982)
Waste water (e.g., from chloroprene production)	Collect waste-water vapour phase	GC/FID	100 mg/l	Geodakyan et al. (1981)
Aqueous solution	Mix with 3-methyl-2-benzothiazolone hydrazone hydrochloride; allow to react (optional heating); add ferric chloride; dilute (acetone)	S	Not given	Sawicki et al. (1961)
Aqueous solution/industrial effluent	Purge with nitrogen gas; trap with Tenax GC sorbent and silica gel; desorb thermally	GC/MS	200 µg/l	Spingarn et al. (1982)
Polyethylene terephthalate resin	Grind under liquid nitrogen; inject into headspace	GC/FID	50 µg/kg	Dong (1981)
Alfalfa	Steam distil; extract (diethyl ether)	GC/MS and GC	Not given	Kami (1983)
Plants (cotton)	Steam distil; collect on Tenax GC	GC/MS	Not given	Berni & Stanley (1982)
Biological fluids	Inject into headspace with internal standard (e.g., n-propanol) solution; heat	GC/FID	0.5-32 mg/l (range of detection)	Suitheimer et al. (1982)
Blood	Mix with perchloric acid to precipitate protein; centrifuge; transfer supernatants to headspace bottles	GC	4 µg (artefactual acetaldehyde is corrected for or blocked with sodium azide)[c]	Eriksson et al. (1982)
Mother's milk	Warm and purge with helium; trap on Tenax cartridge; desorb thermally	GC/MS	Not given	Pellizzari et al. (1982)

Sample matrix	Sample preparation	Assay procedure[a]	Limits of detection	Reference
Wine	Add internal standard (proponaldehyde); extract (ammonium acetate/acetic acid/acetylacetone); dry over anhydrous sodium sulphate	HPLC	0.01 µg	Okamoto et al. (1981)
Beer	Steam distil; extract (pentane/dichloromethane); react to form the p-nitrobenzyloxyamine derivative	HPLC	Not given	Piendl et al. (1981)
Cultures of lactic acid bacteria	Steam distil into 3-methyl-2-benzothiazolone hydrazone reagent mixture	S	0-24 µg/ml (range of detection)	Schmidt et al. (1983)

[a]Abbreviations: HPLC/UV, high-performance liquid chromatography/ultraviolet detection; GC/MS, gas chromatography/mass spectrometry; GC, gas chromatography; GC/FTD, gas chromatography/flame thermoionic detection; GC/FID, gas chromatography/flame ionization detection; HPLC, high-performance liquid chromatography; S, spectrophotometry

[b]According to Eriksson (1983), the blood acetaldehyde levels detected in earlier studies are due mainly to artefactual acetaldehyde formation during analysis.

3. Biological Data Relevant to the Evaluation of Carcinogenic Risk to Humans

3.1 Carcinogenicity studies in animals

(a) Inhalation and/or intratracheal administration

Rat: Four groups of 105 male and 105 female Cpb:WU albino Wistar rats were exposed to 0, 750, 1500 or 3000 (reduced progressively over a period of 11 months to 1000 ppm due to toxicity) ppm [0, 1350, 2700 or 5400-1800 mg/m^3] acetaldehyde vapour [purity unspecified] for six hours per day on five days per week for a maximum of 27 months. Each group comprised five subgroups, three of which were used for interim kills at weeks 13, 26 and 52, respectively. Of the animals killed at these intervals, only one had a tumour of the respiratory tract: a female of the high-dose group killed in week 53, bearing a nasal squamous-cell carcinoma. To date, only results obtained during the first 15 months of the study have been reported. Mortality at month 15 was 60% in the high-dose group and 3-4% in each of the other groups. In animals that died spontaneously or were killed in extremis during the first 15 months of the study, the incidences of nasal carcinomas (squamous-cell carcinomas and adenocarcinomas) were (males and females): 1/7, 1/2, 5/8 and 23/63 in the control, low-, mid- and high-dose groups, respectively. No tumour was found in other segments of the respiratory tract. A variety of compound-related lesions, including hyperplastic and inflammatory changes, occurred in the nose and larynx of acetaldehyde-exposed rats (Woutersen et al., 1984).

Hamster: Groups of 35 male Syrian golden hamsters were exposed to 0 or 1500 ppm [2700 mg/m^3] acetaldehyde vapour for seven hours per day on five days per week for 52 weeks, and to weekly intratracheal instillations of 0, 0.0625, 0.125, 0.25, 0.5 or 1 mg benzo[a]pyrene suspended in saline for the same period. Groups of five animals were killed at the 52nd week and the remainder allowed to survive untreated for an additional 26 weeks. There was no significant difference in mortality between the animals exposed to acetaldehyde and those exposed to air, except for the subgroup treated with the highest dose of benzo[a]pyrene, for which the mortality in the acetaldehyde-exposed animals was increased more rapidly than the mortality in the corresponding benzo[a]pyrene group exposed to air ($p < 0.001$ in both groups). No tumour was found in hamsters exposed to acetaldehyde only; but 3/30, 4/30, 9/30, 25/29 and 26/28 hamsters exposed to benzo[a]pyrene alone developed respiratory-tract tumours and 1/28, 5/29, 8/29, 16/29 and 29/30 hamsters exposed to benzo[a]pyrene and acetaldehyde vapour developed the same type of tumour (Feron, 1979).

Groups of 36 male and 36 female Syrian golden hamsters, six weeks of age, were exposed for seven hours per day on five days per week to room air (chamber controls) or to decreasing concentrations of acetaldehyde (distilled and analysed by gas chromatography) (initial concentration, 2500 ppm [4500 mg/m^3]; final concentration, 1650 ppm [2970 mg/m^3]) for 52 weeks. Six animals killed and examined from each group had no tumour. The remaining animals were observed until 81 weeks and killed. The incidences of respiratory-tract tumours were 0/30, 8/29, 0/28 and 5/29 in control males, exposed males, control females and exposed females, respectively ($p < 0.05$). The acetaldehyde-induced tumours were predominantly laryngeal carcinomas with a few laryngeal polyps, and nasal polyps and carcinomas (Feron et al., 1982).

Seven groups of 35 male and 35 female Syrian golden hamsters, 11 weeks old, were given the following treatments weekly or biweekly for 52 weeks: intratracheal instillations of saline (0.2 ml weekly; vehicle controls); acetaldehyde alone (two groups: one receiving 0.2 ml 2% acetaldehyde weekly, and one receiving 0.2 ml 4% acetaldehyde weekly); acetaldehyde plus benzo[a]pyrene or N-nitrosodiethylamine; or benzo[a]pyrene or N-nitrosodiethylamine (positive controls). After 13, 26 and 52 weeks, three animals per sex and per group were killed and autopsied. The experiment was terminated at week 104, when the survivors were killed. At that time, overall mortality was 83% in males and 97% in females. Intratracheal instillation of acetaldehyde alone produced extensive 'peribronchiolar adenomatoid lesions' in the lungs, but no tumour. Both benzo[a]pyrene and N-nitrosodiethylamine induced a variety of tumours in several segments of the respiratory tract. This tumour response was not influenced by simultaneous treatment with acetaldehyde (Feron, 1979).

In a similar experiment to that described above, groups of 30 male and 30 female Syrian golden hamsters were exposed to air or to 2500-1650 ppm [4500-2870 mg/m^3] acetaldehyde vapour for seven hours per day on five days per week for 52 weeks and were given simultaneous weekly intratracheal instillations of 0.2 ml 0.175% or 0.35% benzo[a]pyrene (total amounts instilled, 18.2 and 36.4 mg/hamster) suspended in saline or subcutaneous injections once every three weeks of 0.2 ml 0.0625% (total volume injected, 2.1 µl) N-nitrosodiethylamine in saline. The experiment was terminated at 81 weeks. The types and incidences of benign and malignant respiratory-tract tumours in the various groups are summarized in Table 7. The main difference between the air- an acetaldehyde-exposed groups was a three- to five-fold increase in malignant respiratory-tract tumours in the acetaldehyde/high-dose benzo[a]pyrene group (Feron et al., 1982).

(b) Subcutaneous administration

Rat: A group of 20 rats [strain, sex and age unspecified], weighing 80-140 g, received 26-41 subcutaneous injections of 0.5-1 ml 0.5% acetaldehyde [purity unspecified] solution [solvent unspecified] followed by 40-50 subcutaneous injections of 1-1.5 ml acetaldehyde solution, the concentration of which was gradually increased from 1% to 5% (total number of injections ranged from 76-81). The injections were given twice a week during the first three weeks and once a week thereafter. The study lasted 489-554 days. Four of 14 rats that received injections until the end of the experimental period [no further specification given] developed sarcomas at the site of injection. There was no metastasis; three of the four tumours were transplantable (Watanabe & Sugimoto, 1956). [The Working Group noted that the experiment was inadequate for evaluation because of the small number of animals and the lack of a control group.]

Table 7. Types and incidences of respiratory-tract tumours in hamsters exposed to air or acetaldehyde and treated intratracheally with benzo[a]pyrene or subcutaneously with N-nitrosodiethylamine[a]

Carcinogen	Type of tumour	Treatment with air		Treatment with acetaldehyde	
		Males	Females	Males	Females
Benzo[a]pyrene, low dose	Benign	3 papillomas 1 adenoma 1 adeno-squamous adenoma	3 papillomas 1 adeno-squamous adenoma	4 papillomas 1 adeno-squamous adenoma	5 papillomas 1 adenoma
	Malignant	None	None	1 carcinoma in situ 9 squamous-cell carcinomas	7 squamous-cell carcinomas
Total[b]		5 in 4/29	3 in 3/27	15 in 12/29	13 in 11/29
Benzo[a]pyrene, high dose	Benign	8 papillomas 9 adenomas 2 adeno-squamous adenomas	1 papilloma 5 adenomas	3 papillomas 1 adenoma 2 adeno-squamous adenomas	
	Malignant	2 squamous-cell carcinomas 3 adenocarcinomas 2 anaplastic carcinomas 1 sarcoma	2 squamous-cell carcinomas 1 adenocarcinoma	3 carcinomas in situ 17 squamous-cell carcinomas 4 adenocarcinomas 2 sarcomas	2 carcinomas in situ 9 squamous-cell carcinomas 1 adeno-squamous carcinoma 1 anaplastic carcinoma
Total[b]		27 in 19/30	9 in 7/24	32 in 22/27	19 in 16/29
N-Nitrosodiethylamine	Benign	14 papillomas	13 papillomas	8 papillomas	2 papillomas
	Malignant	1 adenocarcinoma	None	3 carcinomas in situ	3 carcinomas in situ
Total[b]		15 in 12/29	13 in 11/27	15 in 11/30	9 in 8/20

[a]From Feron et al. (1982)
[b]Number of tumours in number of animals with tumours over number of animals examined

3.2 Other relevant biological data

(a) *Experimental systems*

The pharmacology of acetaldehyde in experimental animals has been reviewed (Brien & Loomis, 1983).

Toxic effects

In a four-hour inhalation study, the LC_{50} of acetaldehyde in hamsters was reported to be 17 000 ppm (31 000 mg/m^3) (Kruysse, 1970). The four-hour LC_{50} of acetaldehyde in rats was reported to be 13 300 ppm (24 000 mg/m^3) (Appelman et al., 1982). Subcutaneous LD_{50}s in rats and mice were found to be 640 and 560 mg/kg bw, respectively (Skog, 1950). The intraperitoneal LD_{50} in mice was found to be 500 mg/kg bw (Truitt & Walsh, 1971). The LD_{50} of acetaldehyde administered intratracheally to hamsters was reported to be 96.1 mg/kg bw (Feron & de Jong, 1971), and the intravenous LD_{50} in mice, 165 mg/kg bw (O'Shea & Kaufman, 1979).

Groups of 20 Syrian golden hamsters were exposed to 390-4560 ppm (700-8200 mg/m^3) acetaldehyde vapours for six hours per day, on five days per week for 90 days. The highest level (4560 ppm) induced growth retardation, ocular and nasal irritation, increased numbers of erythrocytes, increased heart and kidney weights and severe histopathological changes in the respiratory tract. At a level of 390 ppm, no toxic effect was observed (Kruysse et al., 1975).

In a four-week inhalation study, groups of male and female albino Wistar rats were exposed to 400-5000 ppm (790-9000 mg/m^3) acetaldehyde for six hours per day, on five days per week. The major changes reported with doses of 1000 and 2200 ppm (1800-3960 mg/m^3) were growth retardation, an increase in the production of urine in males, and slight to moderate degeneration with or without hyper- and metaplasia of the nasal epithelium. The only change seen at the lowest concentration (400 ppm) was slight degeneration of the nasal olfactory epithelium (Appelman et al., 1982).

In an isolated, perfused guinea-pig heart, 1 mM [44 µg/ml] acetaldehyde increased heart rate and coronary flow (Gailis, 1975). It also increased heart rate and oxygen consumption and decreased coronary vascular resistance of isolated, perfused hearts of rats and guinea-pigs (Gailis & Verdy, 1971). A dose-dependent biphasic response to acetaldehyde (4.4-1320 µg/ml) was found for left atrial contractility in guinea-pigs (Truitt & Walsh, 1971). Similar observations were reported in dogs (James & Bear, 1967; Stratton et al., 1981).

Green and Egle (1983) reported that intravenous administration of 2.0-40.0 mg/kg bw acetaldehyde to hypertensive SHR Wistar rats caused a blood pressor response; a depressor response was seen at higher (20-40 mg/kg bw) doses. A similar pressor response has been observed in dogs and cats (Handovsky & Heymans, 1936; Akabane et al., 1964).

Intravenous administration of 1-40 mg/kg bw acetaldehyde to anaesthetized male Wistar rats had opposing effects on the cardiovascular system: a sympathomimetic effect, resulting in a rise in blood pressure at doses below 20 mg/kg bw, and, at higher doses, bradycardia and hypotension (Egle et al., 1973).

Incubation of acetaldehyde with rat-liver mitochondria has been shown to cause several biochemical changes (at relatively high concentrations of 1-3 mM [44-132 µg/ml]), including alteration of mitochondrial respiration and hepatic oxidation of fatty acids, and increases in hepatic triglyceride levels (Brien & Loomis, 1983); incubation of [1,2-^{14}C]-acetaldehyde for 60 min at 37°C with liver microsomes from male Sprague-Dawley rats resulted in binding with microsomal protein (Nomura & Lieber, 1981).

In an in-vitro study, a suppressive effect on rat testicular steroidogenesis was reported at concentrations as low as 50 μM [2.2 μg/ml] acetaldehyde (Cicero et al., 1980).

Incubation of acetaldehyde (100 mg %) with isolated hepatocytes for 60 min significantly increased lipid peroxide formation, which was inhibited by prior addition of antioxidants such as vitamin E or glutathione (Stege, 1982).

Effects on reproduction and prenatal toxicity

Ethanol is metabolized to acetaldehyde, a recognized teratogen. Much experimental work has been carried out to develop a suitable animal model for ethanol-induced teratogenicity, and some studies have focused on the role of metabolism in this effect. Reviews of the research up to the early 1980s are available (Véghelyi et al., 1978; Obe & Ristow, 1979; Kumar, 1982).

Acetaldehyde [purity unspecified] dissolved in saline was administered intraperitoneally to groups of five to ten CF rats at single doses of 50, 75 or 100 mg/kg bw either on day 10, 11 or 12 or on each of days 10, 11 and 12 of gestation. Foetuses were examined on day 21. Foetal resorptions, malformations (including oedema, microcephaly, micrognathia, micromelia, hydrocephaly and exencephaly) and growth retardation were found. In general, these effects were greatest in the high-dose groups and in the multiple-treatment groups, although there was no strong dose-dependency, as shown by a wide variation in foetal response. No resorption or malformation was found in 13 control litters. Maternal effects were not reported (Sreenathan et al., 1982).

Rats received intra-amniotic injections of 0.02 ml of a 1% or 10% solution of acetaldehyde on day 13 of gestation. Foetal mortality was high in the control groups, but even higher in the treated groups (reaching 100% in the high-dose group). Surviving low-dose embryos had an increased incidence of malformations (80% compared to 14% in controls) (Bariliak & Kozachuk, 1983).

Groups of four to eleven pregnant CFLP mice were injected intravenously with either 0.1 ml/25 g bw of 0.9% saline or a similar volume of saline containing either 2% (v/v) or 1% (v/v) (approx. 62 and 31 mg/kg bw) acetaldehyde on days 7, 8 and 9 of gestation. Embryos were examined either on day 10 or day 19 of gestation. Dose-dependent embryolethality was observed (2.3%, 18.0% and 31.0% in the control, low- and high-dose groups, respectively, in day-19 embryos; and 9.8%, 31.7% and 46.3% in day-10 embryos); malformations that involved non-closure of the anterior or posterior neuropore were seen in day-10 embryos [incidence not specified] but not in day-19 embryos. Acetaldehyde-treated embryos were significantly smaller than controls on day 19 (O'Shea & Kaufman, 1979). Using the same general protocol, but examining the effects on day-10 or -12 embryos after single (days 6, 7, or 8) compared to those after multiple (days 6-8, 7-8 or 7-9) exposures to 0.1 ml of 2% acetaldehyde in saline, a high incidence of neural tube defects was observed after a single exposure (15/56 embryos after day-6 injections, 10/31 after day-7 and 0/71 after day-8 compared to a saline control group of 0/163), but embryonic mortality was more prevalent after multiple exposures (as high as 50/106 embryos after exposure on days 7-9 compared to 6/52 in the appropriate control) (O'Shea & Kaufman, 1981).

Groups of 5-14 C57Bl/6J mice received a single intraperitoneal dose of 0.32 g/kg bw of 4% acetaldehyde in arachis oil (v/v) on days 7, 8, 9 or 10 of gestation; or two injections, 30 min or six hours apart, on the same days. Embryos were examined on day 18. A low incidence of stage-specific teratogenic effects, with facial defects (exencephaly, and mandibular and maxillary hypoplasia) occurred after treatment on day 7 or 8, and limb defects (polydactyly, club foot) occurred after treatment on day 9 or 10. While the types of malformation observed in the acetaldehyde-treated foetuses were similar to those found in alcohol-treated animals, the overall incidence in the latter case was much greater (Webster et al., 1983).

A group of eight pregnant CD-1 mice received intraperitoneal injections of 200 mg/kg bw of a 0.69% solution (v/v) of acetaldehyde in sterile saline (volume of injection, 0.037 ml/g bw) every two hours for a total of five doses (total dose, 1000 mg/kg bw) on day 10 of gestation. Foetuses were examined on day 18 and compared with a variety of ethanol-treated groups. In general, ethanol-treated litters showed dose-related decreases in viability and growth, as well as teratogenic effects, including malformations of the head, and cardiovascular, urogenital and skeletal systems. Co-administration of 4-methylpyrazole, an inhibitor of alcohol dehydrogenase, dramatically increased embryotoxicity. The acetaldehyde treatment did not increase the percentage of resorptions and malformed foetuses or decrease foetal weight over that in saline or untreated controls (Blakley & Scott, 1984a). Acetaldehyde was detected in the embryos of CD-1 mice within five minutes of intraperitoneal treatment with 200 mg/kg bw acetaldehyde on day 10 of gestation (Blakley & Scott, 1984b).

The effects of acetaldehyde on the embryonic development of mice and rats have also been investigated in whole embryo cultures (Popov et al., 1981; Thompson & Folb, 1982; Campbell & Fantel, 1983). [In evaluating these findings, it is important to remember that acetaldehyde has a boiling temperature of 20.16°C.] In all studies, acetaldehyde adversely affected embryonic growth and differentiation.

Absorption, distribution, excretion and metabolism

Acetaldehyde is a toxic metabolite formed in the mammalian liver during the oxidation of ethanol (Truitt & Walsh, 1971).

Intravenous infusion of 0.5-5% solutions of acetaldehyde to rabbits resulted in rapid elimination of acetaldehyde, at a rate of 7-10 mg/min (Hald & Larsen, 1949).

Acetaldehyde is metabolized to acetic acid by NAD^+-dependent aldehyde dehydrogenase (Brien & Loomis, 1983). It has been reported that rat liver contains several aldehyde dehydrogenases with different properties and subcellular locations (Marjanen, 1973; Tottmar et al., 1973). The rate of acetaldehyde metabolism to acetic acid varies with the form of the enzyme, as does the site and kinetics of acetaldehyde oxidation in liver cells (Deitrich & Siew, 1974). Acetaldehyde is oxidized mainly in rat-liver mitochondria by an aldehyde dehydrogenase with a low Km (Tottmar et al., 1973; Parilla et al., 1974). Acetaldehyde is also oxidized, in rat nasal mucosal homogenates, by NAD^+-dependent dehydrogenases (Casanova-Schmitz et al., 1984).

Administration to rats of a single intraperitoneal dose of 6.2 mmol (273 mg) acetaldehyde caused a significant increase (about 100%) in the amount of a urinary, sulphur-containing metabolite that was found in the alkali-hydrolysable sulphydryl fraction (Hemminki, 1982). A thiazolidine 4-carboxylic acid derivative has also been shown to be produced when acetaldehyde reacts with L-cysteine (Nagasawa et al., 1975; Sprince et al., 1974; Nagasawa et al., 1984).

Concurrent administration of acetaldehyde, L-cysteine and nitrite to rats yielded *N*-nitroso-2-methylthiazolidine-4-carboxylic acid (*cis*- and *trans*-isomers), >90% of which was excreted in the urine (Ohshima et al., 1984).

The metabolism of acetaldehyde is influenced by a variety of factors, such as diet (Marchner & Tottmar, 1976), genetic factors (Eriksson, 1973), pregnancy (Kesäniemi, 1974), chronic alcohol consumption (Teschke et al., 1977) and various drugs (Truitt & Walsh, 1971; Brien & Loomis, 1983).

Metabolism of [1,2-^{14}C]-acetaldehyde by isolated perfused rat heart was demonstrated by measuring quantitatively the ensuing $^{14}CO_2$ production (Forsyth et al., 1973).

Incubation of acetaldehyde (purity, 99%) with ribonucleosides and deoxyribonucleosides has been shown to form adducts with cytosine- and purine-containing nucleotides. After sodium borohydrate reduction, one of the acetaldehyde-guanosine adducts was shown to be N^2-ethylguanosine (Hemminki & Suni, 1984).

Mutagenicity and other short-term tests (see also 'Appendix: Activity Profiles for Short-Term Tests', p. 334)

Reviews of the genetic effects of aldehydes, including acetaldehyde, are available (Auerbach et al., 1977; Obe & Ristow, 1979; Obe, 1981).

Acetaldehyde has been reported to induce cross-links in isolated DNA (Ristow & Obe, 1978) and to be positive in an *Escherichia coli* polA$^+$/polA$^-$ assay (Rosenkranz, 1977).

Acetaldehyde was not mutagenic to *Salmonella typhimurium* TA1535, TA1538 (Rosenkranz, 1977), TA98 or TA100 (Sasaki & Endo, 1978) [abstract, details not given], with or without liver microsomes from polychlorinated biphenyl-treated rats, or to strain TA104 (Marnett et al., 1984) in the absence of an exogenous metabolic system. It was mutagenic to *E. coli* when treatment was carried out in stoppered test tubes at 0°C (Véghelyi et al., 1978; Igali & Gazsó, 1980 [abstract, details not given]). Acetaldehyde induced mutations in mitochondria of *Saccharomyces cerevisiae* (Bandas, 1982).

Acetaldehyde caused a dose-dependent induction of chromosomal aberrations in *Vicia faba* root-tip meristems (Rieger & Michaelis, 1960).

In rat fibroblasts, acetaldehyde induced micronuclei, chromosomal aberrations and aneuploidy (Bird & Draper, 1980; Bird et al., 1982)

Chromosomal aberrations (Au & Badr, 1979 [abstract, details not given]) and sister chromatid exchanges (Obe & Ristow, 1977; Obe & Beek, 1979; de Raat et al., 1983) were obtained after treating Chinese hamster ovary cells with acetaldehyde.

In human peripheral lymphocytes *in vitro*, acetaldehyde induced a dose-dependent increase in the incidence of sister chromatid exchanges (Ristow & Obe, 1978; Véghelyi & Osztovics, 1978; Jansson, 1982; Böhlke et al., 1983) and chromosomal aberrations (Badr & Hussain, 1977; Obe et al., 1978; Véghelyi & Osztovics, 1978; Böhlke et al., 1983; Obe et al., 1984).

Abernethy et al., (1982) reported in an abstract that acetaldehyde alone (10-100 µg/ml) did not induce morphological transformation in C3H/10T$\frac{1}{2}$ Cl8 cells, but showed initiating activity at 10 and 25 µg/ml when assayed in the presence of 0.25 µg/ml 12-*O*-tetradecanoylphorbol 13-acetate.

Treatment of mice (Obe et al., 1979) and Chinese hamsters (Korte et al., 1981) with acetaldehyde led to an increase in the incidence of sister chromatid exchanges in bone-marrow cells.

Intra-amniotic administration of acetaldehyde solutions to rat embryos on day 13 of pregnancy led to structural chromosomal aberrations in embryonic cells (Bariliak & Kozachuk, 1983).

(b) Humans

Toxic effects

Human volunteers exposed for 15 min to 50 ppm (90 mg/m^3) acetaldehyde vapour experienced mild eye irritation (Silverman *et al.*, 1946). Men exposed to 200 ppm (360 mg/m^3) for 15 min developed transient conjunctivitis (Proctor & Hughes, 1978), whereas all of 14 men exposed to 134 ppm (241 mg/m^3) acetaldehyde for 30 min developed mild upper-respiratory irritation (Sim & Pattle, 1957).

The irritant effect of acetaldehyde vapour, which is reported to cause coughing and a burning sensation in the nose, throat and eyes, usually prevents exposure to a level sufficient to cause depression of the central nervous system. A splash of liquid acetaldehyde was reported to cause a burning sensation, lachrymation and blurred vision. Prolonged periods of contact with the skin result in erythema and burns; repeated contact may result in dermatitis, due either to primary irritation or to sensitization (Proctor & Hughes, 1978).

Intravenous infusion of 5% acetaldehyde [purity unspecified] at a rate of 20.6-82.4 mg/min for up to 36 min into normal human subjects caused an increase in heart rate, ventilation and dead space, and a decrease in alveolar carbon dioxide levels. These symptoms are qualitatively and quantitatively similar to those seen after ethanol intake in subjects previously treated with disulfiram (Antabuse), a known inhibitor of aldehyde dehydrogenase (Asmussen *et al.*, 1948).

Effects on reproduction and prenatal toxicity

It is not known whether acetaldehyde, the primary metabolite of ethanol, is involved in the etiology of the human foetal alcohol syndrome (Clarren & Smith, 1978; Kumar, 1982).

Absorption, distribution, excretion and metabolism

The apparent in-vivo concentration of acetaldehyde in blood was found to be mainly artefactual and seems to be due to oxidation of ethanol during the analytical procedure (Eriksson, 1983) (see section 2.3).

The retention of 45-70% inhaled acetaldehyde from a recording respirometer was reported in human subjects during both oral and nasal breathing; there was no difference in total retention between oral and nasal breathing. The percentage of acetaldehyde retention was primarily dependent upon the duration of the ventilatory cycle (Egle, 1970).

N-Nitroso-2-methylthiazolidine 4-carboxylic acid (*cis*- and *trans*-isomers) was frequently detected in the urine of human subjects; a fraction of this may be formed as a two-step synthesis *in vivo* from acetaldehyde and L-cysteine to yield 2-methylthiazolidine 4-carboxylic acid, which is easily nitrosated (Tsuda *et al.*, 1983; Ohshima *et al.*, 1984).

Mutagenicity and chromosomal effects

No data on acetaldehyde-exposed persons were available to the Working Group.

3.3 Case reports and epidemiological studies on carcinogenicity to humans

Nine cases of malignant neoplasm were reported among an unspecified number of workers in an aldehyde factory in the German Democratic Republic between 1967 and 1972 (Bittersohl, 1974, 1975). The factory was reported on because, among a number of chemical plants studied simultaneously, it had one of the higher cancer rates. Of the cancers found, five were bronchial tumours and two were carcinomas of the oral cavity; all nine patients were cigarette smokers. The authors state that the relative frequencies of these tumours were higher than the expected frequency in the German Democratic Republic. The main process at the factory was the dimerization of acetaldehyde; the product generated consisted of about 70% acetaldol with varying portions of acetaldehyde, butyraldehyde, crotonaldehyde and other higher, condensed aldehydes, as well as traces of acrolein. [The Working Group noted that the factory was studied without an a-priori hypothesis.]

4. Summary of Data Reported and Evaluation

4.1 Exposure data

Acetaldehyde was first produced commercially in 1916. It is produced in large quantities in many countries, where it is used principally as a chemical intermediate; a minor use is as a food additive. It occurs as a metabolic intermediate in higher plants and as a metabolite of ethanol in humans. It is found as a combustion product of various fuels and in tobacco smoke.

4.2 Experimental data

Acetaldehyde was tested for carcinogenicity in rats by inhalation exposure and in hamsters by inhalation exposure and intratracheal instillation. Following inhalation exposure, an increased incidence of carcinomas was induced in the nasal mucosa of rats, and laryngeal carcinomas were induced in hamsters. In another inhalation study in hamsters, using a lower exposure level, and in an intratracheal instillation study, no increased incidence of tumours was observed. In hamsters, inhalation of acetaldehyde enhanced the incidence of respiratory-tract tumours produced by intratracheal instillation of benzo[a]pyrene.

Foetal malformations were found in mice and rats treated with acetaldehyde *in vivo* and *in vitro*, and resorptions were observed in both species *in vivo*.

Acetaldehyde induced DNA damage and mutations in bacteria and mutations in yeast mitochondria. It induced chromosomal aberrations in plants. It induced chromosomal aberrations, aneuploidy, micronuclei and sister chromatid exchanges in cultured mammalian cells. An increased incidence of sister chromatid exchanges in bone-marrow cells was observed in mice and hamsters treated *in vivo*.

4.3 Human data

The one study of workers in an aldehyde plant was inadequate for evaluation of the carcinogenicity of acetaldehyde to humans.

Overall assessment of data from short-term tests: acetaldehyde[a]

	Genetic activity			Cell transformation
	DNA damage	Mutation	Chromosomal effects	
Prokaryotes	+	+		
Fungi/green plants		+	+	
Insects		+	+	
Mammalian cells (in vitro)	+	+	+	
Mammals (in vivo)	+		+	
Humans (in vivo)			+	
Degree of evidence in short-term tests for genetic activity: Sufficient				Cell transformation: No data

[a]The groups into which the table is divided and the symbol + are defined on pp. 17-18 of the Preamble; the degrees of evidence are defined on p. 18.

4.4 Evaluation[1]

There is *sufficient evidence*[2] for the carcinogenicity of acetaldehyde to experimental animals.

There is *inadequate evidence* for the carcinogenicity of acetaldehyde to humans.

5. References

Abernethy, D.J., Frazelle, J.H. & Boreiko, C.J. (1982) Effects of ethanol, acetaldehyde and acetic acid in the C3H/10T$\frac{1}{2}$ Cl 8 cell transformation system (Abstract No. Bf-1). *Environ. Mutagenesis, 4*, 331

Ahrenholz, S. & Gorman, R. (1980) *Health Hazard Evaluation Determination Report No. HE 80-2-727, Continental Plastic Containers, Springdale, Ohio*, Cincinnati, OH, National Institute for Occupational Safety and Health

Akabane, J., Nakanishi, S., Kohei, H., Matsumura, R. & Ogata, H. (1964) Studies on sympathomimetic action of acetaldehyde. I. Experiments with blood pressure and nictitating membrane responses. *Jpn. J. Pharmacol., 14*, 295-307

---

[1]For definitions of the italicized terms, see the Preamble, pp. 15-16.
[2]In the absence of adequate data on humans, it is reasonable, for practical purposes, to regard chemicals for which there is *sufficient evidence* of carcinogenicity in animals as if they presented a carcinogenic risk to humans.

American Conference of Governmental Industrial Hygienists (1984) *TLVs Threshold Limit Values for Chemical Substances and Physical Agents in the Work Environment with Intended Changes for 1984-85*, Cincinnati, OH, p. 9

Anderson, G.E., Liu, C.S., Holman, H.Y. & Killus, J.P. (1980) *Human Exposure to Atmospheric Concentrations of Selected Chemicals*, Environmental Protection Agency, Contract No. 68-02-3066, San Rafael, CA, Systems Applications, Inc., pp. 20-32

Aoyama, T. & Yashiro, T. (1983) Analytical study of low-concentration gases. II. Collection of acetaldehyde at low concentrations in air in a solid reaction tube and its analysis using a flame thermionic detector. *J. Chromatogr.*, 265, 45-55

Appelman, L.M., Woutersen, R.A. & Feron, V.J. (1982) Inhalation toxicity of acetaldehyde in rats. I. Acute and subacute studies. *Toxicology*, 23, 293-307

Arnts, R.R. & Meeks, S.A. (1981) Biogenic hydrocarbon contribution to the ambient air of selected areas. *Atmos. Environ.*, 15, 1643-1651

Asmussen, E., Hald, J. & Larsen, V. (1948) The pharmacological action of acetaldehyde on the human organism. *Acta pharmacol.*, 4, 311-320

Au, W. & Badr, F.M. (1979) Does ethanol induce chromosomal damage ? (Abstract No. 243). *In Vitro*, 15, 221

Auerbach, C., Moutschen-Dahmen, M. & Moutschen, J. (1977) Genetic and cytogenetical effects of formaldehyde and related compounds. *Mutat. Res.*, 39, 317-362

Badr, F.M. & Hussain, F. (1977) Action of ethanol and its metabolite acetaldehyde in human lymphocytes. In vivo and in vitro study (Abstract). *Genetics*, 86, s2-s3

Bandas, E.L. (1982) Studies on the role of metabolites and contaminants in the mutagenic action of ethanol on the yeast mitochondria (Russ.). *Genetika*, 18, 1056-1061

Barilliak, I.R. & Kozachuk, S.I. (1983) Embryotoxic and mutagenic activity of ethanol and acetaldehyde in intra-amniotic exposure (Russ.). *Tsitol. Genet.*, 17, 57-60

Berni, R.J. & Stanley, J.B. (1982) Volatiles from cotton plant parts. *Text. Res. J.*, 52, 539-541

Bird, R.P. & Draper, H.H. (1980) Effect of malonaldehyde and acetaldehyde on cultured mammalian cells: Growth, morphology, and synthesis of macromolecules. *J. Toxicol. environ. Health*, 6, 811-823

Bird, R.P., Draper, H.H. & Basrur, P.K. (1982) Effect of malonaldehyde and acetaldehyde on cultured mammalian cells. Production of micronuclei and chromosomal aberrations. *Mutat. Res.*, 101, 237-246

Bittersohl, G. (1974) Epidemiologic investigations on cancer incidence in workers contacted by acetaldol and other aliphatic aldehydes (Ger.). *Arch. Geschwulstforsch.*, 43, 172-176

Bittersohl, G. (1975) Epidemiological research on cancer risk by aldol and aliphatic aldehydes. *Environ. Qual. Saf.*, 4, 235-238

Blakley, P.M. & Scott, W.J., Jr (1984a) Determination of the proximate teratogen of the mouse fetal alcohol syndrome. I. Teratogenicity of ethanol and acetaldehyde. *Toxicol. appl. Pharmacol.*, *72*, 355-363

Blakley, P.M. & Scott, W.J., Jr (1984b) Determination of the proximate teratogen of the mouse fetal alcohol syndrome. 2. Pharmacokinetics of the placental transfer of ethanol and acetaldehyde. *Toxicol. appl. Pharmacol.*, *72*, 364-371

Boettner, E.A, Ball, G.L. & Weiss, B. (1973) *Combustion Products from the Incineration of Plastics (EPA-670/2-73-049, PB-222 001)*, Cincinnati, OH, US Environmental Protection Agency, pp. 1-5, 106-122

Böhlke, J.U., Singh, S. & Goedde, H.W. (1983) Cytogenetic effects of acetaldehyde in lymphocytes of Germans and Japanese: SCE, clastogenic activity, and cell cycle delay. *Human Genet.*, *63*, 285-289

Brien, J.F. & Loomis, C.W. (1983) Pharmacology of acetaldehyde. *Can. J. Physiol. Pharmacol.*, *61*, 1-22

Brodzinsky, R. & Singh, H.B. (1982) *Volatile Organic Chemicals in the Atmosphere: An Assessment of Available Data (Final Report, US Environmental Protection Agency, Contract No. 68-02-3452)*, Menlo Park, CA, SRI International, pp. 3-4, 12-15, 26-27, 37, 120

Campbell, M.A. & Fantel, A.G. (1983) Teratogenicity of acetaldehyde in vitro: Relevance to the fetal alcohol syndrome. *Life Sci.*, *32*, 2641-2647

Carey, P.M. (1981) *Mobile Source Emissions of Formaldehyde and Other Aldehydes (EPA/AA/CTAB/PA/31-11, PA82-118159)*, Springfield, VA, National Technical Information Service, pp. 9-11, 23-26

Casanova-Schmitz, M., David, R.M. & Heck, H.d'A. (1984) Oxidation of formaldehyde and acetaldehyde by NAD^+-dependent dehydrogenases in rat nasal mucosal homogenates. *Biochem. Pharmacol.*, *33*, 1137-1142

Chrostek, W.J. (1981) *Health Hazard Evaluation Report No. HETA-81-098-941, Lab-Crest Scientific Glass Company, Subsidiary of Fischer and Porter Company, Warminster, Pennsylvania*, Cincinnati, OH, National Institute for Occupational Safety and Health

Chrostek, W.J. & Shoemaker, W.E. (1981) *Health Hazard Evaluation Report No. HETA-81-298-944, Hospital of the University of Pennsylvania, Philadelphia, Pennsylvania*, Cincinnati, OH, National Institute for Occupational Safety and Health

Cicero, T.J., Bell, R.D., Meyer, E.R. & Badger, T.M. (1980) Ethanol and acetaldehyde directly inhibit testicular steroidogenesis. *J. Pharmacol. exp. Ther.*, *213*, 228-233

Clarren, S.K. & Smith, D.W. (1978) The fetal alcohol syndrome. *New Engl. J. Med.*, *298*, 1063-1067

Deitrich, R.A. & Siew, C. (1974) *Localization and function of aldehyde dehydrogenases*. In: Therman, R.G., Yonetani, I., Williamson, J.R. & Chance, B., eds, *Alcohol and Aldehyde Metabolizing Systems*, New York, Academic Press, pp. 125-135

Delcour, J.A., Caers, J.M., Dondeyne, P., Delvaux, F. & Robberechts, E. (1982) An enzymic assay for the determination of acetaldehyde in beers. *J. Inst. Brew.*, *88*, 384-386 [*Chem. Abstr.*, *98*, 15399j]

Deutsche Forschungsgemeinschaft (1984) *Maximal Work Place Concentrations and Biological Tolerance Values for Compounds in the Work Place* (Ger.), Part XX, Weinheim, Verlag Chemie GmbH, p. 15

Dong, M.W. (1981) Novel applications for headspace gas chromatography. *Chromatographia*, *14*, 447-451

Eastman Kodak Company (1982) *Acetaldehyde, Specification No. 3535-8*, Kingsport, TN, Eastman Chemical Products, Inc.

Eastman Kodak Company (1983) *Acetaldehyde, Publication No. C-109A*, Kingsport, TN, Eastman Chemical Products, Inc.

Egle, J.L., Jr (1970) Retention of inhaled acetaldehyde in man. *J. Pharmacol. exp. Ther.*, *174*, 14-19

Egle, J.L., Jr, Hudgins, P.M. & Lai, F.M. (1973) Cardiovascular effects of intravenous acetaldehyde and propionaldehyde in the anesthetized rat. *Toxicol. appl. Pharmacol.*, *24*, 636-644

Eimutis, E.C., Quill, R.P. & Rinaldi, G.M. (1978) *Source Assessment Noncriteria Pollutant Emissions (1978 Update)* (*PB-291 747*), Research Triangle Park, NC, Industrial Environmental Research Laboratory, US Environmental Protection Agency, pp. 1-5

Eriksson, C.J.P. (1973) Ethanol and acetaldehyde metabolism in rat strains genetically selected for their ethanol preference. *Biochem. Pharmacol.*, *22*, 2283-2292

Eriksson, C.J. (1983) Human blood acetaldehyde concentration during ethanol oxidation (Update 1982). *Pharmacol. Biochem. Behav.*, *18*, Suppl. 1, 141-150

Eriksson, C.J., Mizoi, Y. & Fukunaga, T. (1982) The determination of acetaldehyde in human blood by the perchloric acid precipitation method: The characterization and elimination of artefactual acetaldehyde formation. *Anal. Biochem.*, *125*, 259-263

Fassett, D.W. (1963) Aldehydes and acetals. In: Fassett, D.W. & Irish, D.D., eds, *Industrial Hygiene and Toxicology*, Vol. 2, *Toxicology*, 2nd ed., New York, Interscience, p. 1966

Feron, V.J. (1979) Effects of exposure to acetaldehyde in Syrian hamsters simultaneously treated with benzo(a)pyrene or diethylnitrosamine. *Prog. exp. Tumor Res.*, *24*, 162-176

Feron, V.J. & de Jong, D. (1971) *Acute Intratracheal Toxicity of Acetaldehyde in Syrian Golden Hamsters (Report No. R 3600)*, Zeist, The Netherlands, Central Institute for Nutrition and Food Research TNO

Feron, V.J., Kruysse, A. & Woutersen, R.A. (1982) Respiratory tract tumours in hamsters exposed to acetaldehyde vapour alone or simultaneously to benzo(a)pyrene or diethylnitrosamine. *Eur. J. Cancer clin. Oncol.*, *18*, 13-31

Forsyth, G.W., Nagasawa, H.T. & Alexander, C.S. (1973) Acetaldehyde metabolism by the rat heart. *Proc. Soc. exp. Biol. Med.*, *144*, 498-500

Furia, T.E. & Bellanca, N., eds (1975) *Fenaroli's Handbook of Flavor Ingredients*, 2nd ed., Vol. 2, Cleveland, OH, CRC Press, Inc., p. 3

Gailis, L. (1975) *Cardiovascular effects of acetaldehyde: Evidence for the involvement of tissue SH groups*. In: Lindros, K.O. & Eriksson, C.J.P., eds, *The Role of Acetaldehyde in the Actions of Ethanol*, Helsinki, The Finnish Foundation for Alcohol Studies, pp. 135-147

Gailis, L. & Verdy, M. (1971) The effect of ethanol and acetaldehyde on the metabolism and vascular resistance of the perfused heart. *Can. J. Biochem.*, *49*, 227-233

Geodakyan, K.T., Khanamiryan, K. M. & Vardanyan, A.M. (1981) Chromatographic determination of some compounds in the wastewater of chloroprene production (Russ.). *Prom-st. Arm.*, *8*, 36-37 [*Chem. Abstr.*, *96*, 109830d]

Green, M.A. & Egle, J.L., Jr (1983) The effects of acetaldehyde and acrolein on blood pressure in guanethidine-pretreated hypertensive rats. *Toxicol. appl. Pharmacol.*, *69*, 29-36

Grosjean, D. (1982) Formaldehyde and other carbonyls in Los Angeles ambient air. *Environ. Sci. Technol.*, *16*, 254-262

Gunderson, E.C. & Anderson, C.C. (1980) *Development and Validation of Methods for Sampling and Analysis of Workplace Toxic Substances (DHHS (NIOSH) Publication No. 80-133)*, Cincinnati, OH, National Institute for Occupational Safety and Health, pp. 1-3, 10-11, A1-A6

Hagemeyer, H.J. (1978) *Acetaldehyde*. In: Grayson, M., ed., *Kirk-Othmer Encyclopedia of Chemical Technology*, 3rd ed., Vol. 1, New York, John Wiley & Sons, pp. 97-112

Hagen, E. (1967) The composition of combustion gases of polyurethane foams and polycarbonates. *Plaste Kautsch.*, *6*, 391-392 (Translation distributed by the Defense Technical Information Center, Alexandria, VA, FTD-HT-23-1474-68)

Hald, J. & Larsen, V. (1949) The rate of acetaldehyde metabolism in rabbits treated with antabuse (tetraethylthiuramdisulphide). *Acta pharmacol. toxicol.*, *5*, 292-297

Handovsky, H. & Heymans, C. (1936) On the effect of acetaldehyde on rhythmicity and involuntary muscular tonus (Fr.). *C.R. Soc. Biol.*, *123*, 1242-1244

Hare, C.T. & Bradow, R.L. (1979) Characterization of heavy-duty diesel gaseous and particulate emissions, and effects of fuel composition. *Soc. Automot. Eng. Tech. Pap. Ser.*, No. 790490, 1-3, 7, 13, 18-20, 24

Hayes, E.R. (1963) *Acetaldehyde*. In: Kirk, R.E. & Othmer, D.F., eds, *Kirk-Othmer Encyclopedia of Chemical Technology*, 2nd ed., Vol. 1, John Wiley & Sons, pp. 92, 95

Hemminki, K. (1982) Urinary sulfur containing metabolites after administration of ethanol, acetaldehyde and formaldehyde to rats. *Toxicol. Lett.*, *11*, 1-6

Hemminki, K. & Suni, R. (1984) Sites of reaction of glutaraldehyde and acetaldehyde with nucleosides. *Arch. Toxicol.*, *55*, 186-190

Herrick, R.F. (1980) *Industrial Hygiene Survey Report on Urea Formaldehyde Foam Insulation Manufacturing (PB83-107797)*, Cincinnati, OH, National Institute for Occupational Safety and Health, pp. 2-5, 9-12

Hoffmann, D., Brunnemann, K.D., Gori, G.B. & Wynder, E.L. (1975) On the carcinogenicity of marijuana smoke. *Rec. Adv. Phytochem.*, *9*, 63-81

Horvat, R.J., Senter, S.D. & Dekazos, E.D. (1983) GLC-MS analysis of volatile constituents in rabbiteye blueberries. *J. Food Sci.*, *48*, 278-279

Hoshika, Y., Nihei, Y. & Muto, G. (1981) Pattern display for characterisation of trace amounts of odorants discharged from nine odour sources. *Analyst*, *106*, 1187-1202

IARC (1979) *IARC Monographs on the Evaluation of the Carcinogenic Risk of Chemicals to Humans*, Vol. 19, *Some Monomers, Plastics and Synthetic Elastomers, and Acrolein*, Lyon, pp. 341-366

IARC (1982) *IARC Monographs on the Evaluation of the Carcinogenic Risk of Chemicals to Humans*, Vol. 28, *The Rubber Industry*, Lyon

Igali, S. & Gazsó, L. (1980) Mutagenic effect of alcohol and acetaldehyde on *Escherichia coli* (Abstract No. 81). *Mutat. Res.*, *74*, 209-210

International Labour Office (1980) *Occupational Exposure Limits for Airborne Toxic Substances*, 2nd (rev.) ed. (*Occupational Safety and Health Series No. 37*), Geneva, pp. 34-35

James, T.N. & Bear, E.S. (1967) Effects of ethanol and acetaldehyde on the heart. *Am. Heart J.*, *74*, 243-255

Jansson, T. (1982) The frequency of sister chromatid exchanges in human lymphocytes treated with ethanol and acetaldehyde. *Hereditas*, *97*, 301-303

Jarke, F.H., Dravnieks, A. & Gordon, S.M. (1981) Organic contaminants in indoor air and their relation to outdoor contaminants. *Am. Soc. Heat. Refrig. Air-Cond. Eng. Trans.*, *87*, 153-166

Kami, T. (1983) Composition of the essential oil of alfalfa. *J. agric. Food Chem.*, *31*, 38-41

Kesäniemi, T.A. (1974) Metabolism of ethanol and acetaldehyde in intact rats during pregnancy. *Biochem. Pharmacol.*, *23*, 1157-1162

Kondakova, L.V., Trakhtengerts, T.Y., Shaposhnikov, Y.K., Sergeeva, V.V. & Lukanina, L.K. (1981) Composition of gaseous products formed during pulping by oxidative ammonolysis (Russ.). *Sb. Tr. Vses. Nauchno-Issled. Inst. Tsellyul.-Bum. Prom-sti.*, 40-42 [*Chem. Abstr.*, *97*, 57326t]

Korte, A., Obe, G., Ingwersen, I. & Rückert, G. (1981) Influence of chronic ethanol uptake and acute acetaldehyde treatment on the chromosomes of bone-marrow cells and peripheral lymphocytes of Chinese hamsters. *Mutat. Res.*, *88*, 389-395

Krost, K.J., Pellizzari, E.D., Walburn, S.G. & Hubbard, S.A. (1982) Collection and analysis of hazardous organic emissions. *Anal. Chem.*, *54*, 810-817

Kruysse, A. (1970) *Acute Inhalation Toxicity of Acetaldehyde in Hamsters (Report No. R 3270)*, Zeist, The Netherlands, Central Institute for Nutrition and Food Research TNO

Kruysse, A., Feron, V.J. & Til, H.P. (1975) Repeated exposure to acetaldehyde vapor. Studies in Syrian golden hamsters. *Arch. environ. Health*, *30*, 449-452

Kumar, S.P. (1982) Fetal alcohol syndrome mechanisms of teratogenesis. *Ann. clin. Lab. Sci.*, *12*, 254-257

Levaggi, D.A. & Feldstein, M. (1970) The determination of formaldehyde, acrolein, and low molecular weight aldehydes in industrial emissions on a single collection sample. *J. Air Pollut. Control Assoc.*, *20*, 312-313

Lipari, F. & Swarin, S.J. (1982) Determination of formaldehyde and other aldehydes in automobile exhaust with an improved 2,4-dinitrophenylhydrazine method. *J. Chromatogr.*, *247*, 297-306

Lipari, F., Dasch, J.M. & Scruggs, W.F. (1984) Aldehyde emissions from wood-burning fireplaces. *Environ. Sci. Technol.*, *18*, 326-330

Lipfert, F.W. & Dungan, J.L. (1982) National estimates of residential firewood and air pollution emissions. *Alternative Energy Sources*, *4*, 379-388

Lynch, C., Lim, C.K., Thomas, M. & Peters, T.J. (1983) Assay of blood and tissue aldehydes by HPLC analysis of their 2,4-dinitrophenylhydrazine adducts. *Clin. chim. Acta*, *130*, 117-122

Manning, D.L., Maskarinec, M.P., Jenkins, R.A. & Marshall, A.H. (1983) High performance liquid chromatographic determination of selected gas phase carbonyls in tobacco smoke. *J. Assoc. off. anal. Chem.*, *66*, 8-12

Marchner, H. & Tottmar, O. (1976) Influence of the diet on the metabolism of acetaldehyde in rats. *Acta pharmacol. toxicol.*, *38*, 59-71

Marjanen, L.A. (1973) Comparison of aldehyde dehydrogenases from cytosol and mitochondria of rat liver. *Biochim. biophys. Acta*, *327*, 238-246

Marnett, L.J., Hurd, H., Hollstein, M., Levin, D.E., Esterbauer, H. & Ames, B.N. (1984) Naturally occurring carbonyl compounds are mutagens in *Salmonella* tester strain TA104. *Mutat. Res.* (in press)

Nagasawa, H.T., Goon, D.J.W., Constantino, N.V. & Alexander, C.S. (1975) Diversion of ethanol metabolism by sulfhydryl amino acids. D-Penicillamine-directed excretion of 2,5,5-trimethyl-D-thiazolidine-4-carboxylic acid in the urine of rats after ethanol administration. *Life Sci.*, *17*, 704-714

Nagasawa, H.T., Goon, D.J.W., Muldoon, W.P. & Zera, R.T. (1984) 2-Substituted thiazolidine-4(R)-carboxylic acids as prodrugs of L-cysteine. Protection of mice against acetaminophen hepatotoxicity. *J. med. Chem.*, *27*, 591-596

National Academy of Sciences/National Research Council (1965) *Chemicals Used in Food Processing* (*Publication 1274*), Washington DC, p. 65

National Finnish Board of Occupational Safety and Health (1981) *Airborne Contaminants in the Workplaces* (*Safety Bulletin No. 3*), Helsinki, p. 7

National Institute for Occupational Safety and Health (1974) *Manual of Analytical Methods*, Rockville, MD (as reported by American Industrial Hygiene Association, *Hygienic Guide Series, Acetaldehyde*, Akron, OH)

National Institute for Occupational Safety and Health (1980) *Projected Number of Occupational Exposures to Chemical and Physical Hazards*, Cincinnati, OH, p. 1

National Institute for Occupational Safety and Health (1981) *National Occupational Hazard Survey* (microfiche), Cincinnati, OH, p. 14 (generics excluded)

National Research Council (1981) *Food Chemicals Codex*, 3rd ed., Washington DC, National Academy Press, pp. 353, 354-355, 583, 615

National Research Council/National Academy of Sciences (1979) *The 1977 Survey of Industry on the Use of Food Additives*, Part 1 of 3 (*PB 80-113418*), Springfield, VA, National Technical Information Service, pp. 310-312

National Swedish Board of Occupational Safety and Health (1981) *Hygiene Limit Values* (*AFS 1981:8*), Stockholm, p. 9

Nomura, F. & Lieber, C.S. (1981) Binding of acetaldehyde to rat liver microsomes: Enhancement after chronic alcohol consumption. *Biochem. biophys. Res. Commun.*, *100*, 131-137

Obe, G. (1981) *Acetaldehyde not ethanol is mutagenic*. In: Kappas, A., ed., *Progress in Mutation Research*, Vol. 2, *Progress in Environmental Mutagenesis and Carcinogenesis*, Amsterdam, Elsevier/North-Holland, pp. 19-23

Obe, G. & Beek, B. (1979) Mutagenic activity of aldehydes. *Drug Alcohol Depend.*, *4*, 91-94

Obe, G. & Ristow, H. (1977) Acetaldehyde, but not ethanol, induces sister chromatid exchanges in Chinese hamster cells *in vitro*. *Mutat. Res.*, *56*, 211-213

Obe, G. & Ristow, H. (1979) Mutagenic, cancerogenic and teratogenic effects of alcohol. *Mutat. Res.*, *65*, 229-259

Obe, G., Ristow, H. & Herha, J. (1978) *Mutagenic activity of alcohol in man*. In: *Mutations: Their Origin, Nature and Potential Relevance to Genetic Risk in Man. Deutsche Forschungsgemeinschaft, Jahreskonferenz 1977*, Boppard, Harald Boldt Verlag, pp. 151-161

Obe, G., Natarajan, A.T., Meyers, M. & Den Hertog, A. (1979) Induction of chromosomal aberrations in peripheral lymphocytes of human blood *in vitro,* and of SCEs in bone-marrow cells of mice *in vivo* by ethanol and its metabolite acetaldehyde. *Mutat. Res.*, *68*, 291-294

Obe, G., Brodmann, R., Fleischer, R., Engeln, H., Göbel, D. & Herha, J. (1984) *Mutagenic and carcinogenic activity of addict products* (Ger.). In: Keup, W., ed., *Biologie der Sucht* (Biology of Addiction), Berlin, Springer-Verlag (in press)

Ohshima, H., O'Neill, I.K., Friesen, M., Béréziat, J.-C. & Bartsch, H. (1984) Occurrence in human urine of new sulphur-containing *N*-nitrosoamino acids, *N*-nitrosothiazolidine 4-carboxylic acid and its 2-methyl derivative, and their formation. *J. Cancer Res. clin. Oncol., 108*, 121-128

Okamoto, M., Ohtsuka, K., Imai, J. & Yamada, F. (1981) High-performance liquid chromatographic determination of acetaldehyde in wine as its lutidine derivatives. *J. Chromatogr., 219*, 175-178

O'Shea, K.S. & Kaufman, M.H. (1979) The teratogenic effect of acetaldehyde: Implications for the study of the fetal alcohol syndrome. *J. Anat., 128*, 65-76

O'Shea, K.S. & Kaufman, M.H. (1981) Effect of acetaldehyde on the neuroepithelium of early mouse embryos. *J. Anat., 132*, 107-118

Parrilla, R., Ohkawa, K., Lindros, K.O., Zimmerman, U.-J.P., Kobayashi, K. & Williamson, J.R. (1974) Functional compartmentation of acetaldehyde oxidation in rat liver. *J. biol. Chem., 249*, 4926-4933

Pellizzari, E.D. (1979) *Information on the Characteristics of Ambient Organic Vapors in Areas of High Chemical Production (Draft Report, Contract No. 68-02-2721)*, Research Triangle Park, NC, US Environmental Protection Agency, pp. 1, 51-52, 56-62, 64-66, 71-73, 79-81, 94-96, 103-104, 119-120

Pellizzari, E.D., Hartwell, T.D., Harris, B.S.H., III, Waddell, R.D., Whitaker, D.A. & Erickson, M.D. (1982) Purgeable organic compounds in mother's milk. *Bull. environ. Contam. Toxicol., 28*, 322-328

Piendl, A., Westner, H. & Geiger, E. (1981) Detection and separation of aldehydes in beer using high pressure liquid chromatography (HPLC) (Ger.). *Brauwisssenschaft, 34*, 30l-307 [*Chem. Abstr., 96*, 4870p]

Pino, J.A. (1982) Analysis of volatile carbonyl components of aqueous condensate obtained from the concentration of sweet orange juice (Span.). *Rev. Cienc. Quím., 13*, 125-132 [*Chem. Abstr., 99*, 69097y]

Popov, V.B., Vaisman, B.L., Puchkov, V.F. & Ignatyeva, T.V. (1981) Embryotoxic effect of ethanol and biotransformation products in the culture of postimplantation of rat embryos (Russ.). *Bull. exp. Biol. Med., 92*, 725-728

Proctor, N.H. & Hughes, J.P. (1978) *Acetaldehyde*. In: Proctor, N.N. & Hughes, J.P., eds, *Chemical Hazards of the Workplace*, Philadelphia, J.B. Lippincott Co., pp. 79-80

de Raat, W.K, Davis, P.B. & Bakker, G.L. (1983) Induction of sister-chromatid exhanges by alcohol and alcoholic beverages after metabolic activation by rat-liver homogenate. *Mutat. Res., 124*, 85-90

Ramdahl, T., Alfheim, I., Rustad, S. & Olsen, T. (1982) Chemical and biological characterization of emissions from small residential stoves burning wood and charcoal. *Chemosphere*, *11*, 601-611

Rieger, R. & Michaelis, A. (1960) Chromosome aberrations induced by acetaldehyde on primary roots of *Vicia faba* (Ger.). *Biol. Zentralbl.*, *79*, 1-5

Ristow, H. & Obe, G. (1978) Acetaldehyde induces cross-links in DNA and causes sister-chromatid exchanges in human cells. *Mutat. Res.*, *58*, 115-119

Rosenkranz, H.S. (1977) Mutagenicity of halogenates, alkanes and their derivatives. *Environ. Health Perspect.*, *21*, 79-84

Rosensteel, R.E. & Tanaka, S. (1976) *Health Hazard Evaluation Determination Report 75-89-344, Rock Hill Printing and Finishing Company, Rock Hill, South Carolina*, Cincinnati, OH, National Institute for Occupational Safety and Health

Rudling, L., Ahling, B. & Löfroth, G. (1982) *Chemical and biological characterization of emissions from combustion of wood and wood-chips in small furnaces and stoves*. In: Cooper, J.A. & Malik, D., eds, *Proceedings of an International Conference on Residues in Solid Fuels: Environmental Impacts Solutions*, Beaverton, OR, Oregon Graduate Center, pp. 34-53

Sasaki, Y. & Endo, R. (1978) Mutagenicity of aldehydes in *Salmonella* (Abstract No. 27). *Mutat. Res.*, *54*, 251-252

Sawicki, E., Hauser, T.R., Stanley, T.W. & Elbert, W. (1961) The 3-methyl-2-benzothiazolone hydrazone test. Sensitive new methods for the detection, rapid estimation, and determination of aliphatic aldehydes. *Anal. Chem.*, *33*, 93-96

Schmidt, R.H., Davidson, S.M. & Lowry, S.P. (1983) Determination of acetaldehyde in *Streptococcus lactis* cultures as 2,4-dinitrophenylhydrazone by high-performance liquid chromatography. *J. agric. Food Chem.*, *31*, 978-980

Serth, R.W., Tierney, D.R. & Hughes, T.W. (1978) *Source Assessment: Acrylic Acid Manufacture, State of the Art (EPA-600/2-78-004w)*, Cincinnati, OH, Industrial Environmental Research Laboratory, US Environmental Protection Agency, pp. vi-xiii, 1-5, 25-38, 46-51

Shackelford, W.M. & Keith, L.H. (1976) *Frequency of Organic Compounds Identified in Water (EPA-600/4-76-062)*, Athens, GA, US Environmental Protection Agency, pp. 7, 37, 46

Silverman, L., Schulte, H.F. & First, M.W. (1946) Further studies on sensory response to certain industrial solvent vapors. *J. ind. Hyg. Toxicol.*, *28*, 262-266

Sim, V.M. & Pattle, R.E. (1957) Effect of possible smog irritations on human subjects. *J. Am. med. Assoc.*, *165*, 1908-1913

Singh, H.B., Salas, L.J., Stiles, R. & Shigeishi, H. (1982) *Measurements of Hazardous Organic Chemicals in the Ambient Atmosphere (SRI International Project 7774)*, Research Triangle Park, NC, Environmental Sciences Research Laboratory, US Environmental Protection Agency, pp. 32-37, 70-79, 84-90

Skog, E. (1950) A toxicological investigation of lower aliphatic aldehydes. I. Toxicity of formaldehyde, acetaldehyde, propionaldehyde and butyraldehyde; as well as of acrolein and crotonaldehyde. *Acta pharmacol., 6*, 299-318

Spingarn, N.E., Northington, D.J. & Pressely, T. (1982) Analysis of volatile hazardous substances by GC/MS. *J. chromatogr. Sci., 20*, 286-288

Sprince, H., Parker, C.M., Smith, G.G. & Gonzales, L.J. (1974) Protection against acetaldehyde toxicity in the rat by L-cysteine, thiamin and L-2-methylthiazolidine-4-carboxylic acid. *Agents Actions, 4*, 125-130

Sreenathan, R.N., Padmanabhan, R. & Singh, S. (1982) Teratogenic effects of acetaldehyde in the rat. *Drug Alcohol Depend., 9*, 339-350

Stege, T.E. (1982) Acetaldehyde-induced lipid peroxidation in isolated hepatocytes. *Chem. Pathol. Pharmacol., 36*, 287-297

Stepanova, E.N. & Tsapalova, I.E. (1982) Flavor-forming components of some kinds of fresh mushrooms (Russ.). *Izv. Vyssh. Uchebn. Zaved., Pishch. Tekhnol., 5*, 154 [*Chem. Abstr., 98*, 33403k]

Stratton, R., Dormer, K.J. & Zeiner, A.R. (1981) The cardiovascular effects of ethanol and acetaldehyde in exercising dogs. *Alcohol. clin. exp. Res., 5*, 56-63

Suitheimer, C., Bost, R. & Sunshine, I. (1982) *Volatiles by headspace chromatography*. In: Sunshine, I. & Jatlow, P.I., eds, *Methodology for Analytical Toxicology*, Vol. 2, Boca Raton, FL, CRC Press, pp. 1-9

Teschke, R., Hasumura, Y. & Lieber, C.S. (1977) Hepatic pathways of ethanol and acetaldehyde metabolism and their role in the pathogenesis of alcohol-induced liver injury. *Nutr. Metab., 21* (Suppl. 1), 144-147

Thibodeaux, L.J., Parker, D.G. & Heck, H.H. (1982) *Measurement of Volatile Chemical Emissions from Wastewater Basins (EPA-600/2-82-095)*, Cincinnati, OH, Industrial Environmental Research Laboratory, US Environmental Protection Agency, pp. iii-vii, 1-4, 24-36

Thompson, P.A.C. & Folb, P.I. (1982) An in vitro model of alcohol and acetaldehyde teratogenicity. *J. appl. Toxicol., 2*, 190-195

Tottmar, S.O.C., Pettersson, H. & Kiessling, K.-H. (1973) The subcellular distribution and properties of aldehyde dehydrogenases in rat liver. *Biochem. J., 135*, 577-586

Truitt, E.B., Jr & Walsh, M.J. (1971) *The role of acetaldehyde in the actions of ethanol*. In: Kissin, B. & Begleiter, H., eds, *The Biology of Alcoholism*, Vol. 1, Biochemistry, New York, Plenum Press, pp. 161-195

Tsuda, M., Hirayama, T. & Sugimura, T. (1983) Presence of *N*-nitroso-L-thioproline and *N*-nitroso-L-methylthioprolines in human urine as major *N*-nitroso compounds. *Gann, 74*, 331-333

US Department of Commerce (1983) *US Imports for Consumption and General Imports (FT246/Annual 1982)*, Bureau of the Census, Washington DC, US Government Printing Office, p. 1-289

US Department of the Treasury (1983) Formulas for denatured alcohol and rum. *US Code Fed. Regul., Title 27*, Parts 21, 212; *Fed. Regist. 48* (No. 107), pp. 24672-24675, 24679, 24682

US Department of Transportation (1982) Performance-oriented packagings standards. *US Code Fed. Regul., Title 49*, Parts 171, 172, 173, 178; *Fed. Regist., 47* (No. 73), pp. 16268, 16273

US Environmental Protection Agency (1975) *Preliminary Assessment of Suspected Carcinogens in Drinking Water: Report to Congress (EPA-560/4-75-005, PB-250-961)*, Washington, DC, pp. 1-15, I-i, I-13, II-1-II-2

US Environmental Protection Agency (1980) Hazardous waste management system: Identification and listing of hazardous wastes. *US Code Fed. Regul., Title 40*, Part 261; *Fed. Regist., 45* (No. 98), pp. 33084, 33122-33127, 33131-33133

US Environmental Protection Agency (1982) Acetaldehyde; exemption from the requirement of a tolerance. *US Code Fed. Regul., Title 40*, Part 180.1031

US Environmental Protection Agency (1983) Notification requirements; reportable quantity adjustments. *US Code Fed. Regul., Title 40*, Part 302; *Fed. Regist., 48* (No. 102), pp. 23552, 23571

US Food and Drug Administration (1980) Food and drugs. *US Code Fed. Regul., Title 21*, Parts 177.2410, 182.60, 582.60

US Food and Drug Administration (1982) Starter distillate and diacetyl; proposed GRAS status as direct human food ingredients. *US Code Fed. Regul., Title 21*, Parts 172, 182, 184; *Fed. Regist. 47* (No. 152), pp. 34155, 34156

Van Langenhove, H.R., Van Acker, M. & Schamp, N.M. (1983) Quantitative determination of carbonyl compounds in rendering emissions by reversed-phase high-performance liquid chromatography of the 2,4-dinitrophenylhydrazones. *Analyst, 108*, 329-334

Véghelyi, P.V. & Osztovics, M. (1978) The alcohol syndromes: The intrarecombigenic effect of acetaldehyde. *Experientia, 34*, 195-196

Véghelyi, P.V. Osztovics, M., Kardos, G., Leisztner, L., Szaszovszky, E., Igali, S. & Imrei, J. (1978) The fetal alcohol syndrome: Symptoms and pathogenesis. *Acta pediat. acad. sci. hung., 19*, 171-189

Versar, Inc. (1975) *Identification of Organic Compounds in Effluents from Industrial Sources (EPA-560/3-75-002, 2B-241 841)*, Washington DC, US Environmental Protection Agency, pp. C-2, 4-14, 4-15, 4-17

Verschueren, K. (1977) *Handbook of Environmental Data on Organic Chemicals*, New York, Van Nostrand Reinhold Co., pp. 57-59

Watanabe, F. & Sugimoto, S. (1956) Study on the carcinogenicity of aldehyde. 3rd Report. Four cases of sarcomas of rats appearing in the areas of repeated subcutaneous injections of acetaldehyde (Jpn.). *Gann, 47*, 599-601

Webster, W.S., Walsh, D.A., McEwen, S.E. & Lipson, A.H. (1983) Some teratogenic properties of ethanol and acetaldehyde in C57BL/6J mice: Implications for the study of the fetal alcohol syndrome. *Teratology, 27*, 231-243

Windholz, M., ed. (1983) *The Merck Index,* 10th ed., Rahway, NJ, Merck & Co., p. 6

Woutersen, R.A., Appelman, L.M., Feron, V.J. & van der Heijden, C.A. (1984) Inhalation toxicity of acetaldehyde in rats. II. Carcinogenicity study: Interim results after 15 months. *Toxicology, 31,* 123-133

ACROLEIN

This substance was considered by a previous working group, in February 1978 (IARC, 1979). Since that time, new data have become available, and these have been incorporated into the monograph and taken into account in the present evaluation.

1. Chemical and Physical Data

1.1 Synonyms and trade names

Chem. Abstr. Services Reg. No.: 107-02-8

Chem. Abstr. Name: 2-Propenal

IUPAC Systematic Name: Acrolein

Synonyms: Acraldehyde; *trans*-acrolein; acrylaldehyde; acrylic aldehyde; allyl aldehyde; ethylene aldehyde; NSC 8819; propenal; prop-2-en-1-al

Trade Names: Acquinite; Aqualin; Aqualine; Crolean; Magnacide; Magnacide H

1.2 Structural and molecular formulae and molecular weight

$$CH_2 = CH - CHO$$

C_3H_4O Mol. wt: 56.1

1.3 Chemical and physical properties of the pure substance

From Hess *et al.* (1978), unless otherwise specified

(a) *Description*: Colourless liquid that causes lachrymation

(b) *Boiling-point*: 53°C

(c) *Melting-point*: -87.0°C

(d) *Density*: Specific gravity (20/20°C), 0.8427

(e) *Refractive index*: n_D^{20} 1.4013

(f) *Spectroscopy data*: λ_{max} 207 nm (E_1^1 = 2000); infrared, Raman, nuclear magnetic resonance and mass spectral data have been tabulated (Grasselli & Ritchey, 1975)

(g) *Solubility*: Soluble in diethyl ether and ethanol (Windholz, 1983); 20.6 wt % dissolves in water at 20°C

(h) *Viscosity*: 0.35 cP at 20°C

(i) *Volatility*: Vapour pressure, 220 mm Hg at 20°C

(j) *Stability*: Flash-point, -26°C (Buckingham, 1982); light and air, or presence of alkali or strong acid, catalyse polymerization, forming disacryl (Windholz, 1983); stability is very pH-dependent.

(k) *Reactivity*: Reactive as both an olefin and an aldehyde and, because of the carbonyl-double bond conjugation, undergoes Diels-Alder reactions such as dimerization

(l) *Conversion factor*: 1 ppm = 2.3 mg/m^3 at 760 mm Hg and 25°C (Verschueren, 1977)

1.4 Technical products and impurities

Acrolein is available in the USA with the following specifications: purity, 95.5 wt % min; water, 3.0 wt % max; total carbonyls other than acrolein, 1.5 wt % max; hydroquinone (stabilizer) (IARC, 1977), 0.10-0.25 wt %; specific gravity (20°/20°C), 0.842-0.846; and pH 6.0 max for a 10% aqueous solution. Propionaldehyde and acetone are the principal carbonyl impurities (Hess *et al.*, 1978).

2. Production, Use, Occurrence and Analysis

2.1 Production and use

(a) *Production*

Acrolein was first prepared in 1843 by Redtenbacher by the dry distillation of fat (Prager *et al.*, 1918). It was produced commercially starting in 1938 by the vapour-phase condensation of acetaldehyde and formaldehyde. In 1959, the direct oxidation of propylene in the presence of a catalyst became the preferred commercial process, and variations of this process are the only methods currently used commercially (Hess *et al.*, 1978). The acetaldehyde-formaldehyde route was last used in the USA in 1970.

Acrolein was first produced commercially in the USA in 1955 (US Tariff Commission, 1956). US production of acrolein other than as an unisolated intermediate for acrylic acid and its esters amounted to an estimated 29 thousand tonnes in 1979. In addition, 180-185 thousand tonnes of acrolein were produced as an unisolated intermediate during the production of acrylic acid and its esters. Currently, only one US company manufactures acrolein for sale but this company and three others produce it as an intermediate in the synthesis of acrylic acid and its esters. Separate data on US imports and exports of acrolein are not published.

Acrolein is produced for sale in western Europe by one company each in France and Germany, and a second company in each of these countries produces acrolein as an intermediate in the synthesis of acrylic acid and its esters. All of these companies make acrolein by the catalytic oxidation of propylene, and their combined annual production is estimated to total 50-100 thousand tonnes.

The commercial production of acrolein started in Japan in 1960. Three Japanese companies currently manufacture it for sale, and 1983 production is estimated to have been 5-6 thousand tonnes. In addition, one of these companies and three others produce acrolein as an intermediate in the synthesis of acrylic acid and its esters. All of these companies make acrolein by the catalytic oxidation of propylene.

Worldwide production of acrolein (probably only isolated material) in 1975 has been estimated to have been about 59 thousand tonnes (Hess *et al.*, 1978).

(b) Use

The principal use for acrolein is as a chemical intermediate for acrylic acid and its esters. It also finds major usage as a biocide and in the synthesis of methionine and its hydroxy analogue. Smaller quantities are used in a variety of other applications.

When propylene is converted to acrylic acid and its esters, acrolein is produced in an initial catalytic oxidation step; it is then further oxidized to acrylic acid, most of which is converted to its lower alkyl esters (see IARC, 1979, for information on the uses of these chemicals).

Acrolein is used directly as an aquatic herbicide and algicide in irrigation canals (to control grass and weeds), as a microbiocide in oil wells, liquid hydrocarbon fuels, cooling-water towers and water-treatment ponds, and as a slimicide in the manufacture of paper and paperboard.

Acrolein is also used as a chemical intermediate for DL-methionine, its hydroxy analogue, and their salts. These compounds are used as protein supplements in poultry feeds, and DL-methionine is also used in human medicine. Combined US production of these products in 1981 is estimated to have been approximately 32 thousand tonnes (99% DL-methionine basis).

Other commercially significant derivatives of acrolein are glutaraldehyde, allyl alcohol and β-picoline. The latter, prepared by the reaction of acrolein and ammonia, is used commercially in the synthesis of the antipellagra vitamins, niacin and niacinamide.

Acrolein is used to make 3-cyclohexene-1-carboxaldehyde (tetrahydrobenzaldehyde), an intermediate for cycloaliphatic epoxy resins and 1,2,6-hexanetriol. It has been used to make its homopolymer and several copolymers (e.g., with acrylic acid, acrylonitrile [see IARC, 1979] and acrylic esters) as well as polymers by reaction with formaldehyde (see IARC, 1982a), guanidine hydrochloride, and aliphatic diamines (e.g., ethylene diamine). It has also been used to make modified food starch (alone or in combination with vinyl acetate [see IARC, 1979]), and it was used in the USA as a chemical intermediate in one method to make synthetic glycerol until 1980 when the only plant using this process was shut down.

Acrolein has reportedly been used in the manufacture of colloidal forms of metals, in the production of perfumes, and as a warning agent in methyl chloride refrigerant (Windholz, 1983). During the First World War, it was used as a poison gas (Izard & Libermann, 1978), and 183 thousand kg were produced between 1914 and 1918 (Champeix & Catilina, 1967).

In Japan, acrolein is used principally as a chemical intermediate in the synthesis of DL-methionine.

Occupational exposure to acrolein has been limited by regulation or recommended guidelines in at least 17 countries. The standards are listed in Table 1.

Acrolein has been approved for use as an etherifying agent, at levels not to exceed 0.6%, in the preparation of modified food starch and, subject to certain limitations, for use as a slimicide in the manufacture of paper and paperboard products in contact with food (US Food and Drug Administration, 1980).

The US Environmental Protection Agency (EPA) (1982) has classified all pesticides containing acrolein for restricted use; these products are limited to use by or under the direct supervision of a certified applicator. The EPA has also identified acrolein as a toxic waste

Table 1. National occupational exposure limits for acrolein[a]

Country	Year	Concentration (mg/m^3)	ppm	Interpretation[b]	Status
Australia	1978	0.25	0.1	TWA	Guideline
Belgium	1978	0.25	0.1	TWA	Regulation
Czechoslovakia	1976	0.5	-	TWA	Regulation
		1	-	Ceiling (10 min)	
Finland	1981	0.25	0.1	Ceiling	Guideline
German Democratic Republic	1979	0.25	-	TWA	Regulation
		0.25	-	Maximum (30 min)	
Germany, Federal Republic of	1984	0.25	0.1	TWA	Guideline
Hungary	1974	0.7	-	TWA[c]	Regulation
Italy	1978	0.25	0.1	TWA	Guideline
		0.75	0.3	STEL	
Japan	1978	0.25	0.1	Ceiling	Guideline
Netherlands	1978	0.25	0.1	TWA	Guideline
Poland	1976	0.5	-	Ceiling	Regulation
Romania	1975	0.3	-	TWA	Guideline
		0.5	-	Maximum	
Sweden	1981	0.7	0.3	STEL	Guideline
		0.2	0.1	TWA	
Switzerland	1978	0.25	0.1	TWA	Regulation
USSR	1977	0.2	-	Maximum	Regulation
USA[d]					
OSHA	1978	0.25	0.1	TWA	Regulation
		-	5	Maximum (30 min)	
ACGIH	1984/85	0.25	0.1	TWA	Guideline
		0.8	0.3	STEL	
Yugoslavia	1971	0.25	0.1	Ceiling	Regulation

[a]From International Labour Office (1980); National Finnish Board of Occupational Safety and Health (1981); National Swedish Board of Occupational Safety and Health (1981); American Conference of Governmental Industrial Hygienists (1984); Deutsche Forschungsgemeinschaft (1984)
[b]TWA, time-weighted average; STEL, short-term exposure limit
[c]Sensitizer notation added
[d]OSHA, Occupational Safety and Health Administration; ACGIH, American Conference of Governmental Hygienists

and requires that persons who generate, transport, treat, store or dispose of it comply with the regulations of a federal hazardous waste management programme (US Environmental Protection Agency, 1980a).

In March 1983, the EPA proposed standards to limit effluent discharges of acrolein into publicly owned treatment works from point sources in the plastics and synthetic fibres industry. The proposed limit is 50 µg/l max per day (US Environmental Protection Agency, 1983a). The EPA also requires that notification be given whenever discharges containing 0.454 kg or more of acrolein are made into waterways (US Environmental Protection Agency, 1983b).

As part of the Hazardous Materials Regulations of the US Department of Transportation (1982), shipments of acrolein are subject to a variety of labelling, packaging, quantity and shipping restrictions consistent with its designation as a hazardous material.

2.2 Occurrence

(a) *Natural occurrence*

Acrolein has been identified as a volatile component of essential oils extracted from the wood of oak trees (Egorov *et al.*, 1976).

(b) *Occupational exposure*

On the basis of the 1974 National Occupational Hazard Survey, the National Institute for Occupational Safety and Health (1980, 1981) estimated that 7.3 thousand US workers in 16 nonagricultural industries were exposed to acrolein. The principal industries in which exposure was found were hospitals, the aircraft industry and the primary metal products industry.

Acrolein was detected in a truck-maintenance shop in the USA at a mean concentration of 4.6 µg/m^3 (Castle & Smith, 1974). The following exposures to acrolein in workplace air have also been reported: (1) levels of 0.44-1.5 mg/m^3 in a rubber-vulcanization plant producing styrene-butadiene rubber-footwear components (Volkova & Dagdinov, 1969); (2) 0.11-1.04 mg/m^3 during the welding of metals coated with anti-corrosion primers (Protsenko *et al.*, 1973); (3) 0.22-0.32 mg/m^3 in pitch-coking plants, 0.004-0.014 in coal-coking plants (Mašek, 1972); and (4) <0.1 mg/m^3 from diesel train-engine exhaust during repair and servicing (Apol, 1973). Acrolein was found at quarries in exhaust gases from diesel engines and in the workplace air at levels of 2.1-7.2 mg/m^3 (Klochkovskii *et al.*, 1981).

(c) *Air*

A review on levels of acrolein reported in ambient air has been published (Carson *et al.*, 1981).

Acrolein has been found at very low concentrations (0.44-32 µg/m^3) in ambient air in urban and suburban areas. Air-monitoring data obtained between 1961 and 1976 show that this compound occurred at mean ambient levels of 16 µg/m^3 in Los Angeles, CA (urban atmosphere, 42 data points) and at mean levels of 0.7 µg/m^3 in Edison, NJ (near emissions source, 19 data points) (Brodzinsky & Singh, 1982). In another study, acrolein was detected in the air of the Los Angeles Basin over a period of 12 weeks in 1968 at levels of none detected-0.04 mg/m^3; most values recorded, however, were between 0.002 and 0.02 mg/m^3 (Scott Research Laboratories, 1969).

Acrolein has been determined to be one product of the photooxidation of 1,3-butadiene in air (Maldotti et al., 1980). This photooxidation reaction may contribute to significant ambient atmospheric levels of acrolein because of the occurrence of 1,3-butadiene at concentrations of approximately 4.6 µg/m^3 in urban ambient air (Brodzinsky & Singh, 1982).

An estimated 46.7 tonnes of acrolein were emitted into the US atmosphere during 1978 (Anderson et al., 1980).

Acrolein has been identified as an emission from plants manufacturing acrylic acid (Serth et al., 1978). Acrolein emissions have also been reported from coffee-roasting operations (none detected-0.6 mg/m^3), from a lithographic plate coater (<0.23-3.9 mg/m^3) and from an automobile-spray booth (1.1-1.6 mg/m^3) (Levaggi & Feldstein, 1970). Additional sources of atmospheric acrolein that have been identified include turbine engines, the manufacture of fish oils, lacquers, plastics and synthetic rubber, forest fires and spray painting (Graedel, 1978).

It has been estimated that acrolein, acetone and low-molecular-weight fatty acids are emitted at the rate of 1 million kg/year during the manufacture of oxidation-hardening enamels in the Netherlands (Doorgeest, 1970). Acrolein was detected in the USSR in air samples from populated areas located in the vicinity of three enamelled wire manufacturing plants (Vorob'eva et al., 1982). It has also been detected in ventilation gases from paint and varnish preparation and distributing shops in the USSR (Stepanova et al., 1983).

Acrolein was detected among other trace odours in air in Japan: (1) in exhaust gas from a metal paint drier (6.1 mg/m^3); (2) in exhaust gas from two poultry-manure dryers (3.1-4.2 mg/m^3); and (3) in exhaust gas from a corn-starch manufacturing works (1.8 mg/m^3) (Hoshika et al., 1981).

(d) *Water and sediments*

Analysis of municipal effluents in Dayton, OH, showed the presence of acrolein in six of 11 samples, with concentrations ranging from 20-200 µg/l (US Environmental Protection Agency, 1980b).

(e) *Food, beverages and animal feeds*

A review on the occurrence of acrolein in various foods and beverages was published in 1980 (US Environmental Protection Agency, 1980b).

Greenhoff and Wheeler (1981) detected acrolein in fresh lager beer at levels of 1.11-2.00 µg/l (mean, 1.60 µg/l) and in lager aged under forced conditions at a mean level of 5.05 µg/l.

Acrolein has been separated from sugar-cane molasses (Hrdlicka & Janicek, 1968). It has also been detected in: (1) souring salted pork (Cantoni et al., 1969); (2) the fish odour of cooked horse mackerel (Shimomura et al., 1971); (3) the aroma volatiles of white bread (Mulders & Dhont, 1972); (4) the volatile components of raw chicken breast muscle (Grey & Shrimpton, 1967); and (5) the aroma volatiles of ripe arctic bramble berries (Kallio & Linko, 1973).

(f) Tobacco, tobacco smoke and marijuana smoke

Reviews on the occurrence of acrolein in tobacco smoke have been published (US Environmental Protection Agency, 1980b; Carson et al., 1981). The results of measurements of acrolein in the smoke of tobacco and marijuana cigarettes are summarized in Table 2.

Table 2. Acrolein in the smoke of tobacco and marijuana cigarettes

Cigarette	Acrolein concentration (μg/cigarette)	Reference
Tobacco cigarettes		
6 Experimental brands	60-220	Manning et al. (1983)
3 Low-tar brands	3-20	Manning et al. (1983)
32 Commercial brands	3-141	Jenkins et al. (1983)
32 Commercial low-tar brands	10-109	Griest et al. (1977)
Standard 85-mm blended cigarettes of the National Cancer Institute	85	Hoffmann et al. (1975)
Commercial filter brand	13-37	Magin (1980)
Commercial low-delivery filter brand	3-4	Magin (1980)
Commercial non-filter brand	20-25	Magin (1980)
Marijuana cigarettes	92	Hoffmann et al. (1975)

(g) Other pyrolysis products

A review on the occurrence of acrolein in various combustion products has been published (US Environmental Protection Agency, 1980b).

Acrolein has been found among the products from heating animal fats or vegetable oils (Bauer et al., 1977; Izard & Libermann, 1978). It was detected (Robles, 1968) among the decomposition products of cellophane (nitrocellulose-coated regenerated cellulose sheet) used in sealing meat packages; however, airborne concentrations of acrolein were below the detection limit (0.02 mg/m^3) during processing of thermoplastics containing polyethylene and styrene (Pfäffli, 1982). A study of the acrolein content of emissions from several coating operations where the coating was dried or baked in an oven to remove solvent found concentrations as high as 25 mg/m^3 in samples taken after the emissions had passed through an afterburner (Stahl, 1969).

Acrolein can be produced during the manufacture of candles as a decomposition product of overheated wax (Tanne, 1983). It has been detected in diesel-engine exhaust at levels of 0.9-19.6 mg/m^3 (Saito et al., 1983), 0.06-0.13 mg/m^3 (Swarin & Lipari, 1983) and 0.5-0.8 mg/m^3 (Smythe & Karasek, 1973). It has also been reported in gasoline-engine exhaust at levels of 0.46-12.2 mg/m^3 (Verschueren, 1977) and in exhaust from a gasoline rotary engine at a level of 0.46 mg/m^3 (Hoshika & Takata, 1976). Acrolein was found by Zinn et al. (1980) among the major volatile components of the combustion products of hydraulic fluid. On the basis of a 20-gallon [76-l] sample burning into a space of 25 000 ft^3 [708 m^3], the acrolein concentration was calculated to be 200 mg/m^3.

Acrolein has been found in the smoke resulting from combustion of wood (115 mg/m^3), kerosene (<2.3 mg/m^3) and cotton (138 mg/m^3) (Einhorn, 1975). In another study, fireplace acrolein emissions ranged from 21-132 mg/kg of wood burned (Lipari et al., 1984).

Acrolein has been identified in cool flame-combustion products (Cohen & Altshuller, 1961).

2.3 Analysis

Reviews of available methods for the measurement of acrolein concentrations were published in 1969 and 1980 (Stahl, 1969; US Environmental Protection Agency, 1980a).

Typical methods for the analysis of acrolein in various matrices are summarized in Table 3.

An analyser for the continuous determination of acrolein in the atmosphere has been developed based on the colorimetric determination of the complex formed by acrolein with 4-hexylresorcinol. The operational range of the analyser is 0.023-39 mg/m^3 (Reddish, 1982).

Table 3. Methods for the analysis of acrolein

Sample matrix	Sample preparation	Assay procedure[a]	Limits of detection	Reference
Air	Collect with molecular sieves; react with o-aminobiphenyl	Fluorimetry	<0.056 µg (60-litre samples)	Suzuki & Imai (1982)
	Collect in cartridge containing Tenax GC or other resin; desorb thermally and collect in a trap cooled by liquid nitrogen	GC/MS	0.1 µg/m^3	Krost et al. (1982)
	Collect with absorption device containing chromotropic acid/sulphuric acid	S (450 nm)	0.5 mg/m^3	Gronsberg (1974)
	Trap using liquid argon	GC	0.1-5 µg/m^3	Hoshika et al. (1981)
	Collect in an impinger containing sodium bisulphide; treat with trichloroacetic acid	C	Not given	Mackison et al. (1981)
		Direct-reading devices	Not given	Mackison et al. (1981)
Workplace air	Collect in molecular sieve tubes; dissolve (water)	GC	0.5-2.3 mg/m^3 (range of operation)	Singal & Love (1981)
Air in mines containing exhaust from vehicles	Collect in an absorber containing thiosemicarbazide solution	S (290 nm)	0.4 µg/sample	Shadrin (1970)
Air and automobile exhaust	Collect in bubblers containing ethanol; add 4-hexylresorcinol, mercuric chloride and trichloroacetic acid solutions	S (605 nm)	0.2 mg/m^3 (10-litre samples)	Cohen & Altshuller (1961)
Automobile exhaust	Collect gases	Chemical ionization MS (using ammonia as a reagent gas)	Not given	Day et al. (1971)
	Collect in impingers containing 2,4-dinitrophenylhydrazine in acidic solution; extract (chloroform); add internal standard (anthracene)	GC	0.44 mg/m^3 (10-litre samples)	Saito et al. (1983)
	Dilute; bubble through impingers containing 2-diphenylacetyl-1,3-indandione-1-hydrazone in acetonitrile and hydrochloric acid catalyst	HPLC/F	1.4 µg/m^3 (20-litre samples)	Swarin & Lipari (1983)
	Dilute; bubble through midget impingers containing 2,4-dinitrophenylhydrazine	HPLC/UV	11 µg/m^3 (20-litre samples)	Lipari & Swarin (1982)
	Adsorb with diatomaceous earth; immerse in an acetone/dry-ice bath with recurrent agitation; desorb thermally	Microwave S	Not given	Tanimoto & Uehara (1975)

Sample matrix	Sample preparation	Assay procedure[a]	Limits of detection	Reference
Aqueous solution	Derivatize with 2,4-dinitrophenylhydrazine	LC/EC	99 pg	Jacobs & Kissinger (1982)
	Mix with 2,4-dinitrophenylhydrazine in hydrochloric acid	GC/FID or GC/EC	10 ng 20 pg	Kallio et al. (1972)
	Mix with Girard-T reagent in t-butanol; filter; concentrate; dilute (t-butanol); evaporate and redissolve; regenerate aldehydes by adding paraformaldehyde or methylolphthalimide and heating	GC	1 ng/μl	Gadbois et al. (1968)
Natural water	Buffer with phosphate; add ethylenediaminetetraacetic acid	Differential pulse polarography	0.05-0.5 mg/l (range of detection)	Howe (1976)
Wastewater	-	GC/FID	0.5 mg/l	Voloshina (1971)
Irrigation water	Add mixed reagent (m-aminophenol, hydroxylamine hydrochloride and hydrochloric acid); heat in a boiling water bath	Fluorescence spectroscopy	4 μg/l-2 mg/l (range of detection)	Hopkins & Hattrup (1974)
Tobacco smoke	Collect mainstream smoke in a trap immersed in a dry ice/acetone bath	GC	Not given	Rathkamp et al. (1973)
Cigarette smoke	Collect gas phase at the head of the GC column at -75°C	GC	Not given	Griest et al. (1977); Jenkins et al. (1983)
	Trap with silica gel; elute with water; treat with benzyloxyamine; extract (diethyl ether)	GC/NSD	Not given	Magin (1980)
Biological systems	Add mixed reagent (m-aminophenol, hydroxylamine hydrochloride and hydrochloric acid); heat in a boiling water bath	Fluorescence spectroscopy	0.003 μg/ml	Alarcon (1968)
Aroma volatiles of arctic bramble berries	Macerate and distil; react with 2,4-dinitrophenylhydrazine; extract (pentane); evaporate; dissolve (ethyl acetate); separate using thin-layer chromatography	GC/FID or GC/MS	Not given	Kallio & Linko (1973)
Beer	Steam distil; extract (pentane/ dichloromethane); react to form the p-nitrobenzyloxyamine derivative	HPLC	Not given (abstract)	Piendl et al. (1981)
	Chill; add sodium chloride; distil trap liquid nitrogen containing 2,4-dinitrophenylhydrazine; extract (chloroform); dry; remove chloroform	HPLC	<0.1 μg/l (carbonyls)	Greenhoff & Wheeler (1981)

[a]Abbreviations: GC/MS, gas chromatography/mass spectrometry; S, spectrophotometry; GC, gas chromatography; C, colorimetry; MS, mass spectrometry; HPLC/F, high-performance liquid chromatography/fluorescence detection; HPLC/UV, high-performance liquid chromatography/ultraviolet detection; LC/EC, liquid chromatography/electrochemistry; GC/FID, gas chromatography/flame ionization detection; GC/EC, gas chromatography/electron capture detection; GC/NSD, gas chromatography/nitrogen selective detection; HPLC, high-performance liquid chromatography

3. Biological Data Relevant to the Evaluation of Carcinogenic Risk to Humans

3.1 Carcinogenicity studies in animals

Acrolein is a metabolite of cyclophosphamide, which was evaluated previously (IARC, 1981, 1982b).

(a) Skin application

Mouse: A group of 15 S strain mice [sex and age unspecified] received 10 weekly skin applications of a 0.5% solution of acrolein [purity unspecified] in acetone (total dose, 12.6 mg acrolein/animal). Starting 25 days after the first acrolein application, the mice received weekly skin applications of 0.17% croton oil for 18 weeks; for the second and third application, the concentration was reduced to 0.085%. When croton oil and acrolein were administered together, each compound was given alternately at three- or four-day intervals. At the end of the croton-oil treatment, all 15 mice were still alive, and two had a total of three skin papillomas, compared with four skin papillomas in 4/19 controls that received the croton-oil treatment alone (Salaman & Roe, 1956). [The Working Group noted the small number of animals and the short duration of the experiment].

(b) Inhalation and/or intratracheal administration

Hamster: Two groups of 18 male and 18 female Syrian golden hamsters, six weeks old, were exposed to 0 or 4 ppm (0 or 9.2 mg/m^3) acrolein vapour [purity unspecified] for seven hours per day on five days per week for 52 weeks. Six animals per group were killed at 52 weeks and the remainder at 81 weeks. Survival was similar in treated and control animals. No tumour of the respiratory tract was found in any group (Feron & Kruysse, 1977).

Groups of 30 male and 30 female Syrian golden hamsters were exposed to 0 or 4 ppm (0 or 9.2 mg/m^3) acrolein vapour [purity unspecified] for seven hours per day on five days per week for 52 weeks and were given, at the same time and also for 52 weeks, weekly intratracheal instillations of two dose levels of benzo[a]pyrene or subcutaneous injections of N-nitrosodiethylamine (once every three weeks). All surviving animals were killed at 81 weeks. Additional exposure to acrolein did not significantly increase the tumour incidence produced by benzo[a]pyrene or N-nitrosodiethylamine (Feron & Kruysse, 1977).

(c) Carcinogenicity of possible metabolites

Acrolein is metabolized *in vitro* by liver and lung microsomes to glycidaldehyde (see IARC, 1976); the carcinogenicity of this compound is discussed below.

Mouse: In a two-stage mouse-skin assay, groups of 30 female Swiss albino mice, 55 days old, received a single *skin application* of 2.5 mg glycidaldehyde [purity unspecified] or 0.125 mg 7,12-dimethylbenz[a]anthracene (DMBA) (positive control) in 0.25 ml acetone. After three weeks, all mice were given skin applications of 0.25 ml 0.1% croton oil in acetone once a day on five days per week for 30 weeks. Controls either received no treatment or were treated with DMBA, croton oil or acetone only. [No information on the treatment schedule of these controls was given. No control group received glycidaldehyde treatment only.] At week 30, when the experiment was terminated, skin tumours, reported only as keratoacanthomas, were found in 40% and 95% of the mice initiated with glycidaldehyde and with DMBA, respectively. No skin tumour occurred in the control groups (Shamberger et al., 1974, 1975).

Skin applications of a 3% solution of glycidaldehyde [purity unspecified] in benzene onto the clipped backs of eight-week-old female ICR/Ha Swiss mice thrice weekly for life resulted in the development of skin tumours in 16/30 animals, with a mean survival time of 496 days; eight had papillomas and eight had carcinomas. A group of 60 control animals were painted with benzene alone. No skin papilloma or carcinoma was observed; median survival time was 498 days (Van Duuren et al., 1965).

A group of 41 eight-week-old female ICR/Ha Swiss mice received thrice weekly *skin applications* on the back of 100 mg glycidaldehyde solution [purity unspecified] as a 10% solution in acetone for life (598 days). Skin papillomas developed in 6/41 mice; three of the papillomas became squamous-cell carcinomas. The median survival time was 445 days. No skin tumour was seen in 300 acetone-treated controls [$p < 0.01$] with a median survival time of more than 526 days (Van Duuren et al., 1967a).

Groups of 110, 50 or 30 female ICR/Ha Swiss mice were given weekly *subcutaneous injections* of tricaprylin alone (vehicle controls), or 0.1 or 3.3 mg glycidaldehyde [purity unspecified] in 0.05 ml tricaprylin for lifetime, beginning at eight weeks of age. Local sarcomas or squamous-cell carcinomas occurred in 0/110, 3/50 and 7/30 animals in the control, low-dose and high-dose groups, respectively [$p < 0.01$] (Van Duuren et al., 1966).

Rat: Two groups of 50 and 20 six-week-old female Sprague-Dawley rats received weekly *subcutaneous injections* of 1 or 33 mg glycidaldehyde [purity unspecified] in 0.1 ml tricaprylin. Local sarcomas occurred in 1/50 and 5/20 rats at the two dose levels, respectively (maximum duration of tests, 558 and 539 days). In two groups of 20 and 50 tricaprylin-injected controls, one local carcinoma was observed (maximum duration of tests, 555 and 565 days). In two groups of 20 and 50 untreated controls, no local tumour was observed (maximum duration of tests, 559 and 563 days) (Van Duuren et al., 1966, 1967b).

3.2 Other relevant biological data

Acrolein is a metabolite of cyclophosphamide, which was evaluated previously (IARC, 1981, 1982b).

(a) Experimental systems

The toxic effects of acrolein in experimental animals have been reviewed (Izard & Libermann, 1978).

Toxic effects

Oral LD_{50}s of acrolein were reported to be 28 mg/kg bw in mice (Safe Drinking Water Committee, 1977), 46 mg/kg bw in rats (Smyth et al., 1951) and 7 mg/kg bw in rabbits (International Technical Information Institute, 1975). Subcutaneous LD_{50}s were found to be 30 mg/kg bw in mice and 50 mg/kg bw in rats (Skog, 1950). The dermal LD_{50} in rabbits was reported to be 562 mg/kg bw (Ben-Dyke et al., 1970).

The LC_{50} in rats for a 30-min exposure to acrolein vapour was reported to be 130 ppm (300 mg/m^3) (Fassett, 1963) and the inhalation LC_{50} in mice to be 66 ppm (152 mg/m^3) for a six-hour exposure period (Philippin et al., 1970). The LC_{50} for hamsters was 25 ppm (58 mg/m^3) for a four-hour exposure period (Kruysse, 1971). The lowest lethal concentrations (LCLO) of acrolein vapour in male and female rats was 8 ppm (18.4 mg/m^3); 2-4/6 animals died after exposure to this dose for four hours (Carpenter et al., 1949). The LCLO for a six-hour exposure to acrolein vapour was reported to be 10 ppm (24 mg/m^3) in mice, guinea-pigs and rabbits; in cats it was found to be 680 ppm (1570 mg/m^3) for an eight-hour exposure period (International Technical Information Institute, 1975).

Acrolein vapour caused a pressor response at lower dose levels and a depressor response at higher dose levels (Green & Egle, 1983). Subchronic exposure to 0.4-4.0 ppm (0.9-9 mg/m^3) acrolein vapour of rats susceptible (S) and resistant (R) to salt-induced hypertension resulted in death of all S rats exposed to the highest dose within the first 11 days, whereas 60% of R rats survived (Kutzman et al., 1982a).

Concentration-dependent changes in pulmonary function (breathing patterns, quasi-static compliance, diffusion capacity, flow-volume dynamics and distribution of ventilation) were reported in male Fischer 344 rats exposed to 0.4-4.0 ppm (0.9-9 mg/m^3) acrolein vapour for six hours per day on five days per week for 62 days (Costa & Kutzman, 1982). In a subchronic inhalation experiment using the same protocol as above, 32/57 male and 0/8 female Fischer 344 rats died after exposure to 4.0 ppm acrolein vapour; the body-weight gain was reduced significantly in both male and female rats; in addition, increased lung weight, bronchiolar epithelial necrosis and sloughing, bronchiolar oedema with macrophages and focal pulmonary oedema were observed in this group. Occasionally, these lesions were associated with oedema of the trachea and peribronchial lymph nodes, and acute rhinitis (Kutzman et al., 1982b). Exposure of Wistar rats, Syrian golden hamsters and Dutch rabbits to 0.4-4.9 ppm (0.9-11 mg/m^3) acrolein vapour for six hours per day on five days per week for 13 weeks produced eye and nasal irritation and hyper- and metaplasia of the epithelial lining of the respiratory tract; and 3/12 male and 3/12 female rats died during the first four weeks of exposure to the highest concentrations (4.9 ppm); exposure to the lowest level (0.4 ppm) produced no toxic effect in hamsters or rabbits (Feron et al., 1978).

Exposure of rats to 0.55 ppm (1.3 mg/m^3) acrolein vapour for 24 hours per day on seven days per week for 180 days caused irritation of the nasal mucosa during days 7-21 of exposure, and thereafter disappeared. In contrast, there was persistent loss of body weight and alterations in some biochemical parameters, such as decrease in liver weight and acid phosphatase activity, throughout the period of exposure (Bouley et al., 1975).

Groups of pure-bred beagle dogs, squirrel monkeys (Saimiri sciurea), guinea-pigs and Sprague-Dawley-derived rats were exposed to 0.7 and 3.7 ppm (1.6 and 8.5 mg/m^3) acrolein vapour for eight hours per day on five days per week for six consecutive weeks; squamous metaplasia and basal-cell hyperplasia in the trachea were observed in dogs and monkeys, and squamous metaplasia of the lung in 7/9 monkeys (Lyon et al., 1970).

Repeated inhalation by chickens of 50 and 200 ppm (115 and 450 mg/m^3) acrolein vapour for five minutes per day for 1-27 days produced concentration-dependent decreases in the numbers of ciliated and goblet cells and mucous glands in the trachea, and lymphocytic inflammatory lesions in the tracheal mucosa (Denine et al., 1971).

Acrolein has also been reported to cause alterations in lung and liver biochemistry, including significant reductions in microsomal mixed-function oxidase activity, in rats given two intraperitoneal injections of 5 mg/kg bw acrolein (Patel & Leibman, 1979). A single intraperitoneal injection of 3 mg/kg bw acrolein to male Holtzman rats also caused a prolongation of both pentobarbital and hexobarbital sleeping time (Jaeger & Murphy, 1973). In in-vitro studies, Patel et al. (1980a) and Leibman et al. (1984) reported total destruction of liver and lung microsomal NADPH-cytochrome c reductase by 1.5-6 mM [0.084-3.3 mg/ml] acrolein, total loss of nonprotein sulphydryl (GSH) content and partial loss of protein sulphydryl content in these organs. Depletion of GSH (21-63%) in the respiratory mucosa of male Fischer 344 rats after inhalation of 0.1-5 ppm (0.23-11.5 mg/m^3) acrolein vapour has also been reported (McNulty et al., 1984).

Acrolein was also implicated as a causative agent of hepatic periportal necrosis and other biochemical toxicity during biotransformation of allyl alcohol and allylamine (Piazza, 1915; Reid, 1972; Nelson & Boor, 1982; Patel et al., 1983). Cyclophosphamide (IARC, 1981), which is metabolized to acrolein, is known to cause urotoxic side-effects, especially haemorrhagic cystitis (Brock et al., 1979; Cox & Abel, 1979), liver and lung enzyme inactivation, and increased lung lipid peroxidation (Patel & Block, 1982).

Intraperitoneal injection of 1.65-2.7 mg/kg bw acrolein to partly-hepatectomized adult rats inhibited DNA and RNA synthesis in liver and lungs (Munsch & Frayssinet, 1971). Acrolein inhibited the transcriptional ability of isolated nuclei of rat-liver cells (Moulé & Frayssinet, 1971).

Effects on reproduction and prenatal toxicity

Acrolein is a metabolite of the antineoplastic agent, cyclophosphamide, a well-recognized animal teratogen (IARC, 1981). The role of metabolism in the teratogenic action of cyclophosphamide has been examined extensively, but considerable variation is observed in different test systems as to whether the parent chemical or one of its metabolites is responsible for these effects.

Acrolein (practical grade), stabilized with 0.2% hydroquinone (IARC, 1977) and dissolved in 25 µl of 0.9% sodium chloride, was injected at doses of 0.001, 0.01, 0.1, 1 and 10 µmol/egg (0.006-56 µg/egg) into either the air space or the yolk sac of three-day-old White Leghorn SK 12 strain chick embryos. On day 14 of incubation, the embryos were examined for both viability and malformations. Dose-related lethality was observed, with an estimated LD$_{50}$ of 0.05 µmol/egg (2.8 µg/egg). No clear evidence of teratogenic potential was found (Kankaanpää et al., 1979). Acrolein (practical grade), dissolved in 5 µl acetone, was injected into three-day-old White Leghorn chicken eggs at doses of 0.025-0.2 µmol/egg (1.4-11.2 µg/egg), and embryos were examined on day 14. In addition to the dose-related mortality seen earlier, some malformed embryos (principally corneal and lid defects and open coeloms) were found. The LD$_{50}$ was reported to be 0.08 µmol/egg (4.5 µg/egg) and the ED$_{50}$ for both malformations and mortality, 0.05 µmol/egg (2.8 µg/egg) (Korhonen et al., 1983).

Acrolein was dissolved in saline and injected into the air space of 48- and 72-hour chick embryos in doses of 0.001-0.1 mg/egg, and embryos were examined on day 13. The LD$_{50}$ for the 48-hour treatment was 0.01 mg/egg. No significant embryonic mortality was found in the group treated at 72 hours, but malformations (narrowed sixth aortic arch, cardiomegaly and atrial hemorrhage [incidence not reported]) similar to those seen in survivors from exposure at 48 hours were observed (Gilani & Chhibber, 1983).

A dose-related increase in embryolethality, but not in malformations (4/69 foetuses at the high dose, compared to 2/121 in the control group were malformed, but this difference was not significant), was found when groups of 16 New Zealand white rabbits were injected intravenously on day 9 of gestation with 3, 4.5 or 6 mg/kg bw acrolein (stabilized with

0.2% hydroquinone). The high dose killed 6/16 rabbits, compared to 0/13 in controls. However, direct injections into the yolk sac of 10, 20 or 40 µl of a 0.84% solution of acrolein in physiological saline into day-9 embryos resulted in a dose-related increase in both resorptions (63% in the high-dose group compared to 21.2 in controls) and malformations (23.3% in the high-dose group compared to 3% in controls). The defects in the high-dose group included hypoplastic and asymmetrical cervical and thoracic vertebrae, shortened extremities and a ventricular septal defect (Claussen et al., 1980).

Acrolein itself is not a teratogen in an in-vitro rat-embryo culture system, as addition of up to 250 µM [14 µg/ml] (purity, 97%) to cultured day-10.5 rat embryos produced only growth retardation and no terata (Schmid et al., 1981). Phosphoramide mustard, another metabolite of cyclophosphamide, does produce malformations in this system (Mirkes et al., 1981).

Direct intra-amniotic injections of cyclophosphamide, 4-hydroperoxycyclophosphamide (which decomposes in aqueous solution to 4-hydroxycyclophosphamide, an intermediate metabolite), phosphoramide mustard and acrolein were tested for teratogenicity in day-13 rat embryos. All were teratogenic, but only 4-hydroperoxycyclophosphamide and acrolein produced the same malformations (oedema, hydrocephaly, open eyes, micrognathia, cleft palate, omphalocele, bent tail, and forelimb and hindlimb defects) as cyclophosphamide, although at 100-fold lower doses (1 µg/foetus) (Hales, 1982).

More recently, the dechlorinated derivative of cyclophosphamide (D-CP) was evaluated for teratogenicity in an in-vitro rat-embryo culture system (Mirkes et al., 1984). This derivative is metabolized to acrolein and dechlorophosphoramide mustard (D-PM), a non-alkylating derivative of phosphoramide mustard. D-PM, D-CP and acrolein were tested in this system. D-CP, but not D-PM, was teratogenic to day-10 embryos. Acrolein caused growth retardation and abnormal flexions, but none of the malformations produced by cyclophosphamide. D-CP could not be shown to be teratogenic *in vivo*.

Absorption, distribution, excretion and metabolism

Draminski et al. (1983) have studied the excretion of acrolein metabolites in the urine of adult female Wistar rats after a single oral administration of 10 mg/kg bw acrolein in corn oil. S-Carboxyethyl-N-acetylcysteine (S-carboxyethylmercapturic acid) and S-(propionic acid methyl ester) mercapturic acid were reported to be the major metabolites.

Male CFE albino rats metabolized 10.5% of a single subcutaneous dose of 1 ml of a 1% (v/v) solution of acrolein in arachis oil to N-acetyl-S-(3-hydroxypropyl)-L-cysteine (3-hydroxypropylmercapturic acid) which was isolated from urine (Kaye, 1973).

Boyland and Chasseaud (1967) have reported that acrolein can conjugate with glutathione non-enzymatically and enzymatically by a glutathione-S-transferase-catalysed enzymic reaction.

Patel et al. (1980b) have reported in-vitro metabolism of acrolein in lung and liver fractions from male Holtzman rats. Acrolein was metabolized to acrylic acid by $9000 \times g$ supernatant, cytosolic and microsomal preparations of rat liver; lung and liver microsomal preparations catalysed the epoxidation of acrolein to glycidaldehyde; and glycidaldehyde was converted to glyceraldehyde by liver epoxide hydrase (see Figure 1 in General Remarks on the Substances Considered, p. 32).

Acrolein, when reacted with deoxyguanosine and DNA at pH 7 *in vitro*, forms two diastereomeric cyclic 1,N^2-propanodeoxy adducts (Chung et al., 1984).

Mutagenicity and other short-term tests (see also 'Appendix: Activity Profiles for Short-Term Tests', p. 335)

Reviews of the genetic effects of aldehydes, including acrolein, are available (Auerbach *et al.*, 1977; Izard & Libermann, 1978).

Acrolein was reported to give a positive result in an *Escherichia coli* pol A^+/pol A^- assay without metabolic activation (Bilimoria, 1975 [abstract, details not given]).

Acrolein was mutagenic to *Salmonella typhimurium* TA104 without an exogenous metabolic system (Marnett *et al.*, 1984). Both positive and negative results have been reported in strains TA98 and TA100 under various test conditions (Florin *et al.*, 1980; Lijinsky & Andrews, 1980; Loquet *et al.*, 1981; Lutz *et al.*, 1982; Haworth *et al.*, 1983). No mutagenic activity was seen in strains TA1535, TA1537 or TA1538 (Florin *et al.*, 1980; Lijinsky & Andrews, 1980; Loquet *et al.*, 1981); however, a weak mutagenic response was reported in strain TA1535 in the presence of a hepatic microsomal fraction from phenobarbital-induced rats (Hales, 1982).

A slight mutagenic effect was reported with *E. coli* WP2 *uvr*A (Hemminki *et al.*, 1980).

Acrolein did not cause DNA cross-links or DNA breaks in *Saccharomyces cerevisiae* (Fleer & Brendel, 1982). It was reported to be mutagenic to *Streptomyces coelicolor* [no data reported] (Ortali *et al.*, 1977) but not to *Aspergillus nidulans* [no data reported] (Bignami *et al.*, 1977) or *Saccharomyces cerevisiae* (Izard, 1973).

Abernethy *et al*, (1983) reported in an abstract that acrolein alone did not induce morphological transformation in C3H 10T$\frac{1}{2}$ cells, but showed some initiating activity when assayed near the LC$_{50}$ concentration of 6.3 μM (0.4 μg/ml) in the presence of 0.25 μg/ml 12-*O*-tetradecanoylphorbol 13-acetate. [The exact concentrations used were not given.]

Acrolein induced sister chromatid exchanges in Chinese hamster ovary cells (Au *et al.*, 1980).

It was negative in the dominant lethal test in male mice given intraperitoneal doses of 1.5 or 2.2 mg/kg bw (Epstein *et al.*, 1972).

Glycidaldehyde, a metabolite of acrolein, induced mutations in *Klebsiella pneumoniae* (Voogd *et al.*, 1981), *S. typhimurium* TA1535 and TA100 (McCann *et al.*, 1975) and *Saccharomyces cerevisiae* (Izard, 1973). It was positive in the mouse-lymphoma L5178Y/TK assay (Amacher & Turner, 1982).

(b) Humans

Toxic effects

Exposure to 1 ppm (2.3 mg/m^3) acrolein vapour in the air causes lachrymation and marked eye, nose and throat irritation within a period of five minutes (Fassett, 1963). Acrolein is a severe pulmonary irritant and powerful lachrymogen at a concentration of 3 ppm (7 mg/m^3) and greatly irritates the conjunctiva and mucous membranes of the upper respiratory tract (Prentiss, 1937). At higher concentrations, it also causes injury to the lung; respiratory insufficiency may persist for at least 18 months after exposure (Champeix & Catilina, 1967). A 10-min exposure to 350 mg/m^3 was lethal (Prentiss, 1937).

Accidental exposure of a human subject to vapours from an overheated frying pan containing fat and food items resulted in symptoms similar to those reported in cases of acrolein intoxication (Bauer et al., 1977). Exposure of another two subjects to vapours of frying oil resulted in death. Autopsy revealed massive cellular desquamation of the bronchial lining and there were multiple pulmonary infarcts. Acrolein was suspected of being among the causative agents (Gosselin et al., 1979).

Case reports of several patients who received prolonged cyclophosphamide treatment showed haemorrhagic cystitis of the bladder (Wall & Clausen, 1975; Beyer-Boon et al., 1978; IARC, 1981). The bladder toxicity of cyclophosphamide appears to be due to the formation of acrolein and may be prevented by co-administration of sodium 2-mercaptoethane sulphonate (Brock et al., 1979).

Effects on reproduction and postnatal toxicity

Reproductive effects have been reported in humans exposed therapeutically to cyclophosphamide (IARC, 1981). However, these studies did not address the role of metabolism in the effects. No data were available on the reproductive effects of acrolein.

No data were available to the Working Group on absorption, distribution, excretion and metabolism or on mutagenicity and chromosomal effects in acrolein-exposed groups.

3.3 Case reports and epidemiological studies of carcinogenicity to humans

Acrolein is a metabolite of cyclophosphamide, which was evaluated previously (IARC, 1981, 1982b).

A study in which exposure to traces of acrolein is mentioned (Bittersohl, 1975) is described in the monograph on acetaldehyde (p. 101).

4. Summary of Data Reported and Evaluation

4.1 Exposure data

Acrolein has been produced commercially since 1955 and is used principally as a chemical intermediate in the production of acrylic acid and its esters. It is employed to a lesser degree as a biocide and in the synthesis of methionine. It is a degradation product of the pyrolysis of animal and vegetable fats and is found in tobacco smoke. It is also present as an urban air pollutant arising from combustion of fossil fuels. Acrolein is a metabolite of many compounds to which humans are exposed, including allyl alcohol and cyclophosphamide.

4.2 Experimental data

Acrolein was tested in mice by skin application and in hamsters by inhalation exposure. The study in mice was inadequate for an evaluation of carcinogenicity. No carcinogenic effect was detected in hamsters.

Under certain conditions of direct embryonic exposure, acrolein was teratogenic to rats and rabbits. After intravenous administration to rabbits of maternally toxic doses, embryonic mortality but no significant malformation was seen in one experiment.

Acrolein was mutagenic to bacteria. It did not induce DNA damage or mutations in fungi. It induced sister chromatid exchanges in mammalian cells *in vitro*. Acrolein did not induce dominant lethal mutations in mice.

Overall assessment of data from short-term tests: acrolein[a]

	Genetic activity			Cell transformation
	DNA damage	Mutation	Chromosomal effects	
Prokaryotes		+		
Fungi/Green plants	−	−		
Insects				
Mammalian cells (*in vitro*)			+	
Mammals (*in vivo*)			−	
Humans (*in vivo*)				
Degree of evidence in short-term tests for genetic activity: *Limited*				Cell transformation: No data

[a]The groups into which the table is divided and the symbols are defined on pp. 17-18 of the Preamble; the degrees of evidence are defined on p. 18.

Glycidaldehyde (see IARC, 1976), a possible metabolite of acrolein, was tested in mice by skin application and subcutaneous injection and in rats by subcutaneous injection. It produced malignant tumours at the site of application in animals of both species. Glycidaldehyde had initiating activity in a two-stage mouse-skin bioassay. It induced mutations in bacteria, yeast and mammalian cells *in vitro*.

4.3 Human data

The one study of workers in an aldehyde plant was inadequate for evaluation of the carcinogenicity of acrolein to humans.

4.4 Evaluation[1]

There is *inadequate evidence* for the carcinogenicity of acrolein to experimental animals.

There is *inadequate evidence* for the carcinogenicity of acrolein to humans.

No evaluation could be made of the carcinogenicity of acrolein to humans.

---

[1]For definitions of the italicized terms, see the Preamble, pp. 15-16.

5. References

Abernethy, D.J., Frazelle, J.H. & Boreiko, C.J. (1983) Relative cytotoxic and transforming potential of respiratory irritants in the C3H/10T$\frac{1}{2}$ cell transformation system (Abstract No. Cd-20). *Environ. Mutagenesis*, *5*, 419

Alarcon, R.A. (1968) Fluorometric determination of acrolein and related compounds with *m*-aminophenol. *Anal. Chem.*, *40*, 1704-1708

Amacher, D.E. & Turner, G.N. (1982) Mutagenic evaluation of carcinogens and non-carcinogens in the L5178Y/TK assay utilizing postmitochondrial fractions (S9) from normal rat liver. *Mutat. Res.*, *97*, 49-65

American Conference of Governmental Industrial Hygienists (1984) *TLVs Threshold Limit Values for Chemical Substances and Physical Agents in the Work Environment with Intended Changes for 1984-1985*, Cincinnati, OH, p. 9

Anderson, G.E., Liu, C.S., Holman, H.Y. & Killus, J.P. (1980) *Human Exposure to Atmospheric Concentrations of Selected Chemicals*, Environmental Protection Agency, Contract No. 68-02-3066, San Rafael, CA, Systems Applications, Inc., pp. 20-32

Apol, A.G. (1973) *Health Hazard Evaluation/Toxicity Determination Report 72-32, Union Pacific Railroad, Pocatello, Idaho*, Cincinnati, OH, National Institute for Occupational Safety and Health

Au, W., Sokova, O.I., Kopnin, B. & Arrighi, F.E. (1980) Cytogenetic toxicity of cyclophosphamide and its metabolites *in vitro*. *Cytogenet. Cell Genet.*, *26*, 108-116

Auerbach, C., Moutschen-Dahmen & Moutschen, J. (1977) Genetic and cytogenetical effects of formaldehyde and related compounds. *Mutat. Res.*, *39*, 317-362

Bauer, K., Czech, K. & Porter, A. (1977) Severe accidental acrolein intoxication at home (Ger.). *Wien. klin. Wochenschr.*, *89*, 243-244

Ben-Dyke, R., Sanderson, D.M. & Noakes, D.N. (1970) Acute toxicity data for pesticides. *World Rev. Pest. Control*, *9*, 119-127

Beyer-Boon, M.E., De Voogt, H.J. & Schaberg, A. (1978) The effects of cyclophosphamide treatment on the epithelium and stroma of the urinary bladder. *Eur. J. Cancer*, *14*, 1029-1035

Bignami, M., Cardamone, G., Comba, P., Ortali, V.A., Morpurgo, G. & Carere, A. (1977) Relationship between chemical structure and mutagenic activity in some pesticides: The use of *Salmonella typhimurium* and *Aspergillus nidulans* (Abstract No. 79). *Mutat. Res.*, *46*, 243-244

Bilimoria, M.H. (1975) The detection of mutagenic activity of chemicals and tobacco smoke in a bacterial system (Abstract No. 39). *Mutat. Res.*, *31*, 328

Bittersohl, G. (1975) Epidemiological research on cancer risk by aldol and aliphatic aldehydes. *Environ. Qual. Saf.*, *4*, 235-238

Bouley, G., Dubreuil, A., Godin, J. & Boudène, C. (1975) Effects of a weak dose of continuously inhaled acrolein in rats (Fr.). *Eur. J. Toxicol.*, *8*, 291-297

Boyland, E. & Chasseaud, L.F. (1967) Enzyme-catalysed conjugations of glutathione with unsaturated compounds. *Biochem. J.*, *104*, 95-102

Brock, N., Stekar, J., Pohl, J., Niemeyer, V. & Scheffler, G. (1979) Acrolein, the causative factor of urotoxic side-effects of cyclophosphamide, ifosfamide, trofosfamide and sufosfamide. *Arzneim.-Forsch./Drug Res.*, *29*, 659-661

Brodzinsky, R. & Singh, H.B. (1982) *Volatile Organic Chemicals in the Atmosphere: An Assessment of Available Data* (*Final Report, US Environmental Protection Agency, Contract No. 68-02-3452*), Menlo Park, CA, SRI International, pp. 3-4, 12-15, 26, 122, 131

Buckingham, J., ed. (1982) *Dictionary of Organic Compounds*, 5th ed., Vol. 1, New York, Chapman & Hall, p. 4784

Cantoni, C., Bianchi, M.A., Renon, P. & Calcinardi, C. (1969) Bacterial and chemical alterations during souring in salted pork (Ital.). *Atti Soc. ital. Sci. vet.*, *23*, 752-756 [*Chem. Abstr.*, *73*, 129686q]

Carpenter, C.P., Smyth, H.F., Jr & Pozzani, U.C. (1949) The assay of acute vapor toxicity, and the grading and interpretation of results on 96 chemical compounds. *J. ind. Hyg. Toxicol.*, *31*, 343-346

Carson, B.L., Beall, C.M., Ellis, H.V., III, Baker, L.H. & Herndon, B.L. (1981) *Acrolein Health Effects* (*EPA-460/3-81-034, PB82-161282*), Springfield, VA, National Technical Information Service, pp. 8-9, 94, 96, 98, 103-104, 107, 111, 113-114, 117-118

Castle, C.N. & Smith, T.J. (1974) *Environmental Sampling at a Copper Smelter* (*PB82-164948*), Cincinnati, OH, National Institute for Occupational Safety and Health, pp. 4-7, 13-14, 22, 29-30

Champeix, J. & Catilina, P. (1967) *Les Intoxications par l'Acroléine* (Acrolein poisoning), Paris, Masson

Chung, F.-L., Young, R. & Hecht, S.S. (1984) Formation of cyclic 1,N^2-propanodeoxyguanosine adducts in DNA upon reaction with acrolein or crotonaldehyde. *Cancer Res.*, *44*, 990-995

Claussen, U., Hellmann, W. & Pache, G. (1980) The embryotoxicity of the cyclophosphamide metabolite acrolein in rabbits, tested *in vivo* by i.v. injection and by the yolk-sac method. *Arzneim.-Forsch./Drug Res.*, *30*, 2080-2083

Cohen, I.R. & Altshuller, A.P. (1961) A new spectrophotometric method for the determination of acrolein in combustion gases and in the atmosphere. *Anal. Chem.*, *33*, 726-733

Costa, D.L. & Kutzman, R.S. (1982) A subchronic acrolein inhalation study in rats. II. Assessment of pulmonary function (Abstract No. 571). *Toxicologist*, *2*, 163

Cox, P.J. & Abel, G. (1979) Cyclophosphamide cystitis. Studies aimed at its minimization. *Biochem. Pharmacol.*, *28*, 3499-3502

Day, A.G., III, Beggs, D.P., Vestal, M.L. & Johnston, W.H. (1971) *Improved Instrumentation for Determination of Exhaust Gas Oxygenate Content (PB-210 151)*, Baltimore, MD, Scientific Research Instruments Corp., pp. 1-3, 19-20, 54-59, 61

Denine, E.P., Robbins, S.L. & Kensler, C.J. (1971) The effects of acrolein inhalation on the tracheal mucosa of the chicken (Abstract No. 144). *Toxicol. appl. Pharmacol.*, *19*, 416

Deutsche Forschungsgemeinschaft (1984) *Maximal Work Place Concentrations and Biological Tolerance Values for Compounds in the Work Place* (Ger.), Part XX, Weinheim, Verlag Chemie GmbH, p. 48

Doorgeest, T. (1970) Paint and air pollution (Dutch). *TNO Nieuws*, *25*, 37-42 [*Chem. Abstr.*, *72*, 136075x]

Draminski, W., Eder, E. & Henschler, D. (1983) A new pathway of acrolein metabolism in rats. *Arch. Toxicol.*, *52*, 243-247

Egorov, I.A., Pisarnitskii, A.F., Zinkevich, E.P. & Gavrilov, A.I. (1976) Study of some volatile components of oak wood (Russ.). *Prikl. Biokhim. Mikrobiol.*, *12*, 108-112 [*Chem. Abstr.*, *85*, 182256y]

Einhorn, I.N. (1975) Physiological and toxicological aspects of smoke produced during the combustion of polymeric materials. *Environ. Health Perspect.*, *11*, 163-189

Epstein, S.S., Arnold, E., Andrea, J., Bass, W. & Bishop, Y. (1972) Detection of chemical mutagens by the dominant lethal assay in the mouse. *Toxicol. appl. Pharmacol.*, *23*, 288-325

Fassett, D.W. (1963) Aldehydes and acetals. In: Patty, F.A., ed., *Industrial Hygiene and Toxicology*, 2nd rev. ed., Vol. 2, *Toxicology*, New York, Interscience, pp. 1978-1979

Feron, V.J. & Kruysse, A. (1977) Effects of exposure to acrolein vapor in hamsters simultaneously treated with benzo[a]pyrene or diethylnitrosamine. *J. Toxicol. environ. Health*, *3*, 379-394

Feron, V.J., Kruysse, A., Til, H.P. & Immel, H.R. (1978) Repeated exposure to acrolein vapour: Subacute studies in hamsters, rats and rabbits. *Toxicology*, *9*, 47-57

Fleer, R. & Brendel, M. (1982) Toxicity, interstrand cross-links and DNA fragmentation induced by 'activated' cyclophosphamide in yeast: Comparative studies on 4-hyperoxy-cyclophosphamide, its monofunctional analogon, acrolein, phosphoramide mustard, and nor-nitrogen mustard. *Chem.-biol. Interactions*, *39*, 1-15

Florin, I., Rutberg, L., Curvall, M. & Enzell, C.R. (1980) Screening of tobacco smoke constituents for mutagenicity using the Ames test. *Toxicology*, *18*, 219-232

Gadbois, D.F., Scheurer, P.G. & King, F.J. (1968) Analysis of saturated aldehydes by gas-liquid chromatography using methylolphthalimide for regeneration of their Girard-T derivatives. *Anal. Chem.*, *40*, 1362-1365

Gilani, S.H. & Chhibber, G. (1983) The teratogenicity of the cyclosphamide metabolite acrolein in chick embryos (Abstract). *Teratology*, *27*, 44A

Gosselin, B., Wattel, F., Chopin, C., Degand, P., Fruchart, J.C., Van der Loo, D. & Crasquin, O. (1979) Acute acrolein poisoning. One observation (Fr.). *Nouv. Presse méd.*, *8*, 2469-2472

Graedel, T.E. (1978) *Chemical Compounds in the Atmosphere*, New York, Academic Press, p. 171

Grasselli, J.G. & Ritchey, W.M., eds (1975) *CRC Atlas of Spectral Data and Physical Constants for Organic Compounds*, 2nd ed., Vol. 4, Cleveland, OH, Chemical Rubber Co., p. 297

Green, M.A. & Egle, J.L., Jr (1983) The effects of acetaldehyde and acrolein on blood pressure in guanethidine-pretreated hypertensive rats. *Toxicol. appl. Pharmacol.*, *69*, 29-36

Greenhoff, K. & Wheeler, R.E. (1981) Analysis of beer carbonyls at the part per billion level by combined liquid chromatography and high pressure liquid chromatography. *J. Inst. Brew.*, *86*, 35-41

Grey, T.C. & Shrimpton, D.H. (1967) Volatile components of raw chicken breast muscle. *Br. Poult. Sci.*, *8*, 23-33

Griest, W.H., Quincy, R.B. & Guerin, M.R. (1977) *Selected Constituents in the Smoke of Domestic Low Tar Cigarettes (ORNL/TM-6144/P1)*, Oak Ridge, TN, Oak Ridge National Laboratory

Gronsberg, Y.S. (1974) Determination of acrolein in air. *Khim. prom. (Moscow)*, *5*, 394 [translation in *Soviet chem. Ind.*, *6*, 337-338]

Hales, B.F. (1982) Comparison of the mutagenicity and teratogenicity of cyclophosphamide and its active metabolites, 4-hydroxycyclophosphamide, phosphoramide mustard, and acrolein. *Cancer Res.*, *42*, 3016-3021

Haworth, S., Lawlor, T., Mortelmans, K., Speck, W. & Zeiger, E. (1983) *Salmonella* mutagenicity test results for 250 chemicals. *Environ. Mutagenesis*, *Suppl. 1*, 3-142

Hemminki, K., Falck, K. & Vainio, H. (1980) Comparison of alkylation rates and mutagenicity of directly acting industrial and laboratory chemicals. Epoxides, glycidyl esters, methylating and ethylating agents, halogenated hydrocarbons, hydrazine derivatives, aldehydes, thiuram and dithiocarbamate derivatives. *Arch. Toxicol.*, *46*, 277-285

Hess, L.G., Kurtz, A.N. & Stanton, D.B. (1978) Acrolein and derivatives. In: Grayson, M., ed., *Kirk-Othmer Encyclopedia of Chemical Technology*, 3rd ed., Vol. 1, New York, John Wiley & Sons, pp. 277-297

Hoffmann, D., Brunnemann, K.D., Gori, G.B. & Wynder, E.L. (1975) On the carcinogenicity of marijuana smoke. *Rec. Adv. Phytochem.*, *9*, 63-81

Hopkins, D.M. & Hattrup, A.R. (1974) *Field Evaluation of a Method to Detect Acrolein in Irrigation Canals (REC-ERC-74-8; PB Rep. No. 234926/4GA)*, Springfield, VA, National Technical Information Service

Hoshika, Y. & Takata, Y. (1976) Gas chromatographic separation of carbonyl compounds as their 2,4-dinitrophenylhydrazones using glass capillary columns. *J. Chromatogr.*, *120*, 379-389

Hoshika, Y., Nihei, Y. & Muto, G. (1981) Pattern display for characterisation of trace amounts of odorants discharged from nine odour sources. *Analyst*, *106*, 1187-1202

Howe, L.H. (1976) Differential pulse polarographic determination of acrolein in water samples. *Anal. Chem.*, *48*, 2167-2169

Hrdlicka, J. & Janicek, G. (1968) Volatile carbonyl compounds isolated from sugar-cane molasses (Czech.). *Sb. Vys. Sk. Chem. Technol. Praze. Potraviny*, *E21*, 77-79 [*Chem. Abstr.*, *71*, 62461a]

IARC (1976) *IARC Monographs on the Evaluation of the Carcinogenic Risk of Chemicals to Man, Vol. 11, Cadmium, Nickel, Some Epoxides, Miscellaneous Industrial Chemicals and General Considerations on Volatile Anaesthetics*, Lyon, pp. 175-181

IARC (1977) *IARC Monographs on the Evaluation of the Carcinogenic Risk of Chemicals to Man, Vol. 15, Some Fumigants, the Herbicides 2,4-D and 2,4,5-T, Chlorinated Dibenzodioxins and Miscellaneous Industrial Chemicals*, Lyon, pp. 155-175

IARC (1979) *IARC Monographs on the Evaluation of the Carcinogenic Risk of Chemicals to Humans, Vol. 19, Some Monomers, Plastics and Synthetic Elastomers, and Acrolein*, Lyon, pp. 47-71, 73-113, 341-366, 479-494

IARC (1981) *IARC Monographs on the Evaluation of the Carcinogenic Risk of Chemicals to Humans, Vol. 26, Some Anticancer and Immunosuppressive Drugs*, Lyon, pp. 165-202

IARC (1982a) *IARC Monographs on the Evaluation of the Carcinogenic Risk of Chemicals to Humans, Vol. 29, Some Industrial Chemicals and Dyestuffs*, Lyon, pp. 345-389

IARC (1982b) *IARC Monographs on the Evaluation of the Carcinogenic Risk of Chemicals to Humans, Supplement 4, Chemicals, Industrial Processes and Industries Associated with Cancer in Humans (IARC Monographs, Volumes 1 to 29)*, Lyon, pp. 99-100

International Labour Office (1980) *Occupational Exposure Limits for Airborne Toxic Substances*, 2nd (rev.) ed. (*Occupational Safety and Health Series No. 37*), Geneva, pp. 36-37

International Technical Information Institute (1975) *Toxic and Hazardous Industrial Chemicals Safety Manual for Handling and Disposal with Toxicity and Hazard Data*, Tokyo, p. AI-35

Izard, C. (1973) Studies on the mutagenic activity of acrolein and of its two epoxides: Glycidol and glycidal on *Saccharomyces cerevisiae* (Fr.). *C.R. Acad. Sci.*, *276* Ser. D, 3037-3040

Izard, C. & Libermann, C. (1978) Acrolein. *Mutat. Res.*, *47*, 115-138

Jacobs, W.A. & Kissinger, P.T. (1982) Determination of carbonyl 2,4-dinitrophenylhydrazones by liquid chromatography/electrochemistry. *J. Liq. Chromatogr.*, *5*, 669-676

Jaeger, R.J. & Murphy, S.D. (1973) Alterations of barbiturate action following 1,1-dichloroethylene, corticosterone, or acrolein. *Arch. int. Pharmacodyn.*, *205*, 281-292

Jenkins, R.A., White, S.K., Griest, W.H. & Guerin, M.R. (1983) *Chemical Characterization of the Smokes of Selected US Commercial Cigarettes: Tar, Nicotine, Carbon Monoxide, Oxides of Nitrogen, Hydrogen Cyanide, and Acrolein (DE83012139; ORNL/TM-8749)*, Oak Ridge, TN, Oak Ridge National Laboratory

Kallio, H. & Linko, R.R. (1973) Volatile monocarbonyl compounds of arctic bramble (*Rubus arcticus* L.) at various stages of ripeness. *Z. Lebensm. Unters.-Forsch.*, *153*, 23-30

Kallio, H., Linko, R.R. & Kaitaranta, J. (1972) Gas-liquid chromatographic analysis of 2,4-dinitrophenylhydrazones of carbonyl compounds. *J. Chromatogr.*, *65*, 355-360

Kankaanpää, J., Elovaara, E., Hemminki, K. & Vainio, H. (1979) Embryotoxicity of acrolein, acrylonitrile and acrylamide in developing chick embryos. *Toxicol. Lett.*, *4*, 93-96

Kaye, C.M. (1973) Biosynthesis of mercapturic acids from allyl alcohol, allyl esters and acrolein. *Biochem. J.*, *134*, 1093-1101

Klochkovskii, S.P., Lukashenko, R.D., Podvysotskii, K.S. & Kagramanyan, N.P. (1981) Acrolein and formaldehyde content in the air of quarries (Russ.). *Bezop. Tr. Prom-sti.*, *12*, 38 [*Chem. Abstr.*, *96*, 128666w]

Korhonen, A., Hemminki, K. & Vainio, H. (1983) Embryotoxic effects of acrolein, methacrylates, guanidines and resorcinol on three day chicken embryos. *Acta pharmacol. toxicol.*, *52*, 95-99

Krost, K.J., Pellizzari, E.D., Walburn, S.G. & Hubbard, S.A. (1982) Collection and analysis of hazardous organic emissions. *Anal. Chem.*, *54*, 810-817

Kruysse, A. (1971) *Acute Inhalation Toxicity of Acrolein in Hamsters (Report R3516)*, Zeist, The Netherlands, Central Institute for Nutrition and Food Research TNO

Kutzman, R.S., Wehner, R.W. & Haber, S.B. (1982a) A subchronic acrolein inhalation study in rats. IV. Impact on hypertension-sensitive and resistant animals (Abstract No. 573). *Toxicologist*, *2*, 163

Kutzman, R.S., Popenoe, E.A., Cockrell, B.Y., Schmaeler, M.A. & Drew, R.T. (1982b) A subchronic acrolein inhalation study in rats. I. Biochemical and pathologic changes (Abstract No. 570). *Toxicologist*, *2*, 162

Leibman, K.C., Kolmstetter, C. & Patel, J.M. (1984) Selective inactivation of rat lung microsomal NADPH-cytochrome *c* reductase by acrolein (Abstract No. 446). *Fed. Proc.*, *43*, 361

Levaggi, D.A. & Feldstein, M. (1970) The determination of formaldehyde, acrolein, and low molecular weight aldehydes in industrial emissions on a single collection sample. *J. Air Poll. Control Assoc.*, *20*, 312-313

Lijinsky, W. & Andrews, A.W. (1980) Mutagenicity of vinyl compounds in *Salmonella typhimurium*. *Teratog. Carcinog. Mutagenesis*, *1*, 259-267

Lipari, F. & Swarin, S.J. (1982) Determination of formaldehyde and other aldehydes in automobile exhaust with an improved 2,4-dinitrophenylhydrazine method. *J. Chromatogr.*, *247*, 297-306

Lipari, F., Dasch, J.M. & Scruggs, W.F. (1984) Aldehyde emissions from wood-burning fireplaces. *Environ. Sci. Technol.*, *18*, 326-330

Loquet, C., Toussaint, G. & LeTalaer, J.Y. (1981) Studies on mutagenic constituents of apple brandy and various alcoholic beverages collected in western France, a high incidence area of oesophageal cancer. *Mutat. Res.*, *88*, 155-164

Lutz, D., Eder, E., Neudecker, T. & Henschler, D. (1982) Structure-mutagenicity relationship in α,β-unsaturated carbonylic compounds and their corresponding allylic alcohols. *Mutat. Res.*, *93*, 305-315

Lyon, J.P., Jenkins, L.J., Jr, Jones, R.A., Coon, R.A. & Siegel, J. (1970) Repeated and continuous exposure of laboratory animals to acrolein. *Toxicol. appl. Pharmacol.*, *17*, 726-732

Mackison, F.W., Stricoff, R.S. & Partridge, L.J., Jr, eds (1981) *NIOSH/OSHA Occupational Health Guidelines for Chemical Hazards (DHHS (NIOSH) Publication No. 81-123)*, Washington DC, National Institute for Occupational Safety and Health/Occupational Safety and Health Administration, pp. 1-5

Magin, D.F. (1980) Gas chromatography of simple monocarbonyls in cigarette whole smoke as the benzyloxime derivatives. *J. Chromatogr.*, *202*, 255-261

Maldotti, A., Chiorboli, C., Bignozzi, C.A., Bartocci, C. & Carassiti, V. (1980) Photooxidation of 1,3-butadiene containing systems: Rate constant determination for the reaction of acrolein with OH radicals. *Int. J. chem. Kinet.*, *12*, 905-913

Manning, D.L., Maskarinec, M.P, Jenkins, R.A. & Marshall, A.H. (1983) High performance liquid chromatographic determination of selected gas phase carbonyls in tobacco smoke. *J. Assoc. off. anal. Chem.*, *66*, 8-12

Marnett, L.J., Hurd, H., Hollstein, M., Levin, D.E., Esterbauer, H. & Ames, B.N. (1984) Naturally occurring carbonyl compounds are mutagens in *Salmonella* tester strain TA104. *Mutat. Res.* (in press)

Mašek, V. (1972) Aldehydes in the air in coal and pitch coking plants *Staub-Reinhalt. Luft*, *32*, 26-28

McCann, J., Choi, E., Yamasaki, E. & Ames, B.N. (1975) Detection of carcinogens as mutagens in the *Salmonella*/microsome test: Assay of 300 chemicals. *Proc. natl Acad. Sci. USA*, *72*, 5135-5139

McNulty, M.J., Heck, H.d'A. & Casanova-Schmitz, M. (1984) Depletion of glutathione in rat respiratory mucosa by inhaled acrolein (Abstract No. 1695). *Fed. Proc.*, *43*, 575

Mirkes, P.E., Fantel, A.G., Greenaway, J.C. & Shepard, T.H. (1981) Teratogenicity of cyclophosphamide metabolites: Phosphoramide mustard, acrolein, and 4-ketocyclophosphamide in rat embryos cultured *in vitro*. *Toxicol. appl. Pharmacol.*, *58*, 322-330

Mirkes, P.E., Greenaway, J.C., Rogers, J.G. & Brundrett, R.B. (1984) Role of acrolein in cyclophosphamide teratogenicity in rat embryos in vitro. Toxicol. appl. Pharmacol., 72, 281-291

Moulé, Y. & Frayssinet, C. (1971) Effects of acrolein on transcription in vitro. FEBS Lett., 16, 216-218

Mulders, E.J. & Dhont, J.H. (1972) Odor of white bread. III. Identification of volatile carbonyl compounds and fatty acids (Ger.). Z. Lebensm. Unters.-Forsch., 150, 228-232 [Chem. Abstr., 78, 83052c]

Munsch, N. & Frayssinet, C. (1971) Action of acrolein on nucleic acid synthesis in vivo (Fr.). Biochimie, 53, 243-248

National Finnish Board of Occupational Safety and Health (1981) Airborne Contaminants in the Workplaces (Safety Bulletin No. 3), Helsinki, p. 7

National Institute for Occupational Safety and Health (1980) Projected Number of Occupational Exposures to Chemical and Physical Hazards, Cincinnati, OH, p. 2

National Institute for Occupational Safety and Health (1981) National Occupational Hazard Survey Data (microfiche), Cincinnati, OH, p. 93 (generics excluded)

National Swedish Board of Occupational Safety and Health (1981) Hygiene Limit Values (AFS 1981:8), Stockholm, p. 9

Nelson, T.J. & Boor, P.J. (1982) Allylamine cardiotoxicity. IV. Metabolism to acrolein by cardiovascular tissues. Biochem. Pharmacol., 31, 509-514

Ortali, V.A., Cardamone, G., Salvini, P., Di Giuseppe, G. & Carere, A. (1977) Relationship between mutagenic activity and chemical structure in some pesticides: Example of Salmonella typhimurium and Streptomyces coelicolor (Ital.). Atti Assoc. genet. ital., 22, 63 64

Patel, J.M. & Block, E.R. (1982) Cyclophosphamide (CP)-induced biochemical toxicity in the lung (Abstract No. 8459). Fed. Proc., 41, 1716

Patel, J.M. & Leibman, K.C. (1979) Biochemical effects of acrolein on rat liver and lung, as influenced by various treatments (Abstract No. 1657). Fed. Proc., 38, 542

Patel, J.M., Ortiz, E. & Leibman, K.C. (1980a) Selective inactivation of rat liver microsomal NADPH-cytochrome c reductase by acrolein (Abstract No. 3156). Fed. Proc., 39, 865

Patel, J.M., Wood, J.C. & Leibman, K.C. (1980b) The biotransformation of allyl alcohol and acrolein in rat liver and lung preparations. Drug Metab. Disposition, 8, 305-308

Patel, J.M., Gordon, W.P., Nelson, S.D. & Leibman, K.C. (1983) Comparison of hepatic biotransformation and toxicity of allyl alcohol and [1,1-^{2}H$_2$]allyl alcohol in rats. Drug Metab. Disposition, 11, 164-166

Pfäffli, P. (1982) Industrial hygiene measurements. Scand. J. Work Environ. Health, 8, Suppl. 2, 27-43

Philippin, C., Gilgen, A. & Grandjean, E. (1970) Toxicological and physiological investigation on acrolein inhalation in the mouse (Fr.). *Int. Arch. Arbeitsmed.*, *26*, 281-305

Piazza, J.G. (1915) Action of allyl compounds (Ger.). *Z. exp. Pathol. Ther.*, *17*, 318-341

Piendl, A., Westner, H. & Geiger, E. (1981) Detection and separation of aldehydes in beer using high pressure liquid chromatography (HPLC) (Ger.). *Brauwissenschaft*, *34*, 301-307 [*Chem. Abstr.*, *96*, 4870p]

Prager, B., Jacobson, P., Schmidt, P. & Stern, D., eds (1918) *Beilsteins Handbuch der Organischen Chemie*, 4th ed., Vol. 1, Syst. No. 90, Berlin, Springer, p. 725

Prentiss, A.M. (1937) *Chemicals in War. A Treatise on Chemical Warfare*, New York, McGraw-Hill, pp. 139-140

Protsenko, G.A., Danilov, V.I., Timchenko, A.N., Nenartovich, A.V., Trubilko, V.I. & Savchenkov, V.A. (1973) Working conditions when metals to which primer has been applied are welded evaluated from the health and hygiene aspect. *Avt. Svarka.*, *2*, 65-68

Rathkamp, G., Tso, T.C. & Hoffmann, D. (1973) Chemical studies on tobacco smoke. XX. Smoke analysis of cigarettes made from Bright tobaccos differing in variety and stalk positions. *Beitr. Tabakforsch.*, *7*, 179-189

Reddish, J.F. (1982) An analyser for the continuous determination of acrolein in the atmosphere. *J. autom. Chem.*, *4*, 116-121

Reid, W.D. (1972) Mechanism of allyl alcohol-induced hepatic necrosis. *Experientia*, *28*, 1058-1061

Robles, E.G., Jr (1968) *Thermal Decomposition Products of Cellophane* (*AD-752 515*), McClellan Air Force Base, CA, US Air Force Environmental Health Laboratory

Safe Drinking Water Committee (1977) *Drinking Water and Health*, Washington DC, National Academy of Sciences, p. 554

Saito, T., Takashina, T., Yanagisawa, S. & Shirai, T. (1983) Determination of trace low molecular weight aliphatic carbonyl compounds in auto exhaust by gas chromatography with a glass capillary column (Jpn.). *Bunseki Kagaku*, *32*, 33-38 [*Chem. Abstr.*, *98*, 112873s]

Salaman, M.H. & Roe, F.J.C. (1956) Further tests for tumour-initiating activity: N,N-Di-(2-chloroethyl)-*p*-aminophenylbutyric acid (CB1348) as an initiator of skin tumour formation in the mouse. *Br. J. Cancer*, *10*, 363-378

Schmid, B.P., Goulding, E., Kitchin, K. & Sanyal, M.K. (1981) Assessment of the teratogenic potential of acrolein and cyclophosphamide in a rat embryo culture system. *Toxicology*, *22*, 235-243

Scott Research Laboratories, Inc. (1969) *Atmospheric Reaction Studies in the Los Angeles Basin, Phase I, Volumes I and II* (*PB-194 058; PB-194 059*), Plumsteadville, PA, pp. 2-9, 2-10, 5-10, 5-11, 5-44 (Vol. I); pp. 143-149, 310-314 (Vol. II)

Serth, R.W., Tierney, D.R. & Hughes, T.W. (1978) *Source Assessment, Acrylic Acid Manufacture, State-of-the-Art (EPA-600/2-78-004w)*, Cincinnati, OH, Industrial Environmental Research Laboratory, US Environmental Protection Agency, pp. vi-xiii, 1-5, 25-38, 46-51

Shadrin, A.S. (1970) Spectrophotometric determination of acrolein in the atmosphere of open-cut mines with truck transportation (Russ.). *Tr. Inst. Gorn. Dela. Min. Chern. Met. SSSR*, *24*, 27-32 [*Chem. Abstr.*, *76*, 131082v]

Shamberger, R.J., Andreone, T.L. & Willis, C.E. (1974) Antioxidants and cancer. IV. Initiating activity of malonaldehyde as a carcinogen. *J. natl Cancer Inst.*, *53*, 1771-1773

Shamberger, R.J., Tytko, S.A. & Willis, C.E. (1975) Malonaldehyde is a carcinogen (Abstract No. 3431). *Fed. Proc.*, *34*, 827

Shimomura, M., Yoshimatsu, F. & Matsumoto, F. (1971) Fish odor of cooked horse mackerel (Jpn.). *Kaseigaku Zasshi*, *22*, 106-112 [*Chem. Abstr.*, *75*, 62292d]

Singal, M. & Love, J.R. (1981) *Health Hazard Evaluation Report, HHE 79-128-806, Bob Gerren Ford, Inc., Manistee, Michigan*, Cincinnati, OH, National Institute for Occupational Safety and Health

Skog, E. (1950) A toxicological investigation of lower aliphatic aldehydes. I. Toxicity of formaldehyde, acetaldehyde, propionaldehyde and butyraldehyde; as well as of acrolein and crotonaldehyde. *Acta pharmacol.*, *6*, 299-318

Smyth, H.F., Jr, Carpenter, C.P. & Weil, C.S. (1951) Range-finding toxicity data: List IV. *Arch. ind. Hyg. occup. Med.*, *4*, 119-122

Smythe, R.J. & Karasek, F.W. (1973) The analysis of diesel engine exhausts for low-molecular-weight carbonyl compounds. *J. Chromatogr.*, *86*, 228-231

Stahl, Q.R. (1969) *Air Pollution Aspects of Aldehydes (PB-188 081)*, Bethesda, MD, Litton Systems, Inc., Clearinghouse for Federal Scientific and Technical Information, pp. 48-49, 55-59, 61-62, 131

Stepanova, M.I., Karpova, N.M. & Shaposhnikov, Y.K. (1983) Composition of volatile components in ventilation waste gases from the production of alkyd resin-based coating materials (Russ.). *Lakokras. Mater. Ikh Primen.*, *3*, 62-64 [*Chem. Abstr.*, *99*, 58111x]

Suzuki, Y. & Imai, S. (1982) Determination of traces of gaseous acrolein by collection on molecular sieves and fluorimetry with o-aminobiphenyl. *Anal. chim. Acta*, *136*, 155-162 [*Chem. Abstr.*, *96*, 204574g]

Swarin, S.J. & Lipari, F. (1983) Determination of formaldehyde and other aldehydes by high performance liquid chromatography with fluorescence detection. *J. liq. Chromatogr.*, *6*, 425-444

Tanimoto, M. & Uehara, H. (1975) Detection of acrolein in engine exhaust with microwave cavity spectrometer of Stark voltage sweep type. *Environ. Sci. Tech.*, *9*, 153-154

Tanne, C. (1983) *Candle manufacture*. In: Parmeggiani, L., ed., *Encyclopaedia of Occupational Health and Safety*, Geneva, International Labour Office, pp. 383-384

US Department of Transportation (1982) Performance-oriented packaging standards. *US Code Fed. Regul., Title 49,* Parts 171, 172, 173, 178; *Fed. Regist., 47* (No. 73), pp. 16268, 16273

US Environmental Protection Agency (1980a) Hazardous waste management system: Identification and listing of hazardous wastes. *US Code Fed. Regul., Title 40,* Part 261; *Fed. Regist., 45* (No. 98), pp. 33084, 33122-33127, 33131-33133

US Environmental Protection Agency (1980b) *Ambient Water Quality Criteria for Acrolein (EPA-440/5-80-016, PB81-117277),* Springfield, VA, National Technical Information Service, pp. A2, C3, C8-C21

US Environmental Protection Agency (1982) Pesticide use classification. *US Code Fed. Regul., Title 40,* Part 162.31

US Environmental Protection Agency (1983a) Organic chemicals and plastics and synthetic fibers category effluent limitations guidelines, pretreatment standards, and new source performance standards. *US Code Fed. Regul., Title 40,* Parts 414, 416; *Fed. Regist., 48* (No. 55), pp. 11828-11867

US Environmental Protection Agency (1983b) Notification requirements; reportable quantity adjustments. *US Code Fed. Regul., Title 40,* Part 302; *Fed. Regist., 48* (No. 102), pp. 23552, 23571

US Food and Drug Administration (1980) Food and drugs. *US Code Fed. Regul., Title 21,* Parts 172.892, 176.300

US Tariff Commission (1956) *Synthetic Organic Chemicals, US Production and Sales, 1955 (Report No. 198, Second Series),* Washington DC, US Government Printing Office, p. 143

Van Duuren, B.L., Orris, L. & Nelson, N. (1965) Carcinogenicity of epoxides, lactones and peroxy compounds. II. *J. natl Cancer Inst., 35,* 707-717

Van Duuren, B.L., Langseth, L., Orris, L., Teebor, G., Nelson, N. & Kuschner, M. (1966) Carcinogenicity of epoxides, lactones, and peroxy compounds. IV. Tumor response in epithelial and connective tissue in mice and rats. *J. natl Cancer Inst., 37,* 825-838

Van Duuren, B.L., Langseth, L., Goldschmidt, B.M. & Orris, L. (1967a) Carcinogenicity of epoxides, lactones, and peroxy compounds. VI. Structure and carcinogenic activity. *J. natl Cancer Inst., 39,* 1217-1228

Van Duuren, B.L., Langseth, L., Orris, L., Baden, M. & Kuschner, M. (1967b) Carcinogenicity of epoxides, lactones, and peroxy compounds. V. Subcutaneous injection in rats. *J. natl Cancer Inst., 39,* 1213-1216

Verschueren, K. (1977) *Handbook of Environmental Data on Organic Chemicals,* New York, Van Nostrand Reinhold Co., pp. 74-76

Volkova, Z.A. & Bagdinov, Z.M. (1969) Industrial hygiene problems in vulcanization processes of rubber production (Russ.). *Gig. Sanit., 34,* 33-40 [*Chem. Abstr., 71,* 128354b]

Voloshina, A. (1971) *Determination of some organic compounds in waste waters by a gas-liquid chromatographic method* (Russ.). In: Dorofeenko, G.N., ed., *Methods of Chemical Analysis of Effluent Waters from Chemical Industries,* Rostov-on-Don, USSR, Rostov University, pp. 6-17 [*Chem. Abstr., 78,* 88404d]

Voogd, C.E., van der Stel, J.J. & Jacobs, J.J.J.A.A. (1981) The mutagenic action of aliphatic epoxides. *Mutat. Res., 89,* 269-282

Vorob'eva, A.I., Volkotrub, L.P., Feoktistova, N.F., Bobin, V.I., Shestakova, N.A. & Ushakova, N.S. (1982) Characteristics of atmospheric pollution from enameled wire manufacturing plants (Russ.). *Gig. Sanit., 6,* 66-67 [*Chem. Abstr., 97,* 58817r]

Wall, R.L. & Clausen, K.P. (1975) Carcinoma of the urinary bladder in patients receiving cyclophosphamide. *New Engl. J. Med., 293,* 271-273

Windholz, M., ed. (1983) *The Merck Index,* 10th ed., Rahway, NJ, Merck & Co., p. 19

Zinn, B.T., Browner, R.F., Powell, E.A., Pasternak, M. & Gardner, R.O. (1980) *The Smoke Hazards Resulting from the Burning of Shipboard Materials Used by the US Navy (NRL Report No. 8414),* Washington DC, Naval Research Laboratory, pp. 48-54

MALONALDEHYDE

1. Chemical and Physical Data

1.1 Synonyms and trade names

Chem. Abstr. Services Reg. No.: 542-78-9

Chem. Abstr. Name: Propanedial

IUPAC Systematic Name: Malonaldehyde

Synonyms: Malondialdehyde; malonic aldehyde; malonic dialdehyde; malonodialdehyde; malonyldialdehyde; MDA; NCI-C54842; 1,3-propanedial; 1,3-propanedialdehyde; 1,3-propanedione

1.2 Structural and molecular formulae and molecular weight

$$CH_2(CHO)_2$$

$C_3H_4O_2$ Mol. wt: 72.1

1.3 Chemical and physical properties of the pure substance

From Buckingham (1982), unless otherwise specified

(a) *Description:* Hygroscopic needles

(b) *Melting-point:* 72-74°C

(c) *Spectroscopy data*: Ultraviolet spectral data (Marnett & Tuttle, 1980) and mass spectral data (Bond *et al.*, 1980) have been reported.

(d) *Stability*: Highly pure malonaldehyde is quite stable under neutral conditions but not under acidic conditions such as those used to prepare it by hydrolysis of its bis(dialkyl)acetal. Since malonaldehyde has a $pK_a = 4.46$, it exists under physiological conditions as its conjugate base (-OCH=CH-CHO), which is relatively stable to self-condensation (Marnett *et al.*, 1979; Marnett & Tuttle, 1980).

(e) *Reactivity*: Reacts with proteins (Apaja, 1980); the conjugate base is much less reactive (Marnett & Tuttle, 1980).

(f) *Conversion factor*: 1 ppm = 2.95 mg/m³ at 760 mm Hg and 25°C [calculated by the Working Group]

1.4 Technical products and impurities

No information was available to the Working Group

2. Production, Use, Occurrence and Analysis

2.1 Production and use

(a) Production

Malonaldehyde was identified in dilute aqueous solution by Claisen in 1903; it was first isolated in 1941 (Apaja, 1980).

No evidence was found that malonaldehyde is produced in commercial quantities in the USA, western Europe or Japan. However, its bis(dialkyl)acetals (e.g., 1,1,3,3-tetraethoxypropane) are available, and free malonaldehyde can be generated from these by acid-catalysed hydrolysis with dilute hydrochloric acid or by shaking them with acidic Dowex 50 ion-exchange resin. The hydrolysis has been reported to proceed through the pathway shown in Fig. 1, and both the dialkoxypropionaldehydes and the β-alkoxyacroleins are present in the crude malonaldehyde so produced. Producing pure malonaldehyde requires special purification steps, e.g., conversion to the sodium salt and repeated column chromatography with Sephadex LH-20 resin (Marnett *et al.*, 1979).

(b) Use

The dimethyl and diethyl acetals of malonaldehyde, 1,1,3,3-tetramethoxypropane and 1,1,3,3-tetraethoxypropane, are used to generate (by hydrolysis) free malonaldehyde for laboratory purposes. No evidence was found that any of these compounds is used for commercial applications.

2.2 Occurrence

(a) Natural occurrence

Malonaldehyde has been detected in lipid-peroxidizing microsomes and is found in animal tissue as an end-product of lipid peroxidation. It is also a side-product of prostaglandin and thromboxane biosynthesis (Marnett *et al.*, 1979; Apaja, 1980; Bond *et al.*, 1980; Esterbauer & Slater, 1981).

(b) Soil and plants

Malonaldehyde has been detected in the leaves of pea and cotton plants (Merzlyak *et al.*, 1979).

(c) Food, beverages and animal feeds

Malonaldehyde is found in many foodstuffs and is present at generally high levels in rancid foods. It has been detected in fish meat, fish oil, rancid salmon oil, rancid nuts, rancid flour, orange juice essence, vegetable oils, fats, fresh frozen green beans, milk, milk fat, rye bread, and in raw, cured and cooked meats (Apaja, 1980; Newburg & Concon, 1980).

Fig. 1. Pathway of hydrolysis of 1,1,3,3-tetraethoxypropane to malonaldehyde (Marnett & Tuttle, 1980)[a]

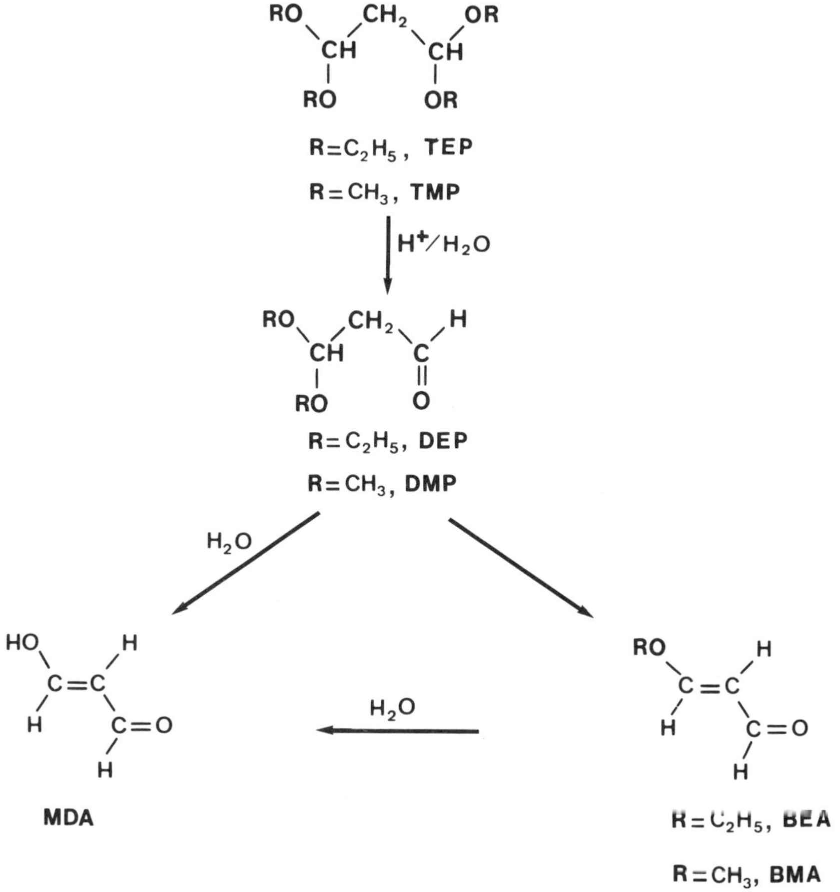

[a]TEP, 1,1,3,3-tetraethoxypropane; TMP, 1,1,3,3-tetramethoxypropane; DEP, 3,3-diethoxypropionaldehyde; DMP, 3,3-dimethoxypropionaldehyde; MDA, malonaldehyde; BEA, β-ethoxyacrolein; BMA, β-methoxyacrolein

Concentrations of free malonaldehyde in commercial samples of refined groundnut oils ranged from 0.04-0.14 mg/kg, and the level in sunflower oil was 0.08 mg/kg. The total malonaldehyde (free and bound) concentrations were 0.53-4.36 mg/kg in groundnut oils and 0.98 mg/kg in sunflower oil (Arya & Nirmala, 1971).

Increased levels of malonaldehyde have been found in hamburger, chicken and beef as a result of cooking under a variety of conditions (e.g., microwave, frying, baking and boiling). Conversely, levels of malonaldehyde in frozen smoked trout and cheddar cheese decreased as a result of cooking. The formation of malonaldehye in foods during cooking seemed to be dependent on many factors, including the degree of unsaturation of the fatty acids and the length of time spent in contact with molecular oxygen. The amount found in cooked food samples may also have depended on loss of free malonaldehyde due to its volatility or reactivity (Newburg & Concon, 1980).

A study conducted in Cleveland, OH, reported that among several meats, beef had the highest content of malonaldehyde (up to 13.7 ± 2.7 µg/g in sirloin steak). (In some meat products the cooking process increased the malonaldehyde content, e.g., the content in a beef roast increased from 9.4 ± 3.1 to 27.0 ± 6.3 µg/g after cooking for two hours at 325°C.) Turkey and cooked chicken also had high levels; but most cheeses, pork and fish had only small amounts, and many vegetables, fruits and canned foods had either minute quantities or no malonaldehyde. Vegetable oil, corn-oil margarine and milk had no detectable malonaldehyde (Shamberger et al., 1977).

(d) Human tissues and secretions

Malonaldehyde is present in blood platelets and in serum. It is formed during the metabolism of prostaglandin endoperoxides and in certain tissues. It appears to be one of the major products of endoperoxide metabolism (Marnett et al., 1979; Roncucci & Lansen, 1980; Shimasaki & Ueta, 1980; Von Voss et al., 1980). In normal subjects, mean levels of malonaldehyde (measured as the 2-thiobarbituric acid derivative) were 3.42 ± 0.94 nmol/ml of serum in males and 3.10 ± 0.62 nmol/ml in females (Yagi, 1982).

2.3 Analysis

The amount of malonaldehyde detected in biological materials is significantly influenced by the method of treatment of the sample. Levels of malonaldehyde were found to range from 2.3-3.6 µg/g and 1.5-2.5 µg/g in two different rodent feeds, from 0.8-2.2 µg/g in pork liver and from 1.1-1.7 µg/g in ground beef samples, depending on the method of treatment (heat, acid extraction, etc.) used (Bird et al., 1983).

Table 1. Methods for the analysis of malonaldehyde

Sample matrix	Sample preparation	Assay procedure[a]	Limits of detection	Reference
Aqueous solutions	React with anthrone	S (510 nm)	2.3 µg/ml	Kwon & Watts (1963a)
	Extract (trichloroacetic acid); heat with 2-thiobarbituric acid	HPLC/visible light detection	1 ng	Bird et al. (1983)
	Distil acidified aqueous slurry	HPLC/UV	0.07 µg/ml	Kakuda et al. (1981)
	Distil acidified aqueous slurry; heat with 2-thiobarbituric acid	S (535 nm)	0.14 µg/ml	Kakuda et al. (1981)
	Distil acidified aqueous slurry; detect absorbance difference between acidified and basified solutions	S (267 nm)	0.36 µg/ml	Kwon & Watts (1963b)
Vegetable oils	React with dansyl hydrazine in acidic medium	HPLC/F	0.01 mg/kg	Hirayama et al. (1983)
Vegetable oils, human saliva, blood plasma and cervical mucus	Extract vegetable oils or the solution in toluene by hydrochloric acid; add hydrochloric acid to saliva, plasma, or mucous; degas with nitrogen	Polarography	0.7 mg/ml	Bond et al. (1980)
Pea and cotton leaves	Heat with 2-thiobarbituric acid; centrifuge; extract (butanol)	F	0.7 ng	Merzlyak et al. (1979)
Lipid-peroxidizing microsomes	Not given	HPLC/UV	≃ 5 ng	Esterbauer & Slater (1981)
Blood platelets	Centrifuge and resuspend; add thrombin	S	Not given	Von Voss et al. (1980)
Serum	React with 2-thiobarbituric acid	S (547 nm)	Not given	Shimasaki & Ueta (1980)

[a]Abbreviations: S, spectrophotometry; HPLC, high-performance liquid chromatography; HPLC/UV, high-performance liquid chromatography/ultraviolet detection; HPLC/F, high-performance liquid chromatography with fluorescence monitor; F, fluorimetry

Heating may result in artefactual formation of malonaldehyde as a product of the decomposition of lipid hydroperoxides. Acid extraction of samples yields higher concentrations than extraction with neutral solutions, presumably due to release of bound malonaldehyde (Bird et al., 1983).

Typical methods for the analysis of malonaldehyde in various matrices are summarized in Table 1. (It has been reported (e.g., Bird et al., 1983) that the spectrophotometric determination of the 2-thiobarbituric acid derivative is not specific, since various other compounds produce interfering derivatives.)

3. Biological Data Relevant to the Evaluation of Carcinogenic Risk to Humans

3.1 Carcinogenicity studies in animals[1]

Since malonaldehyde is unstable and highly reactive, it is generally not available as a free compound, and other forms were tested in the studies described below. Some of the biological effects of malonaldehyde may be due to impurities such as β-alkoxyacroleins (Marnett & Tuttle, 1980) (see also section 2.1(a)).

(a) Oral administration

Mouse: Five groups of 50 female ICR Swiss mice, six to eight weeks old, were given sodium malonaldehyde (purity, >98%; impurities unspecified, but not including the impurities mentioned above) in the drinking-water (pH 4.0) at concentrations providing 0 (two control groups), 0.1, 1 or 10 mg/kg bw per day for 12 months, at which time the surviving animals were killed. Mortality rates at termination were 13, 12, 12 and 28% ($p < 0.05$) in the 0, 0.1, 1 and 10 mg/kg bw-dose groups, respectively. The incidences of hyperplastic liver nodules were 0/97, 1/49, 2/50 and 2/48 in the same groups, respectively. The incidences of hepatomas were 0/97, 0/49, 2/50 and 0/48, respectively; and the incidences of liver haemangiomas were 1/97, 1/49, 0/50 and 4/48. Tumours were found in the stomach in 3/48 high-dose animals (one lymphoma, one squamous-cell carcinoma and one adenocarcinoma) and in 1/49 low-dose animals (one squamous-cell carcinoma); whereas no gastric tumour occurred in control or mid-dose animals (Bird et al., 1982a). [The Working Group noted the short duration of the experiment and that the incidences of individual tumours in the treated groups were not statistically different from those in controls.]

Three groups of 25 male and 25 female random-bred Swiss mice, eight weeks old, were given malonaldehyde bis(dimethylacetal) in the drinking-water, on six days per week for life, at levels that resulted, following hydrolysis, in concentrations of malonaldehyde of 0.125, 0.25 and 0.5%. Since methanol is also produced on hydrolysis of the bis(dimethylacetal), three control groups were given appropriate levels of methanol in the drinking-water. At week 80, mortality in the 0.5% malonaldehyde group was 25/25 males and 22/25 females, compared with 20/25 males and 14/25 females in the corresponding methanol-control group. The overall mortality in high-dose treated animals was statistically significantly different from that in controls ($p < 0.01$). There was no statistically significant difference in mortality between the lower-dose malonaldehyde groups and corresponding controls. Pulmonary tumours (adenomas and one carcinoma), blood-vessel tumours (haemangiomas and

---

[1]The Working Group was aware of a study in progress in mice and rats administered malonaldehyde in the drinking-water (IARC, 1982)

haemangioendotheliomas) and malignant lymphomas were the predominant types of tumour in the control and treated animals. None of the tumour types was significantly increased in incidence in malonaldehyde-treated animals compared with methanol-treated controls (Apaja, 1980). [The Working Group noted the high mortality and the lack of untreated controls.]

(b) *Skin application*

Mouse: One group of 30 female Swiss mice, 55 days old, was given daily skin applications of 12 mg malonaldehyde [purity unspecified] in 0.25 ml acetone for nine weeks. After this treatment proved toxic, the remaining mice received daily skin applications of 0.36 mg malonaldehyde in 0.25 ml acetone for 39 weeks. Another group of mice of the same strain and sex received daily skin applications of 0.36 mg malonaldehyde in 0.25 ml acetone for 48 weeks. A control group of 30 female mice was used [whether the group was treated with vehicle or was untreated was not specified]. Of the 30 mice treated with 12 mg malonaldehyde, 12 died in weeks 4-6; none had a tumour. In weeks 7-9, six more mice of this group died; five had tumours, four with a liver carcinoma, three of which had metastasized, and one with a carcinoma of the rectum. In the subsequent 39 weeks, no mortality occurred and no further tumour was found in this group. In the group treated with 0.36 mg malonaldehyde throughout the experiment, one skin tumour, classified as a keratoacanthoma, occurred at week 43. At termination (week 48), two animals of the control group had died; no skin tumour occurred in controls (Shamberger *et al.*, 1974). [The Working Group noted the short duration of the experiment and found it difficult to relate the early occurrence of liver carcinoma to the skin application of malonaldehyde.]

Two groups of 40 female random-bred Swiss Webster mice, eight weeks old, were given skin applications of 0 (controls) or 0.6 mg free malonaldehyde [purity unspecified] in 0.05 ml methanol per animal thrice weekly for life. The solution in methanol also contained unhydrolysed malonaldehyde bis(dimethylacetal), resulting in test solutions containing 0.6 mg malonaldehyde + 21.5 mg malonaldehyde bis(methylacetal) per 0.05 ml. Four intermediate groups were started, but, due to an unfavourable shift in the equilibrium reaction between malonaldehyde and malonaldehyde bis(methylacetal) in the test solutions, these groups were terminated at weeks 36-39. The control group and the high-dose group were treated for life. Mortality at week 52 was 4/40 in the high-dose group and 11/40 in the control group, and at week 80, 17/40 in the high-dose group and 25/46 in the control group; by week 120 all mice were dead. The predominant types of tumour in control and treated animals were pulmonary adenomas, haemangiomas, granulosa-cell tumours of the ovaries and malignant lymphomas. Only one skin tumour was found, a papilloma in an intermediate group. After adjusting for survival, the incidence of malignant lymphomas (17/40) in the high-dose group appeared to be statistically significantly higher ($p < 0.01$) than that (6/40) in the control group. However, the authors stated that both lymphoma incidences were within the normal range for malignant lymphomas in the strain of mice used (Apaja, 1980). [No data on the incidence of malignant lymphomas in historical controls were available to the Working Group.]

Six groups of 40 male and 40 female SENCAR mice, seven weeks old, were given skin applications of 20, 50, 100, 200 or 500 µg sodium malonaldehyde [stated to be pure and prepared by a method avoiding the presence of mutagenic or carcinogenic impurities; purity and impurities unspecified] or 50.5 µg benzo[a]pyrene (positive control) in 0.2 ml solvent (20% dimethylsulphoxide, 80% acetone) per animal twice weekly for 52 weeks. Mortality at week 52 was 3/80 mice in the low- and high-dose groups and ranged from 4-6/80 mice in the mid-dose groups, whereas all except two female mice of the positive-control group had died. None of the animals treated with sodium malonaldehyde developed a skin tumour, whereas 296 skin tumours (papillomas and carcinomas) were counted in 77 positive controls (Fischer et al., 1983).

In a two-stage mouse-skin assay, groups of 30 female Swiss mice, 55 days old, received a single skin application of 6 or 12 mg malonaldehyde [purity unspecified] or 0.125 mg 7,12-dimethylbenz[a]anthracene (DMBA) (positive control) in 0.25 ml acetone. After three weeks, all mice were given skin applications of 0.25 ml 0.1% croton oil in acetone daily on five days a week for 30 weeks. Controls included groups of mice treated only with DMBA, malonaldehyde, croton oil or acetone, or receiving no treatment. [No further information on the treatment of these control groups was given.] At week 30, when the experiment was terminated, skin tumours classified as keratoacanthomas occurred in 16/30 animals treated with 6 mg and in 16/30 treated with 12 mg malonaldehyde and in 29/30 of the mice treated with DMBA. No tumour occurred in the control groups (Shamberger et al., 1974, 1975).

In a two-stage mouse-skin assay, six groups of 40 male and 40 female SENCAR mice, seven weeks old, received a single skin application of 20, 50, 100, 200 or 500 µg sodium malonaldehyde [stated to be pure and prepared by a method avoiding the presence of carcinogenic or mutagenic impurities; purity and impurities unspecified] or 50.5 µg benzo[a]pyrene (positive control) in 0.2 ml solvent (20% dimethylsulphoxide, 80% acetone). One week after this treatment, application of 2 µg 12-O-tetradecanoylphorbol 13-acetate (TPA) in 0.2 ml solvent (20% dimethylsulphoxide, 80% acetone) was begun twice weekly for a period of at most 41 weeks. At week 42 the incidence of skin papillomas was 22/74, 22/78, 14/74, 15/78 and 18/79 in the groups receiving 20, 50, 100, 200 and 500 µg sodium malonaldehyde, respectively, and 229/65 in the positive-control group. [The number of tumour-bearing animals was not given.] It was stated that TPA had been shown previously to produce skin-tumour yields in uninitiated SENCAR mice that were similar to those found in animals treated with sodium malonaldehyde (Fischer et al., 1983). [The Working Group noted the short duration of the experiment and the use of historical controls rather than concomitant TPA controls.]

In a two-stage mouse-skin assay, six groups of 40 female and 40 male SENCAR mice, seven weeks old, were given skin applications of 20, 50, 100, 200 or 500 µg sodium malonaldehyde [stated to be pure and prepared by a method avoiding the presence of mutagenic or carcinogenic impurities; purity and impurities unspecified] or 2 µg TPA (positive controls) in 0.2 ml solvent (20% dimethylsulphoxide, 80% acetone) twice weekly for 28 weeks. This treatment was begun one week after initiation with a single application of 50.5 µg benzo[a]pyrene in 0.2 ml solvent (20% dimethylsulphoxide, 80% acetone) to the skin. [No mouse treated with the benzo[a]pyrene solution alone was used.] At week 28, the incidences of skin papillomas were 2/78, 3/78, 6/77, 1/78 and 2/80 in the groups receiving 20, 50, 100, 200 and 500 µg sodium malonaldehyde, respectively, and 360/78 in the positive-control group (Fischer et al., 1983). [The Working Group noted that the number of tumour-bearing animals was not given.]

3.2 Other relevant biological data

(a) Experimental systems

Toxic effects

Malonaldehyde, because of its instability and high reactivity, is not available as a free compound, and other forms were tested in the studies described below. Some of the biological effects of malonaldehyde may be due to impurities such as β-alkoxyacroleins (Marnett & Tuttle, 1980) (see also section 2.1(*a*)).

The oral LD_{50} in mice for malonaldehyde [purity unspecified] was 606 mg/kg bw (Apaja, 1980).

The oral LD_{50} of the sodium salt of enolic malonaldehyde (sodium β-oxyacrolein) [purity not determined] in albino Wistar rats was reported to be 824 mg/kg bw (equivalent to 632 mg/kg bw for malonaldehyde). The LD_{50} of the ethoxy derivative of malonaldehyde, 1,1,3,3-tetraethoxypropane, was found to be 1610 mg/kg bw (equivalent to 527 mg/kg bw malonaldehyde) in the same strain. Absorption experiments led the authors to conclude that the predominant toxic effect is due to the action of a compound other than malonaldehyde (Crawford et al., 1965).

The sodium salt of enolic malonaldehyde (shown to be free of β-methoxyacrolein) was administered in the drinking-water to groups of female ICR Swiss mice, eight weeks old, to give daily doses of 2-500 mg/kg bw (as malonaldehyde) for 90 days. No mortality was reported. Animals in the group receiving 500 mg/kg bw lost weight after day 50. Histopathological examination of 27 different tissue samples indicated that the liver underwent dose-dependent changes; all doses of malonaldehyde tested induced irregularities in the size and chromatin distribution of liver nuclei. Pancreatic lesions, consisting primarily of atrophy of the exocrine cells with loss of zymogen granulation, occurred in animals receiving the highest dose of malonaldehyde. Mild dysplasia of the urinary-bladder epithelium was found in all treated groups (Siu et al., 1983).

Acute ulcerative gastritis and fibrosis of glandular mucosa was significantly increased in 50 mice, eight weeks old, given malonaldehyde bis(dimethylacetal) in the drinking-water on six days per week for life at levels of 0.25% and 0.5% malonaldehyde (Apaja, 1980).

Recent studies have demonstrated that the addition to erythrocytes *in vitro* of malonaldehyde prepared from its bis(dimethylacetal) derivative caused a decrease in spectrin, producing higher-molecular-weight protein polymers and a marked decrease in cellular deformability; the proteins in the membranes of older erythrocytes were similarly altered (Jain & Hochstein, 1980; Shohet & Jain, 1982). It decreased erythrocyte survival (Jain et al., 1983). Malonaldehyde prepared from the bis(dimethylacetal) derivative was also found to react with normal haemoglobin A to form a number of less cationic components (Kikugawa et al., 1984). Malonaldehyde prepared from 1,1,3,3-tetramethoxypropane has been shown to inhibit a variety of enzymes (Chio & Tappel, 1969; Shin et al., 1972).

Effects on reproduction and prenatal toxicity

No data were available to the Working Group.

Absorption, distribution, excretion and metabolism

Bird and Draper (1982) have followed the uptake and oxidation of ^{14}C-malonaldehyde by cultured mammalian cells. There was a limited, concentration-dependent uptake of malonaldehyde (4% at concentrations of 0.1-1000 μM [0.007-72 μg/ml]) by 24 hours. They suggested that the limited uptake was due in part to the interaction of malonaldehyde with constitutents of the culture medium. However 83-89% of the ^{14}C-malonaldehyde was oxidized to $^{14}CO_2$ by 24 hours, and approximately 5% was recovered in the major lipids.

Two aldehyde dehydrogenases in the rat-liver cytosol fraction, with apparent Km for malonaldehyde of 16 μM and 128 μM, respectively, accounted for virtually all of the metabolizing activity for 50 μM [3.6 μg/ml] malonaldehyde in the postnuclear supernatant fraction (Hjelle

& Petersen, 1983). A similar low-Km aldehydrogenase has been reported in beef-liver cytosol (Sugimoto et al., 1976). An aldehyde dehydrogenase in the mitochondria with a Km of 7.3 mM could account for the low metabolizing activity of mitochondria (Hjelle & Petersen, 1983) and could result in oxygen uptake (Horton & Packer, 1970).

Malonaldehyde is one of the aldehydes responsible for the 'thiobarbituric acid (TBA)-reactive' material found in animal sera and tissues (Fig. 2).

Fig. 2. Lipid hydroperoxides and the thiobarbituric acid (TBA) reaction (Hayaishi & Shimizu, 1982)

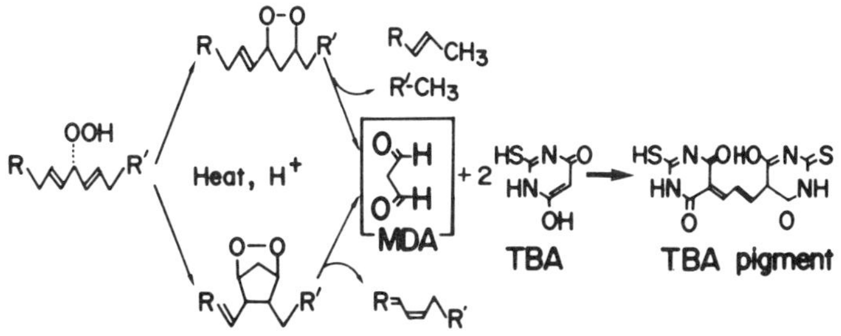

It is produced in a wide range of mammalian organisms as a by-product of prostaglandin biosynthesis and as an end-product of polyunsaturated lipid peroxidation (Bernheim et al., 1948; Diczfalusy et al., 1977; Shimizu et al., 1981; Yagi, 1982). The level of TBA-reacting substances in rabbit serum (expressed in terms of malonaldehyde) was 2 nmol/ml (Yagi, 1982). Platelet prostaglandin endoperoxides seem to be responsible for the serum TBA reaction in rabbits (Shimizu et al., 1981). Increased serum levels of TBA-reactive material have been reported following burn injury and retinopathy (Yagi, 1982).

Malonaldehyde binds to DNA in vitro (Brooks & Klamerth, 1968; Summerfield & Tappel, 1981) and forms covalent adducts with nucleosides, such as cyclic $1,N^6$-adenosine, $3,N^1$-cytidine, and $1,N^2$-guanosine derivatives (Seto et al., 1983; Nair et al., 1984). The extent of DNA cross-linking in vitro correlates with the formation of fluorescent adducts (Reiss et al., 1972; Seto et al., 1983; Summerfield & Tappel, 1983). These adducts are likely to be Schiff bases. A loss of DNA-template activity also occurred in vivo (Klamerth & Levinsky, 1969; Summerfield & Tappel, 1981). Malonaldehyde forms a variety of fluorescent and nonfluorescent adducts, such as amino-immunopropene derivatives, with the amino groups of some amino acids, proteins and phospholipids. Protein adducts often involve cross-links (Tappel, 1978); and malonaldehyde may also cross-link membrane proteins to each other (Jain & Hochstein, 1980), cross-link haemoglobin to the cell membrane (Goldstein et al., 1980) and can probably cross-link membrane lipids (Jain et al., 1983).

Mutagenicity and other short-term tests (see also 'Appendix: Activity Profiles for Short-term Tests', p. 336)

Since malonaldehyde is unstable and highly reactive, it is not available as a free compound, and other forms were tested in the studies described below. Some of the biological effects of malonaldehyde may be due to impurities such as β-alkoxyacroleins (Marnett & Tuttle, 1980) (see also section 2.1(a)).

Highly purified malonaldehyde is weakly mutagenic to *Salmonella typhimurium* TA102, TA104, TA2638 and *his* D3052 (Levin *et al.*, 1982; Basu & Marnett, 1983; Marnett *et al.*, 1984). Negative results were reported in *S. typhimurium* TA1535, TA1537 and TA1538 (Mukai & Goldstein, 1976; Marnett & Tuttle, 1980); and the activity of malonaldehyde in strain *his* D3052 as well as several other *S. typhimurium* strains (Mukai & Goldstein, 1976; Shamberger *et al.*, 1979) may be attributable partly or entirely to mutagenic impurities (Marnett & Tuttle, 1980). Malonaldehyde [purity unspecified] was also reported to be mutagenic to *Escherichia coli* (Yonei & Furui, 1981).

Malonaldehyde induced somatic mutations, but not sex-linked recessive lethal mutations, in *Drosophila melanogaster* (Szabad *et al.*, 1983).

In cultured rat-skin fibroblasts, the compound induced micronuclei, chromosomal aberrations and aneuploidies (Bird & Draper, 1980; Bird *et al.*, 1982b). It induced resistance to thymidine and methotrexate in mouse L5178Y lymphoma cells in the absence of an exogenous metabolic system (Yau, 1979).

(b) Humans

Toxic effects

Malonaldehyde is one of the aldehydes responsible for the 'thiobarbituric acid (TBA)-reactive' material found in human serum (Zlatkis *et al.*, 1981) (Fig. 2). Increased levels of serum TBA-reactive materials have been reported following a myocardial infarction (Aznar *et al.*, 1983; Dousset *et al.*, 1983) and angiopathy (Yagi, 1982).

No data were available to the Working Group on effects on reproduction and prenatal toxicity, on absorption, distribution, excretion and metabolism or on mutagenicity and chromosomal effects.

3.3 Case reports and epidemiological studies of carcinogenicity to humans

No data were available to the Working Group.

4. Summary of Data Reported and Evaluation

4.1 Exposure data

Malonaldehyde is not produced in commercial quantities. It occurs at low levels in the bloodstream as a product of lipid peroxidation and prostaglandin synthesis and in varying concentrations in a wide variety of foods, depending on many factors such as sources and preparation methods.

4.2 Experimental data

Malonaldehyde and its bis(dimethylacetal) and sodium salt were tested in mice by skin application. Its bis(dimethylacetal) and sodium salts were tested in mice by oral administration in drinking-water. The two studies by oral administration were inadequate for evaluation.

After topical application, no increase in the incidence of skin tumours was observed in one study. In one two-stage mouse-skin assay, a high dose of malonaldehyde (possibly containing impurities) showed initiating activity. In two other two-stage assays using lower doses, no initiating or promoting activity was observed.

Malonaldehyde was mutagenic to bacteria and induced somatic mutations in insects. It was mutagenic to mammalian cells *in vitro* and induced micronuclei, chromosomal aberrations and aneuploidy in these cells.

Overall assessment of data from short-term tests: malonaldehyde[a]

	Genetic activity			Cell transformation
	DNA damage	Mutation	Chromosomal effects	
Prokaryotes		+		
Fungi/green plants				
Insects		+[b]		
Mammalian cells (*in vitro*)		+	+	
Mammals (*in vivo*)				
Humans (*in vivo*)				
Degree of evidence in short-term tests for genetic activity: *Sufficient*				Cell transformation: No data

[a]The groups into which the table is divided and the symbol + are defined on pp. 17-18 of the Preamble; the degrees of evidence are defined on p. 18.
[b]Somatic mutations.

4.3 Human data

No case report or epidemiological study of the carcinogenicity of malonaldehyde to humans was available to the Working Group.

4.4 Evaluation[1]

There is *inadequate evidence* for the carcinogenicity of malonaldehyde to experimental animals.

In the absence of epidemiological data, no evaluation could be made of the carcinogenicity of malonaldehyde to humans.

---

[1]For definitions of the italicized terms, see the Preamble, pp. 15-16.

5. References

Apaja, M. (1980) Evaluation of toxicity and carcinogenicity of malonaldehyde. An experimental study in Swiss mice. *Anat. Pathol. Microbiol.*, *8*, Series D, 1-61

Arya, S.S. & Nirmala, N. (1971) Determination of free malonaldehyde in vegetable oils. *J. Food Sci. Technol.*, *8*, 177-180

Aznar, J., Santos, M.T., Valles, J. & Sala, J. (1983) Serum malonaldehyde-like material (MDA-LM) in acute myocardial infarction. *J. clin. Pathol.*, *36*, 712-715

Basu, A.K. & Marnett, L.J. (1983) Unequivocal demonstration that malondialdehyde is a mutagen. *Carcinogenesis*, *4*, 331-333

Bernheim, F., Bernheim, M.L.C. & Wilbur, K.M. (1948) The reaction between thiobarbituric acid and the oxidation products of certain lipids. *J. biol. Chem.*, *174*, 257-264

Bird, R.P. & Draper, H.H. (1980) Effect of malonaldehyde and acetaldehyde on cultured mammalian cells: Growth, morphology, and synthesis of macromolecules. *J. Toxicol. environ. Health*, *6*, 811-823

Bird, R.P. & Draper, H.H. (1982) Uptake and oxidation of malonaldehyde by cultured mammalian cells. *Lipids*, *17*, 519-523

Bird, R.P., Draper, H.H. & Valli, V.E.O. (1982a) Toxicological evaluation of malonaldehyde: A 12-month study of mice. *J. Toxicol. environ. Health*, *10*, 897-905

Bird, R.P., Draper, H.H. & Basrur, P.K. (1982b) Effect of malonaldehyde and acetaldehyde on cultured mammalian cells. Production of micronuclei and chromosomal aberrations. *Mutat. Res.*, *101*, 237-246

Bird, R.P., Hung, S.S.O., Hadley, M. & Draper, H.H. (1983) Determination of malonaldehyde in biological materials by high-pressure liquid chromatography. *Anal. Biochem.*, *128*, 240-244

Bond, A.M., Deprez, P.P., Jones, R.D., Wallace, G.G. & Briggs, M.H. (1980) Polarographic method for the determination of propanedial (malonaldehyde). *Anal. Chem.*, *52*, 2211-2213

Brooks, B.R. & Klamerth, O.L. (1968) Interaction of DNA with bifunctional aldehydes. *Eur. J. Biochem.*, *5*, 178-182

Buckingham, J., ed. (1982) *Dictionary of Organic Compounds*, 5th ed., Vol. 1, New York, Chapman & Hall, p. 4778

Chio, K.S. & Tappel, A.L. (1969) Inactivation of ribonuclease and other enzymes by peroxidizing lipids and by malonaldehyde. *Biochemistry*, *8*, 2827-2832

Crawford, D.L., Sinnhuber, R.O., Stout, F.M., Oldfield, J.E. & Kaufmes, J. (1965) Acute toxicity of malonaldehyde. *Toxicol. appl. Pharmacol.*, *7*, 826-832

Diczfalusy, U., Falardeau, P. & Hammarström, S. (1977) Conversion of prostaglandin endoperoxides to C_{17}-hydroxy acids catalyzed by human platelet thromboxane synthase. *FEBS Lett., 84*, 271-274

Dousset, J.-C., Trouilh, M. & Foglietti, M.-J. (1983) Plasma malonaldehyde levels during myocardial infarction. *Clin. chim. Acta, 129*, 319-322

Esterbauer, H. & Slater, T.P. (1981) The quantitative estimation by high performance liquid chromatography of free malonaldehyde produced by peroxidizing microsomes. *IRCS (Int. Res. Comm. Syst.) med. Sci., 9*, 749-750 [*Chem. Abstr., 95*, 128583w]

Fischer, S.M., Ogle, S., Marnett, L.J., Nesnow, S. & Slaga, T.J. (1983) The lack of initiating and/or promoting activity of sodium malondialdehyde on SENCAR mouse skin. *Cancer Lett., 19*, 61-66

Goldstein, B.D., Rozen, M.G. & Kunis, R.L. (1980) Role of red cell membrane lipid peroxidation in hemolysis due to phenylhydrazine. *Biochem. Pharmacol., 29*, 1355-1359

Hayaishi, O. & Shimizu, T. (1982) *Metabolic and functional significance of prostaglandins in lipid peroxide research.* In: Yagi, K., ed., *Lipid Peroxides in Biology and Medicine*, New York, Academic Press, p. 41

Hirayama, T., Yamada, N., Nohara, M. & Fukui, S. (1983) High performance liquid chromatographic determination of malondialdehyde in vegetable oils. *J. Assoc. off. anal. Chem., 66*, 304-308

Hjelle, J.J. & Petersen, D.R. (1983) Metabolism of malonaldehyde by rat liver aldehyde dehydrogenase. *Toxicol. appl. Pharmacol., 70*, 57-66

Horton, A.A. & Packer, L. (1970) Interactions between malondialdehyde and rat liver mitochondria. *J. Gerontol., 25*, 199-204

IARC (1982) *Information Bulletin on the Survey of Chemicals Being Tested for Carcinogenicity*, No. 10, Lyon, p. 216

Jain, S.K. & Hochstein, P. (1980) Polymerization of membrane components in aging red blood cells. *Biochem. biophys. Res. Commun., 92*, 247-254

Jain, S.K., Mohandas, N., Clark, M.R. & Shohet, S.B. (1983) The effects of malonyldialdehyde, a product of lipid peroxidation, on the deformability, dehydration and ^{51}Cr-survival of erythrocytes. *Br. J. Hematol., 53*, 247-255

Kakuda, Y., Stanley, D.W. & van de Voort, F.R. (1981) Determination of TBA number by high performance liquid chromatography. *J. Am. Oil Chem. Soc., 58*, 773-775

Kikugawa, K., Kosugi, H. & Asakura, T. (1984) Effect of malondialdehyde, a product of lipid peroxidation, on the function and stability of hemoglobin. *Arch. Biochem. Biophys., 229*, 7-14

Klamerth, O.L. & Levinsky, H. (1969) Template activity in liver DNA from rats fed with malondialdehyde. *FEBS Lett., 3*, 205-207

Kwon, T.-W. & Watts, B.M. (1963a) A new color reaction of anthrone with malonaldehyde and other aliphatic aldehydes. *Anal. Chem., 35*, 733-735

Kwon, T.-W. & Watts, B.M. (1963b) Determination of malonaldehyde by ultraviolet spectrophotometry. *J. Food Sci.*, *28*, 627-630

Levin, D.E., Hollstein, M., Christman, M.F., Schwiers, E.A. & Ames, B.N. (1982) A new *Salmonella* tester strain (TA102) with AT base pairs at the site of mutation detects oxidative mutagens. *Proc. natl Acad. Sci. USA*, *79*, 7445-7449

Marnett, L.J. & Tuttle, M.A. (1980) Comparison of the mutagenicities of malondialdehyde and the side products formed during its chemical synthesis. *Cancer Res.*, *40*, 276-282

Marnett, L.J., Bienkowski, M.J., Raban, M. & Tuttle, M.A. (1979) Studies of the hydrolysis of ^{14}C-tetraethoxypropane to malondialdehyde. *Anal. Biochem.*, *99*, 458-463

Marnett, L.J., Hurd, H., Hollstein, M., Levin, D.E., Esterbauer, H. & Ames, B.N. (1984) Naturally occurring carbonyl compounds are mutagenic in *Salmonella* tester strain TA104. *Mutat. Res.* (in press)

Merzlyak, M.N., Shevyreva, V.V., Rumyantseva, V.B. & Zhirov, V.K. (1979) Fluorimetric determination of malonaldehyde in plant tissues (Russ.). *Biol. Nauki (Moscow)*, *10*, 70-74 [*Chem. Abstr.*, *92*, 18273a]

Mukai, F.H. & Goldstein, B.D. (1976) Mutagenicity of malonaldehyde, a decomposition product of peroxidized polyunsaturated fatty acids. *Science*, *191*, 868-869

Nair, V., Turner, G.A. & Offerman, R.J. (1984) Novel adducts from the modification of nucleic acid bases by malondialdehyde. *J. Am. chem. Soc.*, *106*, 3370-3371

Newburg, D.S & Concon, J.M. (1980) Malonaldehyde concentrations in food are affected by cooking conditions. *J. Food Sci.*, *45*, 1681-1683

Reiss, U., Tappel, A.L. & Chio, K.S. (1972) DNA-malonaldehyde reaction: Formation of fluorescent products. *Biochem. biophys. Res. Commun.*, *48*, 921-926

Roncucci, R. & Lansen, J. (1980) Quantitative determination of malonic dialdehyde. *German Patent 2,942,617* (to Continental Pharma) [*Chem. Abstr.*, *93*, 163975f]

Seto, H., Okuda, T., Takesue, T. & Ikemura, T. (1983) Reaction of malonaldehyde with nucleic acid. I. Formation of fluorescent pyrimido[1,2-a]purin-10(3H)-one nucleosides. *Bull. chem. Soc. Jpn*, *56*, 1799-1802

Shamberger, R.J., Andreone, T.L. & Willis, C.E. (1974) Antioxidants and cancer. IV. Initiating activity of malonaldehyde as a carcinogen. *J. natl Cancer Inst.*, *53*, 1771-1773

Shamberger, R.J., Tytko, S.A. & Willis, C.E. (1975) Malonaldehyde is a carcinogen (Abstract No. 3431). *Fed. Proc.*, *34*, 827

Shamberger, R.J., Shamberger, B.A. & Willis, C.E. (1977) Malonaldehyde content of food. *J. Nutr.*, *107*, 1404-1409

Shamberger, R.J., Corlett, C.L., Beaman, K.D. & Kasten, B.L. (1979) Antioxidants reduce the mutagenic effect of malonaldehyde and β-propiolactone. Part IX. Antioxidants and cancer. *Mutat. Res.*, *66*, 349-355

Shimasaki, H. & Ueta, N. (1980) Studies on the fluorescent spectrophotometrical method for the detection of oxidized lipids in serum (Jpn.). *Teikyo Igaku Zasshi*, *1*, 45-51 [*Chem. Abstr.*, *93*, 65152r]

Shimizu, T., Kondo, K. & Hayaishi, O. (1981) Role of prostaglandin endoperoxides in the serum thiobarbituric acid reaction. *Arch. Biochem. Biophys.*, *206*, 271-276

Shin, B.C., Huggins, J.W. & Carraway, K.L. (1972) Effect of pH, concentration and aging on the malonaldehyde reaction with proteins. *Lipids*, *7*, 229-233

Shohet, S.B. & Jain, S.K. (1982) Introduction: Vitamin E and blood cell function. *Ann. N.Y. Acad. Sci.*, *393*, 229-236

Siu, G.M., Draper, H.H. & Valli, V.E.O. (1983) Oral toxicity of malonaldehyde: A 90-day study on mice. *J. Toxicol. environ. Health*, *11*, 105-119

Sugimoto, E., Takahashi, N., Kitagawa, Y. & Chiba, H. (1976) Intracellular localization and characterization of beef liver aldehyde dehydrogenase isozymes. *Agric. biol. Chem.*, *40*, 2063-2070

Summerfield, F.W. & Tappel, A.L. (1981) Determination of malondialdehyde-DNA crosslinks by fluorescence and incorporation of tritium. *Anal. Biochem.*, *111*, 77-82

Summerfield, F.W. & Tappel, A.L. (1983) Determination by fluorescence quenching of the environment of DNA crosslinks made by malondialdehyde. *Biochim. biophys. Acta*, *740*, 185-189

Szabad, J., Soós, I., Polgár, G. & Héjja, G. (1983) Testing the mutagenicity of malondialdehyde and formaldehyde by the *Drosophila* mosaic and the sex-linked recessive lethal tests. *Mutat. Res.*, *113*, 117-133

Tappel, A.L. (1978) *Measurement of and protection from in vivo lipid peroxidation*. In: Pryor, W.A., ed., *Free Radicals in Biology*, Vol. IV, New York, Academic Press, pp. 1-47

Von Voss, H., Bergmann, F., Richter, O., Wegener, M. & Goebel, U. (1980) *Spectrophotometric micromethod for the determination of malonyldialdehyde production of thrombocytes* (Ger.). In: Schimpf, K., ed., *Fibrinogene, Fibrine, Fibrine Adhesive, Proceedings of the 23rd Germany Study Group Meeting on Blood Coagulation* (Ger.), Stuttgart, Schattauer, pp. 555-561 [*Chem. Abstr.*, *94*, 1725t]

Yagi, K. (1982) *Assay for serum lipid peroxide level and its clinical significance*. In: Yagi, K., ed., *Lipid Peroxides in Biology and Medicine*, New York, Academic Press, pp. 223-242

Yau, T.M. (1979) Mutagenicity and cytotoxicity of malonaldehyde in mammalian cells. *Mech. Ageing Dev.*, *11*, 137-144

Yonei, S. & Furui, H. (1981) Lethal and mutagenic effects of malondialdehyde, a decomposition product of peroxidized lipids, on *Escherichia coli* with different DNA-repair capacities. *Mutat. Res.*, *88*, 23-32

Zlatkis, A., Brazell, R.S. & Poole, C.F. (1981) The role of organic volatile profiles in clinical diagnosis. *Clin. Chem.*, *27*, 789-797

EPOXIDES

DIGLYCIDYL RESORCINOL ETHER

This substance was considered by a previous working group, in February 1976 (IARC, 1976a). Since that time, new data have become available, and these have been incorporated into the monograph and taken into account in the present evaluation.

1. Chemical and Physical Data

1.1 Synonyms and trade names

Chem. Abstr. Services Reg. No.: 101-90-6

Chem. Abstr. Name: Oxirane, 2,2'-[1,3-phenylenebis(oxymethylene)]bis-

IUPAC Name: meta-Bis(2,3-epoxypropoxy)benzene

Synonyms: 1,3-Bis(2,3-epoxypropoxy)benzene; *meta*-Bis(glycidyloxy)benzene; diglycidyl ether of resorcinol; 1,0-diglycidyloxybenzene; diglycidyl resorcinol; NCI-C54966; 2,2'-[1,3-phenylenebis(oxymethylene)]bisoxirane; RDGE; resorcinol bis(2,3-epoxypropyl)ether; resorcinol diglycidyl ether; resorcinol glycidyl ether; resorcinyl diglycidyl ether

Trade Name: Araldite ERE 1359

1.2 Structural and molecular formulae and molecular weight

$C_{12}H_{14}O_4$
Mol. wt. 222.2

— 181 —

1.3 Chemical and physical properties of the pure substance

From Hawley (1981), unless otherwise specified

(a) *Description*: Straw-yellow liquid

(b) *Boiling-point*: 172°C at 0.8 mm Hg

(c) *Density*: Specific gravity (25°C), 1.21

(d) *Refractive index*: n_D^{25} 1.541

(e) *Spectroscopy data*: Infrared and nuclear magnetic resonance spectral data have been reported (National Toxicology Program, 1983).

(f) *Solubility*: Miscible with acetone, chloroform, methanol (National Toxicology Program, 1983), benzene (Van Duuren et al., 1965) and most organic resins

(g) *Viscosity*: 500 cP at 25°C

(h) *Stability*: Flash-point (Cleveland open-cup), 176°C; combustible

(i) *Reactivity*: Reacts with compounds having labile hydrogen. As an epoxide, it can react as an alkylating agent.

(j) *Conversion factor*: 1 ppm in air = 9.09 mg/m^3 at 760 mm Hg and 4°C (National Institute for Occupational Safety and Health, 1978)

1.4 Technical products and impurities

The only product for which specifications were available was one formerly available commercially from one US producer, which contained 98.5% min non-volatile matter.

2. Production, Use, Occurrence and Analysis

2.1 Production and use

(a) *Production*

Diglycidyl resorcinol ether was first made in 1948 by the reaction of excess epichlorohydrin (see IARC, 1976b, 1982) with resorcinol (see IARC, 1977) in alkaline solution (Werner & Farenhorst, 1948); this is probably the method used for commercial production.

Diglycidyl resorcinol ether has been produced commercially in the USA since at least 1974. In 1977, one US company reported production in the range of 4.5-45.4 thousand kg and another company produced an unspecified quantity (NIH/EPA Chemical Information System, 1984). Currently only one US company produces diglycidyl resorcinol ether in commercial quantities. Separate data on US imports and exports of diglycidyl resorcinol ether are not published.

No evidence was found that diglycidyl resorcinol ether is produced commercially in western Europe.

Diglycidyl resorcinol ether was first produced commercially in Japan by one company in 1974, and its production in 1975 is estimated to have been five thousand kg. The company stopped production of this compound in 1978.

(b) *Use*

Diglycidyl resorcinol ether is used in the USA as a liquid epoxy resin (Hawley, 1981) and as a reactive diluent in the production of other epoxy resins (Lee & Neville, 1967). It is believed to be used principally in the USA as a diluent to impart special properties to cured epoxy resins (e.g., as in aerospace applications). It has also been used to cure polysulphide rubber, but the current commercial status of this application is not known.

No evidence was found that any country has limited occupational exposure to diglycidyl resorcinol ether by regulation or recommended guideline.

2.2 Occurrence

(a) *Natural occurrence*

Diglycidyl resorcinol ether is not known to occur as a natural product.

(b) *Occupational exposure*

On the basis of the 1974 National Occupational Hazard Survey, the National Institute for Occupational Safety and Health (1980, 1981) estimated that in the USA about 3000 workers in four industries were exposed to diglycidyl resorcinol ether annually at that time. The principal industry in which exposure was found was the aircraft equipment industry.

2.3 Analysis

No information on analytical methods specifically for diglycidyl resorcinol ether was available to the Working Group.

3. Biological Data Relevant to the Evaluation of Carcinogenic Risk to Humans

3.1 Carcinogenicity studies in animals

(a) *Oral administration*

Mouse: Groups of 50 male and 50 female B6C3F$_1$ mice, eight to nine weeks old, were administered 50 or 100 mg/kg bw diglycidyl resorcinol ether (purity, approximately 88% by gas chromatography[1] with 30 unspecified impurities) in corn oil or corn oil alone (vehicle controls) by intragastric intubation, five times per week for 103 weeks. In male mice no significant difference in survival was observed between the treated and control groups. Mortality in treated and control female mice was high, with only 40% of controls, 26% of low-dose and 20% of high-dose animals still alive at the end of two years; the main cause of death was suppurative and necrotizing inflammation of the reproductive tract. Diglycidyl resorcinol ether induced hyperkeratosis, hyperplasia, papillomas and squamous-cell carcinomas of the forestomach in both sexes. The incidences of squamous-cell carcinomas in the control, low-dose and high-dose groups were: males, 0/47, 14/49 and 25/50; and females, 0/47, 12/49 and 23/49, respectively ($p < 0.001$ for both low- and high-dose groups). The incidences of papillomas of the forestomach were: males, 0/47, 4/49 and 10/50 ($p = 0.06$ and 0.001 for low- and high-dose groups, respectively); and females, 0/47, 5/49 and 10/49 ($p = 0.03$ and 0.001 for low- and high-dose groups, respectively). In female mice, the incidences of hepatocellular carcinomas were: 0/48, 1/50 and 3/49 in the control, low-dose and high-dose groups, respectively. The difference between the high-dose group and the controls was statistically significant ($p = 0.04$ by life-table analysis). Increases in the combined incidences of hepatocellular adenomas and carcinomas showed a similar statistical trend. Tumours at other sites could not be related to treatment (National Toxicology Program, 1985).

Rat: Groups of 50 male and 50 female Fischer 344/N rats, eight to nine weeks old, received 25 or 50 mg/kg bw diglycidyl resorcinol ether (purity, approximately 88% by gas chromatography with 30 unspecified impurities, as in the experiment described above) in corn oil or corn oil alone (vehicle controls) by gastric intubation five times per week for 103 weeks. Due to excessive mortality in the high-dose group, supplemental groups of 50 male and 50 female rats were treated with 12 mg/kg bw diglycidyl resorcinol ether in corn oil or with corn oil alone. Mortality among treated rats of each sex in the primary study was dose-related and was significantly greater ($p < 0.001$) than that among the vehicle controls; no high-dose male rat and only 1/50 high-dose female rats survived to the end of the study. Bronchopneumonia, the incidence of which was dose-related, was the most frequent cause of early death. In the supplemental study, there was no significant difference in survival between the treated and control female rats, but survival of the treated male rats was significantly reduced ($p = 0.003$) as compared to their controls. Diglycidyl resorcinol ether induced hyperkeratosis, hyperplasia, papillomas and squamous-cell carcinomas of the forestomach in animals of both sexes. The incidences of squamous-cell carcinoma in the combined control, low-, medium- and high-dose groups were: in males, 0/100, 39/50, 38/50 and 4/49; and in females, 0/99, 27/50, 34/50 and 3/50. The numbers of animals in each group still alive at 104-105 weeks were 81, 23, 5 and 0 males and 76, 35, 16 and 1 females, respectively. The incidences of papilloma of the forestomach were: in males, 0/100, 16/50, 17/50 and 6/49, respectively; and in females, 0/99,, 19/50, 7/50 and 1/50, respectively. The first stomach tumour occurred in a high-dose female at week 42, when only 13 rats were still alive. Similarly, in high-dose males, only 9 animals survived beyond week 60. Since the first squamous-cell carcinoma of the forestomach in this group occurred at week 79, again only 9 animals were at risk for this tumour (National Toxicology Program, 1985).

(b) *Skin application*

Mouse: No skin tumour occurred in 30 female Swiss ICR/Ha mice that received (beginning at eight weeks of age) thrice-weekly skin applications of a 1% solution of diglycidyl resorcinol ether [purity unspecified] in benzene (approximately 100 mg per application) for life; the median survival time was 491 days (Van Duuren *et al.*, 1965).

---

[1]Major impurities identified in the test material were: 1.9% 3-methylbenzoic acid ethyl ester, 1.6% 3-chloropropoxybenzene and 2.8% dihydroxypropoxybenzene. All others were present at levels of <0.8% each.

3.2 Other relevant biological data

(a) *Experimental systems*

Toxic effects

The oral LD_{50}s of diglycidyl resorcinol ether in mice, rabbits and rats are 980, 1240 and 2570 mg/kg bw, respectively. The intraperitoneal LD_{50}s in mice and rats are 243 and 178 mg/kg bw [purity unspecified]. An undefined concentration of diglycidyl resorcinol ether vapour was not lethal to 10 rats after 50 seven-hour exposures by inhalation. It caused irritation to the eyes and skin even after a single application to the eyes or to the skin (Hine *et al.*, 1958).

In monkeys, once-monthly intravenous injection of 100-200 mg/kg bw diglycidyl resorcinol ether produced a progressive lowering of the leucocyte count (Hine *et al.*, 1981). The compound produced inhibition of the growth of Walker carcinoma in rats (Hendry *et al.*, 1951).

Diglycidyl resorcinol ether caused hyperkeratosis and basal-cell hyperplasia of the stomach in rats exposed daily to intragastric doses of 12.5 mg/kg bw in corn oil and to higher concentrations for 13 weeks. Similar findings were observed in mice. In a two-year study of groups of 50 male and 50 female Fischer 344 rats exposed by gastric intubation, bronchopneumonia occurred in 2/50, 17/50, 26/50, 0/50, 10/50 and 17/50 control, low-dose and high-dose males and females, respectively. The bronchopneumonia was not consistent with chemical pneumonitis but was characterized by polymorphonuclear leucocytes in the centriacinar alveoli (National Toxicology Program, 1985).

No data were available to the Working Group on effects on reproduction and prenatal toxicity or on absorption, distribution, excretion and metabolism.

Mutagenicity and other short-term tests (see also 'Appendix: Activity Profiles for Short-term Tests', p. 337)

Diglycidyl resorcinol ether (purity, approximately 88% by gas chromatography with 30 unspecified impurities, as described in section 3.1(*a*)) was mutagenic to *Salmonella typhimurium* TA1535 and TA100 but not TA1537 or TA98 in a preincubation assay, both in the presence and absence of an Aroclor-induced rat- or hamster-liver metabolic system (National Toxicology Program, 1985).

(b) *Humans*

Toxic effects

Diglycidyl resorcinol ether produces severe burns on contact with the skin, and skin sensitization has occurred in a limited number of cases (Hine & Rowe, 1963). [The source of the data was not given.]

No data were available to the Working Group on effects on reproduction and prenatal toxicity, on absorption, distribution, excretion and metabolism or on mutagenicity and chromosomal effects.

3.3 Case reports and epidemiological studies of carcinogenicity to humans

No data were available to the Working Group.

4. Summary of Data Reported and Evaluation

4.1 Exposure data

Diglycidyl resorcinol ether has been produced since at least 1974. It has only limited application, principally in the aerospace industry.

4.2 Experimental data

Diglycidyl resorcinol ether (of technical grade) was tested for carcinogenicity by intragastric intubation in mice of one strain and in rats of one strain. It induced squamous-cell carcinomas and papillomas of the forestomach in animals of both species. In female mice, an increased incidence of hepatocellular tumours was observed. In one experiment in mice, no skin tumour was observed after skin application.

Diglycidyl resorcinol ether (of technical grade) was mutagenic to bacteria.

Overall assessment of data from short term tests: diglycidyl resorcinol ether (of technical grade)[a]

	Genetic activity			Cell transformation
	DNA damage	Mutation	Chromosomal effects	
Prokaryotes		+		
Fungi/green plants				
Insects				
Mammalian cells (in vitro)				
Mammals (in vivo)				
Humans (in vivo)				
Degree of evidence in short-term tests for genetic activity: Inadequate				Cell transformation: No data

[a]The groups into which the table is divided and the symbol + are defined on pp. 17-18 of the Preamble; the degrees of evidence are defined on p. 18.

4.3 Human data

No case report or epidemiological study of the carcinogenicity of diglycidyl resorcinol ether to humans was available to the Working Group.

4.4 Evaluation[1]

There is *sufficient evidence*[2] for the carcinogenicity of a technical grade of diglycidyl resorcinol ether to experimental animals.

No data on the carcinogenicity of diglycidyl resorcinol ether to humans were available to the Working Group.

5. References

Hawley, G.G., ed. (1981) *The Condensed Chemical Dictionary*, 10th ed., New York, Van Nostrand Reinhold Co., p. 892

Hendry, J.A., Homer, R.F., Rose, F.L. & Walpole, A.L. (1951) Cytotoxic agents. II. Bis-epoxides and related compounds. *Br. J. Pharmacol.*, 6, 235-255

Hine, C.H. & Rowe, V.K. (1963) *Resorcinol diglycidyl ether*. In: Patty, F.A., ed., *Industrial Hygiene and Toxicology*, 2nd rev. ed., Vol. 2, New York, Interscience, pp. 1648-1649

Hine, C.H., Kodama, J.K., Anderson, H.H., Simonson, D.W. & Wellington, J.S. (1958) The toxicology of epoxy resins. *Arch. ind. Health*, 17, 129-144

Hine, C., Rowe, V.K., White, E.R., Darmer, K.I., Jr & Youngblood, G.T. (1981) *Epoxy compounds*. In: Clayton, G.D. & Clayton, F.E., eds, *Patty's Industrial Hygiene and Toxicology*, Vol. 2A, *Toxicology*, 3rd rev. ed., New York, John Wiley & Sons, pp. 2232-2233

IARC (1976a) *IARC Monographs on the Evaluation of Carcinogenic Risk of Chemicals to Man, Vol. 11, Cadmium, Nickel, Some Epoxides, Miscellaneous Industrial Chemicals and General Considerations on Volatile Anaesthetics*, Lyon, pp. 125-129

IARC (1976b) *IARC Monographs on the Evaluation of Carcinogenic Risk of Chemicals to Man, Vol. 11, Cadmium, Nickel, Some Epoxides, Miscellaneous Industrial Chemicals and General Considerations on Volatile Anaesthetics*, Lyon, pp. 131-139

IARC (1977) *IARC Monographs on the Evaluation of the Carcinogenic Risk of Chemicals to Man, Vol. 15, Some Fumigants, the Herbicides 2,4-D and 2,4,5-T, Chlorinated Dibenzodioxins and Miscellaneous Industrial Chemicals*, Lyon, pp. 155-175

IARC (1982) *IARC Monographs on the Evaluation of the Carcinogenic Risk of Chemicals to Humans, Suppl. 4, Chemicals, Industrial Processes and Industries Associated with Cancer in Humans (IARC Monographs, Volumes 1-29)*, Lyon, pp. 122-124

---

[1] For definitions of the italicized terms, see the Preamble, pp. 15-16.
[2] In the absence of adequate data on humans, it is reasonable, for practical purposes, to regard chemicals for which there is *sufficient evidence* of carcinogenicity in animals as if they presented a carcinogenic risk to humans.

Lee, H. & Neville, K. (1967) *Handbook of Epoxy Resins*, San Francisco, CA, McGraw-Hill, p. 4-59

National Institute for Occupational Safety and Health (1978) *Criteria for a Recommended Standard ... Occupational Exposure to Glycidyl Ethers (DHEW (NIOSH) Publ. No. 78-166)*, Washington DC, US Department of Health, Education, and Welfare, p. 190

National Institute for Occupational Safety and Health (1980) *Projected Number of Occupational Exposures to Chemical and Physical Hazards*, Cincinnati, OH, p. 119

National Institute for Occupational Safety and Health (1981) *National Occupational Hazard Survey*, microfiche, Cincinnati, OH, p. 7889 (generics excluded)

National Toxicology Program (1985) *NTP Technical Report on the Carcinogenesis Studies of Diglycidyl Resorcinol Ether (Technical Grade) (CAS No. 101-90-6) in F344/N Rats and B6C3F$_1$ Mice (Gavage Study) (NIH Publ. No. 84-2513; NTP-82-064)*, Research Triangle Park, NC, US Department of Health and Human Services

NIH/EPA Chemical Information System (1984) *Toxic Substances Control Act (TSCA) Plant and Production*, Washington DC, Information Sciences Corporation

Van Duuren, B.L., Orris, L. & Nelson, N. (1965) Carcinogenicity of epoxides, lactones, and peroxy compounds. Part II. *J. natl Cancer Inst.*, 35, 707-717

Werner, E.G.G. & Farenhorst, E. (1948) Investigations on glycidyl ethers. I. The synthesis of aromatic bisglycidyl ethers. *Rec. Trav. chim.*, 67, 438-441

ETHYLENE OXIDE

This substance was considered by previous working groups, in February 1976 (IARC, 1976) and in February 1982 (IARC, 1982a). Since that time, new data have become available, and these have been incorporated into the monograph and taken into account in the present evaluation.

1. Chemical and Physical Data

1.1 Synonyms and trade names

Chem. Abstr. Services Reg. No.: 75-21-8

Chem. Abstr. Name: Oxirane

IUPAC Systematic Name: Ethylene oxide

Synonyms: Dihydrooxirene; dimethylene oxide; ENT-26263; EO; epoxyethane; 1,2-epoxyethane; ethene oxide; ethylene (oxide D); ETO; ETOX; FEMA No. 2433; NCI-C50088; oxacyclopropane; oxane; oxidoethane; α,β-oxidoethane; oxiran

Trade Names: Amprolene; Anprolene; Anproline; Oxyfume; Oxyfume 12; Sterilizing Gas Ethylene Oxide 100%; T-Gas

1.2 Structural and molecular formulae and molecular weight

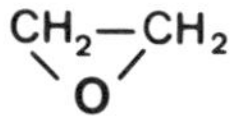

$$CH_2-CH_2 \atop \diagdown O \diagup$$

C_2H_4O Mol. wt: 44.1

1.3 Chemical and physical properties of the pure substance

From Cawse *et al.* (1980), unless otherwise specified

(a) *Description*: Colourless gas

(b) *Boiling-point*: 10.4°C

(c) *Freezing-point*: -112.5°C

(d) *Density*: 0.8969 (liquid at 0°C)

(e) *Refractive index*: n_D^7 1.3597

(f) *Solubility*: Miscible with water, diethyl ether, ethanol and most organic solvents

(g) *Viscosity*: 0.32 cP at 0°C

(h) *Volatility*: Vapour pressure, 494 mm Hg at 0°C

(i) *Stability*: Flash-point (tag open-cup), <-18°C; vapours are inflammable and explosive

(j) *Reactivity*: Very reactive (e.g., reacts with many compounds having labile hydrogen). May undergo slow polymerization during storage. Excessive temperatures or contamination with impurities, such as water, alkalis, acids, metal oxides, and iron and aluminium salts, can cause rapid polymerization or reaction

(k) *Conversion factor*: 1 ppm = 1.83 mg/m^3 at 760 mm Hg and 20°C (Verschueren, 1977)

1.4 Technical products and impurities

Ethylene oxide is available in the USA as a high-purity chemical with the following specifications: water, 0.03% max; aldehydes (as acetaldehyde), 0.003% max; acidity (as acetic acid), 0.002% max; residue, 0.005 g/100 ml max; and acetylene, none (Cawse et al., 1980).

Ethylene oxide is available in western Europe with the following specifications: purity, 99.5% min; water, 0.01% max; and carbon dioxide, 100 mg/kg max.

It has been reported that ethylene oxide made by the chlorohydrin process [see section 2.1 (a)] may contain chlorine-containing C$_2$ chemicals (vinyl chloride, ethylene chloride, chloroethane and ethylene chlorohydrin) at levels of 1-10 mg/kg (Ethylene Oxide Industry Council, 1983).

2. Production, Use, Occurrence and Analysis

2.1 Production and use

(a) *Production*

Ethylene oxide was first prepared in 1859 by Wurtz by the reaction of ethylene chlorohydrin with potassium hydroxide. Commercial production of ethylene oxide was started in Germany during the First World War. Until 1937, essentially all the ethylene oxide produced in the USA was made by the so-called chlorohydrin process, in which ethylene is treated with hypochlorous acid (chlorine and water) to produce ethylene chlorohydrin, and this is converted to ethylene oxide using calcium hydroxide or sodium hydroxide. Since 1937, the chlorohydrin process has gradually been replaced in the USA by the direct oxidation process, in which ethylene (IARC, 1979a) is oxidized to ethylene oxide using either air or oxygen and a silver catalyst (Cawse et al., 1980). The chlorohydrin process was last used to produce ethylene oxide commercially in the USA in about 1973, although one plant that currently produces propylene oxide (see p. 227 of this volume) by the chlorohydrin process has the capacity to produce ethylene oxide by this process also. Currently, all US plants use oxygen (rather than air) in the direct oxidation process to produce ethylene oxide.

US production of ethylene oxide, which started in 1921 (US Tariff Commission, 1922), reached a peak in 1979 when an estimated 2440 thousand tonnes were produced by 12 companies. Total production by the 15 plants of the 12 US producing companies (including one in Puerto Rico) amounted to 2264 thousand tonnes in 1982 (US International Trade Commission, 1983). US imports of ethylene oxide (mostly from Canada) amounted to 4.3 thousand tonnes in 1982 (US Department of Commerce, 1983). US exports (mostly to Canada) in 1983 were 6.3 thousand tonnes (US Department of Commerce, 1984).

Ethylene oxide is produced by three companies in Canada (with an estimated production of 282.5 thousand tonnes in 1981) and by one company each in Mexico and Brazil.

It is produced by four companies in Germany, two companies each in Belgium, France, Italy, the Netherlands and the UK, and by one company each in Spain and Sweden. Production in western Europe is estimated to have been approximately 1370 thousand tonnes in 1981. Ethylene oxide is also produced at two plants each in the Democratic Republic of Germany, Romania and the USSR, and at one plant each in Bulgaria, Czechoslovakia and Poland.

Commercial production of ethylene oxide in Japan started in 1934. Five Japanese companies currently manufacture it at seven plants by the direct oxidation process, and 1982 production is estimated to have been 471 thousand tonnes. Japanese imports were negligible, and exports totalled only 17 tonnes in 1982.

Ethylene oxide is also produced at three plants in China, two plants each in India and Taiwan, and one plant each in Australia, the Democratic People's Republic of Korea and the Republic of Korea.

(b) Use

Almost all ethylene oxide is used as a chemical intermediate. The production of ethylene glycol is the largest market for ethylene oxide in the USA, and the next most important is its use as a chemical intermediate for a variety of nonionic surfactants. The use pattern for the estimated 2330 thousand tonnes of ethylene oxide used in the USA in 1981 was as follows: ethylene glycol, 60%; nonionic surfactants, 12%; glycol ethers, 7%; ethanolamines, 7%; and other uses (including quantities used in nonintermediate applications), 14%.

Ethylene glycol, made by the reaction of ethylene oxide with water, is used principally as an intermediate for terephthalate polyester resins for fibres, film and bottles. Another major use for ethylene glycol is in automotive antifreeze; it is also used in many smaller applications (e.g., solvent, heat-transfer fluid, de-icing fluid) (Cawse et al., 1980).

Nonionic surfactants made by the addition of ethylene oxide to a variety of chemicals having labile hydrogen atoms (principally long-chain alcohols and higher alkylphenols) and the random and block copolymers of ethylene oxide and propylene oxide are used in a variety of household and industrial products as low-foam detergents.

Glycol ethers, made by the addition of ethylene oxide to short-chain alcohols, find use as solvents (e.g., in surface coatings and a variety of consumer products) and as intermediates for glycol ether acetates, which are also important solvents. Additional uses are as components of hydraulic fluids and jet fuel de-icers.

Ethanolamines are made by the reaction of ethylene oxide with ammonia. They are used to remove acidic components (e.g., hydrogen sulphide) from various gaseous hydrocarbon products, as chemical intermediates for fatty alkanolamides (used as foam stabilizers in liquid household detergents), and for the production of fatty acid soaps used in detergents and cosmetics.

Other uses for ethylene oxide are as a chemical intermediate in the manufacture of diethylene glycol (largely used for unsaturated polyester resin production); triethylene glycol (used principally for natural-gas dehydration); tetraethylene glycol (primarily a solvent for extracting aromatic hydrocarbons from mixed hydrocarbon streams); polyethylene glycol (lower-molecular-weight products used in surfactant synthesis and for numerous other purposes, and higher polymers used in fire-fighting, agricultural and other uses); polyether polyols for flexible polyurethane foams (see IARC, 1979b); ethylene chlorohydrin (a chemical intermediate); choline and its derivatives (used as therapeutic agents and dietary supplements); hydroxyethylated cellulose and starch (which have been used as plasma expanders) and similar products; arylethanolamines (dye intermediates); acetal copolymer resins; crown ethers; ethylene carbonate; and cationic surfactants. Until 1969, ethylene oxide was also used in the USA as a chemical intermediate for the manufacture of acrylonitrile (see IARC, 1979c).

Ethylene oxide itself is used (alone or as part of a mixture with carbon dioxide or a fluorochlorocarbon) as a fumigant and sterilant in a variety of applications, and a review article on the use of ethylene oxide in these applications is available (Glaser, 1977). The US Department of Labor (1983) has estimated that approximately 2% of all ethylene oxide produced is used to sterilize or fumigate products, such as: bread, cocoa, desiccated coconut (copra), dried egg powder, fish, flour, dried fruits, meat, spices, dehydrated vegetables and walnut meats; clothing, furs, leather and textiles; cosmetics and drugs; cigarette tobacco; dental, medical, pharmaceutical and other scientific equipment and supplies including disposable and reusable medical items; packaging materials (e.g., for dairy products), paper and books; railway passenger- and freight-cars and buses; motor oil; and other miscellaneous products including experimental animals, beehives, bone meal, furniture, museum artefacts and soils. A review article has been published on the use of ethylene oxide in the sterilization of spices (Coretti, 1978).

In 1978, it was reported that less than 0.24% of the annual US production of ethylene oxide was used in health-care and medical products (US Department of Labor, 1982); in 1977, it was estimated that only about 0.02% of production was used for sterilization in hospitals (Glaser, 1977).

Annual ethylene oxide usage in the US food industry in 1970-1971 was approximately 1060 kg (Flavor and Extract Manufacturers' Association of the United States, 1978). Two companies reported use of ethylene oxide as a food additive in 1977, and use in 1976 amounted to 72.6 thousand kg (National Research Council/National Academy of Sciences, 1979). In 1983, an estimated 27 US spice manufacturers used ethylene oxide to fumigate spices (US Department of Labor, 1983).

The use pattern for the estimated 1340 thousand tonnes of ethylene oxide used in western Europe in 1981 was as follows: ethylene glycol, 46%; nonionic surfactants, 21%; glycol ethers, 10%; polyols, 10%; ethanolamides, 9%; and other applications, 4%.

An estimated 478 thousand tonnes of ethylene oxide were used in Japan in 1981 with the following use pattern: ethylene glycol, 64%; nonionic surfactants, 17%; ethanolamines, 7%; glycol ethers, 5%; polyols, 3%; and other applications, 4%.

Occupational exposure to ethylene oxide has been limited by regulations or recommended guidelines in at least 16 countries. The standards are listed in Table 1.

On 21 April 1983, the Occupational Safety and Health Administration of the US Department of Labor (1983) proposed a reduction in the permissible exposure limit for ethylene oxide of 50 ppm to 1 ppm, which was effective from 22 June 1984 (US Department of Labor, 1984), and established an 'action level' of 0.5 ppm (time-weighted average, TWA).

The US Food and Drug Administration (1980) has approved use of ethylene oxide as a direct and indirect food additive for the following purposes: (1) as a fumigant in sizing used as a component of paper and paperboard in contact with dry foods; (2) as an etherifying agent in the production of modified industrial starch, provided the level of reacted ethylene oxide in the finished product does not exceed 3%; and (3) as a fumigant for spices and other processed natural seasoning materials, except mixtures to which salt has been added, in accordance with prescribed conditions, including a maximum residue of 50 mg/kg in the treated material.

Table 1. National occupational exposure limits for ethylene oxide[a]

Country	Year	Concentration mg/m^3	ppm	Interpretation[b]	Status
Australia	1978	90	50	TWA	Guideline
Belgium	1978	90	50	TWA	Regulation
Bulgaria	1971	1	-	Maximum	Regulation
Finland	1981	20	10	TWA	Guideline
		40	20	STEL	
German Democratic Republic	1977	20	-	TWA	Regulation
		50	-	Maximum (30 min)	
Germany, Federal Republic of	1984	18	10	TWA	Guideline
Italy	1978	60	30	TWA[c]	Guideline
Japan	1978	90	50	Ceiling	Guideline
Netherlands	1978	90	50	TWA	Guideline
Poland	1976	1	-	Ceiling	Regulation
Romania	1975	30	-	TWA	Regulation
		60	-	Maximum	
Sweden	1981	9	5	TWA[d]	Guideline
		18	10	STEL	
Switzerland	1978	90	50	TWA	Regulation
USSR	1977	1	-	Maximum	Regulation
USA[e]					
OSHA	1978	90	50	TWA	Regulation
ACGIH	1984/85	2	1	TWA	Guideline
NIOSH	1983	1.96	0.1	TWA	Guideline
		9	5	Ceiling (10 min)	
Yugoslavia	1971	18	10	Ceiling	Regulation

[a]International Labour Office (1980); National Finnish Board of Occupational Safety and Health (1981); National Swedish Board of Occupational Safety and Health (1981); National Institute for Occupational Safety and Health (1983); American Conference of Governmental Industrial Hygienists (1984); Deutsche Forschungsgemeinschaft (1984)
[b]TWA, time-weighted average; STEL, short-term exposure limit
[c]Sensitizer notation added
[d]Skin penetration and carcinogenicity notation added
[e]OSHA, Occupational Safety and Health Administration; ACGIH, American Conference of Governmental Industrial Hygienists; NIOSH, National Institute for Occupational Safety and Health

The US Food and Drug Administration (1978) has also proposed the tolerances for residues of ethylene oxide in drug products and medical devices shown in Table 2.

Table 2. Tolerances for residues of ethylene oxide in drug products and medical devices[a]

	mg/kg
Drug products	
Ophthalmics (for topical use)	10
Injectables (including veterinary intramammary infusions)	10
Intrauterine devices (containing a drug)	5
Surgical scrub sponges (containing a drug)	25
Hard gelatin capsule shells	35
Medical devices	
Implants:	
Small (<10 g)	250
Medium (10-100 g)	100
Large (>100 g)	25
Intrauterine devices	5
Intraocular lenses	25
Devices in contact with mucosa	250
Devices in contact with blood (*ex vivo*)	25
Devices in contact with skin	250
Surgical scrub sponges	25

[a]From US Food and Drug Administration (1978)

The US Environmental Protection Agency (1978) issued a notice of rebuttable presumption against registration and continued registration of pesticide products containing ethylene oxide. No final decision has yet been reached.

The US Environmental Protection Agency (1982) has established a tolerance of 50 mg/kg for residues of ethylene oxide when used as a postharvest fumigant in or on black walnut meats, copra and whole spices. That Agency has also identified ethylene oxide as a toxic waste and requires that persons who generate, transport, treat, store or dispose of it comply with the regulations of a federal hazardous waste management programme (US Environmental Protection Agency, 1980). The US Environmental Protection Agency (1983) requires that notification be given whenever discharges containing 0.454 kg or more of ethylene oxide are made into waterways.

As part of the Hazardous Materials Regulations of the US Department of Transportation (1982), shipments of ethylene oxide are subject to a variety of labelling, packaging, quantity and shipping restrictions consistent with its designation as a hazardous material.

2.2 Occurrence

(a) *Natural occurrence*

Ethylene oxide is produced endogenously in animals without previous exposure to ethylene (Filser & Bolt, 1983)

(b) *Occupational exposure*

On the basis of the 1974 National Occupational Hazard Survey, the National Institute for Occupational Safety and Health (1980, 1981a) estimated that 141 thousand US workers in 67 nonagricultural industries were exposed to ethylene oxide. The principal industries in which exposure was found were the pumps and compressors industry, hospitals, and the miscellaneous plastics products industry. In 1983, the Occupational Safety and Health Administration estimated that 80 thousand US workers are directly exposed to ethylene oxide and that another 144 thousand workers are incidentally exposed (US Department of Labor, 1983).

In 1977, the National Institute for Occupational Safety and Health (1981b) estimated that approximately 75 thousand US health-care workers employed in sterilization areas were potentially exposed to ethylene oxide and that 25 thousand others may have been incidentally exposed.

In a survey in 1979 of US plants both producing and using ethylene oxide, worker exposure to this chemical was measured. Typical average daily exposures were 0.5-7.3 mg/m^3; worst-case peak exposures were 16-17 500 mg/m^3, the highest exposure being that of maintenance workers changing 0-rings (Flores, 1983).

As part of an industry-wide study by the National Institute for Occupational Safety and Health on the health effects of occupational exposure to ethylene oxide during the years 1977 and 1978, workplace air samples were collected in five US plants producing ethylene oxide and its derivatives. The results of these surveys are given in Table 3.

In an ethylene oxide manufacturing plant in the Netherlands, eight-hour TWAs during the period 1974-1981 were found to be generally below the limit of detection of 0.09 mg/m^3, with occasional transient concentrations of up to 14.6 mg/m^3 (van Sittert et al., 1984).

In a study at a Swedish ethylene oxide production facility, the concentrations of airborne ethylene oxide were estimated as follows: during the 1940s, probably below 25 mg/m^3 with

Table 3. Airborne ethylene oxide (EtO) concentrations observed in five US plants producing this compound and its derivatives

Plant no.	Year of survey	Sample location	Concentration (mg/m^3)	Reference
1	1977	EtO production areas	<18	Koketsu & Alli (1977)
		Derivative unit areas	<18	
		Tank car loading operation (a leak on the slip tube used to gauge the EtO level)	11 000	
2	1977	EtO production areas	<1.8-2.7	Lovegren & Koketsu (1977a)
3	1977	Derivative unit areas	<1.8	Lovegren & Koketsu (1977b)
3	1978	44 Personal samples taken during the production of EtO derivatives	<0.9-<8.3	Oser et al. (1978a)
		Laboratory technician (short-term sample)	150	
		Unloader at an oxide rack	14.6	
3	1978	41 Personal samples taken during derivative production	<0.9-42	Oser et al. (1979)
4	1977	Derivative production:		Lovegren & Koketsu (1977c)
		near a pump	1.8	
		between two reactors	1.8	
5	1978	EtO and mono-, di- and triethylene glycol production:		Oser et al. (1978b)
		supervisor	<0.38	
		control room operator	<0.13	
		technicians (sampling)	0.65-2.07	
		engineer	<0.17	
		loading operator	0.80-113[a]	

[a]A parallel sample analysed by the company was reported to contain 10 mg EtO/m^3

occasional exposures up to 1300 mg/m^3; during the 1950s and early 1960s, 10-50 mg/m^3 with occasional peaks above 1300 mg/m^3; and during the 1970s, 1-10 mg/m^3 with occasional higher values. Ethylene oxide production ceased in 1963, but it was still used in the factory (Hogstedt et al., 1979a).

Potential occupational exposure to ethylene oxide exists in many countries in a wide variety of industries and work settings where this compound is used as a sterilant or fumigant (US Department of Labor, 1982). Worker exposure to ethylene oxide in 12 different types of facilities where it is used as a sterilant/fumigant has been reviewed (Goldgraben & Zank, 1981).

In a limited field survey of hospitals, the National Institute for Occupational Safety and Health (1981b) found that ethylene oxide concentrations near malfunctioning or improperly designed equipment may reach transitory levels of hundreds or even a few thousand mg/m^3. TWA ambient and breathing-zone concentrations were generally below 90 mg/m^3. Emissions of ethylene oxide were reported to occur mainly during discharge of the gas into floor drains following the opening of the door of the sterilization equipment and during the changing of gas cylinders (Anon., 1984).

Exposure of workers to ethylene oxide during its use as a sterilant has been reported in a number of US hospitals and other health-care facilities (US Department of Labor, 1983). The eight-hour TWA for one hospital sterilizing three loads per week was reported to be 4.6 mg/m^3; for another hospital, the TWA was reported to be 5.5-11 mg/m^3; and for Veterans' Administration medical centres, the TWA was reported to be <9 mg/m^3. The TWA for 114 of 121 sites monitored in southern California hospitals from 1978-1982 was <9 mg/m^3. In 27 other hospitals, TWAs were <1.8 mg/m^3 for nine and >18 mg/m^3 for five; levels of ethylene oxide in the breathing zone were in the range of 0-18 mg/m^3, being <7.3 mg/m^3 for 16 of the 27 hospitals and as high as 34.8 mg/m^3 for one exposure. It was suggested that workers probably received much higher exposures for brief periods, as illustrated by the data for one US sterilizer operator (described as typical): a TWA of 5.7 mg/m^3 was generated from a two-minute peak of 1100 mg/m^3 followed by some residual exposure and seven hours at 0 mg/m^3.

Data from other studies of worker exposure to ethylene oxide during its use as a sterilant in hospitals and other health-care facilities are summarized in Table 4. Data on exposures during the sterilization of pharmaceutical and medical products are summarized in Table 5.

A major US producer of pharmaceutical products that were sterilized with ethylene oxide reported that the levels of ethylene oxide to which workers were exposed both directly and indirectly ranged from 0.4-1.8 mg/m^3 (US Department of Labor, 1983).

In a study of employees at two Swedish factories manufacturing medical equipment, packers in one factory were found to have been exposed before the time of the study to an average level of about 8 mg/m^3 (with peaks of about 100 mg/m^3) ethylene oxide; after working routines were changed, the exposure levels decreased to 0.6-1.8 mg/m^3. In the second factory, the exposure level was about 0.2 mg/m^3 (Högstedt et al., 1983).

Occupational exposure to ethylene oxide in a West Virginia museum laboratory ranged from the limit of detection (0.08 mg/sample) to 3 mg/m^3 (Ruhe, 1977).

In an Ohio company manufacturing animal feed, concentrations of ethylene oxide ranging from <1.6-8.2 mg/m^3 were detected in area-air samples; it was not detected in personal breathing-zone samples (Gorman & Horan, 1981).

Table 4. Airborne ethylene oxide concentrations observed in hospitals and other health-care facilities using ethylene oxide as a sterilant

Facility/year	Sample location	Concentration (mg/m^3) [sampling time]	Reference
US (Arizona) hospital/1980	Sterilizer operator At a distance of 8 m from sterilizers during unloading	7200 [3-min peak] 460 [peak] (peaks occurred up to 7 times daily)	Anon. (1982a)
US (Wisconsin) hospital/1978	By the side of a sterilizer (one unloading period included) 8 other sites in a sterilizing unit area	40 [6-h sample] Less than ca 0.5	Johnson et al. (1979)
US (New York) hospital/1981	Sterile room (one unloading period included) Sterile room during unloading At face of sterilizer immediately upon opening Decontamination room	4-6 [2-h sample] 7-13 [12-min samples] 137 [peak] <0.9 (limit of detection)	Burroughs (1981)
US (Washington) hospital/1978	Sterilizer operator during unloading and loading Near sterilizer during unloading and loading General working area (one unloading period included) General working area (sterilizer not in operation)	69-104 [15-min samples] 68-75 [15-min samples] 23 [3-h sample] <7 [4-h sample]	Apol (1978)
US (Connecticut) hospital/1979	Sterilizer operators after charging Sterilizer operators opening sterilizer Near sterilizer Sterilizer tank room	8-16 [1-1.5-h samples] 8-32 [24-min samples] 10 [1.5-h samples] 4 [27-min sample]	Moseley (1979)
US (California) hospital - 1979 - 1981 (after exposure control measures, including better seal, exhaust ventilation, and internal purging prior to opening of doors)	Sterilizer operations Sterilizer operations	Detected (no level given) Not detected	Coye & Belanger (1981)
US (Kentucky) hospital/1980	Central supply room maintenance closet Adjacent to sterilizer	<0.6 [5.5-h sample] 1.5 [5-h sample]	Stephenson et al. (1980)
US (Maryland) hospital/1980	Sterilizer operators	ca 2-15 [15-21-min samples]	Manoff et al. (1982)
Four French sterilization and disinfection facilities (three in hospitals)/1979-1980	During and near sterilizer unloading	0.9-420 [several-minute samples] 0.09-9 [6-8-h samples]	Mouilleseaux et al. (1983)
Finnish hospitals/1976-1981	Sterilizing chamber open Other conditions Sterilizing units	9-18 [20-min sample] <1.8 0.2-0.9 with peak of up to 450	Hemminki et al. (1983) Hemminki et al. (1982)

Table 5. Airborne ethylene oxide concentrations during the sterilization of pharmaceutical and medical products

Facility/year	Sample location	Concentration (mg/m^3) [sampling time]	Reference
US company manufacturing sterilized pulmonary function equipment and artificial kidney filtration systems/1978	Sterilization operator Sterilization area	<2.7-36 [1.5- to 3-h samples] <3.6 [3-h sample]	Ruhe (1978)
US company manufacturing medical therapeutic systems/1979	Sterilization operator Sterilization area	5.7-8 [3-h sample] 5.3 [3-h sample]	Tharr & Donohue (1980)
Three US medical products plants/1980	Plant No. 1 Plant No. 2 Plant No. 3	9-366 1.8-18 1.8	Anon. (1982b)
US plant manufacturing surgical products	Around the front of sterilizer during the early stages of sterilization At breathing height, in front of sterilizer during exhaust cycle Behind sterilizer during exhaust cycle Top of sterilizer interior, 45 min after exhaust cycle (opening sterilizer door)	<0.2-1.8 9.7 >450 220	Collins & Barker (1983)
US plant manufacturing disposable microscope drapes/1977	Personal samples with sterilizer door open Personal samples with sterilizer door cracked Personal sample with sterilizer door closed Area sample near sterilizer as the door was opened	<7.8-94 [1- to 2-h samples] <6.6-16 [1- to 3-h samples] <13.2 140	Schutte (1977)
Two Swedish sterilization plants			Hogstedt et al. (1983)
June 1975	Sterilization room (personal samples)	53 (sterilizer open) 16 (sterilizer closed)	
December 1975	Sterilization room (personal samples)	14 (sterilizer open) 2.4 (sterilizer closed)	
May 1978	Sterilization room (personal samples)	4.4 (sterilizer open and closed together)	

(c) Air

Ethylene oxide has been tentatively identified in atmospheric air samples in the USA (Sawicki, 1976). It has also been observed as a product of the combustion of hydrocarbon fuels and in automobile exhausts. It is a known product of atmospheric oxidations (Bogyo et al., 1980).

It has been estimated that less than 2.3 million kg of ethylene oxide are released into the air annually in the USA during its production and processing, and that all of the ethylene oxide used for fumigant purposes (estimated to be 0.045-4.5 million kg per year) enters the environment (soil, food products, air) (Bogyo et al., 1980). In another report, total nationwide atmospheric emissions of ethylene oxide in 1978 from all US sources were estimated to be about 0.9 million kg (Systems Applications, Inc., 1981). In 1981, the Ethylene Oxide Industry Council estimated that about 1.4 million kg of ethylene oxide are released annually to the air in the USA (US Environmental Protection Agency, 1984).

(d) Water

In 1981, the Ethylene Oxide Industry Council estimated that the amount of ethylene oxide lost to water in the USA during production and processing was 363 thousand kg annually; however, it was reported that this waste-water is usually treated before being discharged from plants (US Environmental Protection Agency, 1984).

(e) Food

Residues of ethylene oxide fumigant were detected in spices (16-41 mg/kg) intended for use in sausage manufacture; but no residue was found in the finished sausages (Jordy, 1981).

Concentrations of ethylene oxide found in seasonings at various times after fumigation are shown in Table 6.

Table 6. Concentrations of ethylene oxide in ground seasonings at various times after fumigation[a]

Spice product	Time after fumigation (days)	Concentration (mg/m^3)
Mustard	7	15.9
	9	0.9
Black pepper	4	5.6
Cassia	4	18.4
	7	4.2
Paprika, Spanish	4	26
	9	0
Red peppers	6	2.9
Ginger	1	41.8
Caraway seed	4	8.0
	11	2.0
Nutmeg	4	46.5
	7	40.5

[a]From Flavor and Extract Manufacturers' Association of the United States (1978)

Residues of ethylene oxide have been measured in the following food and medicinal plants (mg/kg): *Plantago psylium* cuticle, 3.6; red poppy leaves, 0.9; origano leaves, 2.2; liquorice root, 0.6; *Rhamnus purshiana* bark, 1.1; and *Fucus vesiculosus* thallus, 0.4 (Bicchi & Frattini, 1981).

(f) Tobacco

The ethylene oxide concentration in unfumigated tobacco was reported to be 0.02 µg/ml, while fumigated tobacco contained 0.05 µg/ml and extensively fumigated tobacco contained 0.30 µg/ml. The ethylene oxide content of smoke from unfumigated tobacco was reported to be 1 µg/g (Binder & Linder, 1972; Bogyo *et al.*, 1980).

(g) Other

Ethylene oxide residues have been measured on a variety of sterilized surgical equipment. Concentrations of ethylene oxide were not detectable (<25 ng) for most samples (e.g., catheters and tubing); one transfusion unit contained 1.8 mg/kg ethylene oxide and two surgeons' glove samples contained 2.4 and 3.1 mg/kg ethylene oxide (Brown, 1970).

Collins and Barker (1983) reported levels of ethylene oxide at 1.8-2163 mg/m³ in air entrapped in closed packages of surgical products sterilized the previous day, the highest level was found in a humidifier bottle.

2.3 Analysis

A review of analytical methods for the determination of ethylene oxide in air, surgical materials, foods and gasoline combustion products has been published (Bogyo et al., 1980).

Methods used for the analysis of ethylene oxide in a variety of matrices are listed in Table 7.

Table 7. Methods for the analysis of ethylene oxide

Sample matrix	Sample preparation	Assay procedure[a]	Limits of detection	Reference
Air (workplace)	Adsorb (charcoal); desorb (carbon disulphide)	GC/FID	0.27 mg/m^3	National Institute for Occupational Safety and Health (1977); Qazi & Ketcham (1977)
	Adsorb (charcoal); desorb (1% carbon disulphide in benzene); react with hydrobromic acid; treat with sodium carbonate	GC/ECD	0.024 mg/m^3	US Department of Labor (1983)
	-	GC/PID	0.18 mg/m^3	Collins & Barker (1983)
	Adsorb (activated carbon)	GC	Not given	Blome (1982)
	-	IR	Not given	Vanell (1982)
	React with periodic acid; react with xylene; react with sulphuric acid	Colorimetry	9-1800 mg/m^3 (range)	Pritts et al. (1982)
Air (after field fumigation)	-	GC/PID	0.002-183 mg/m^3 (range)	Bond & Dumas (1982)
Ambient air	Preconcentrate; desorb	GC/FID	0.5-1 ng	Dmitriev & Mishchikhin (1982)
Aqueous solution	-	React with sodium sulphite and titrate with hydrochloric acid	Not given	Swan (1954)
Plastic medical devices	Extract (dimethyl formamide); inject into headspace	GC	Not given	Bellenger et al. (1983)
	Extract (ethanol)	GC	1 mg/kg	Tsuge & Senba (1981)
Surgical silk sutures	Inject into headspace	GC/FID	Not given	Kolb (1982)
Penicillin powder and injection	-	GC/FID	Not given	Kiss & Kovacs (1982)
Food and medicinal plants	Extract; inject into headspace	GC/FID	1-10 mg/kg (range)	Bicchi & Frattini (1981)
Food products in heat-sealed packages	Inject into headspace	GC	0.5 mg/kg	Ricottilli et al. (1981)
Sterilized material	Distil into water; hydrolyse (sulphuric acid); react with 3-methyl-2,3-dihydrobenzothiazole hydrazone	Spectrophotometry (630 nm)	Not given	Falcao (1981)

[a]Abbreviations: GC/FID, gas chromatography/flame ionization detection; GC/ECD, gas chromatography/electron capture detection; GC/PID, gas chromatography/photoionization detection; GC, gas chromatography; IR, infrared detection

3. Biological Data Relevant to the Evaluation of Carcinogenic Risk to Humans

3.1 Carcinogenicity studies in animals[1]

(a) *Oral administration*

Rat: A group of 25 male and 25 female rats of an unspecified laboratory strain, weighing 100-150 g at the beginning of the experiment, was fed for two years with a standard laboratory diet (Altromin R) fumigated with ethylene oxide. Groups of 25 male and 25 female control rats were given the untreated diet. The food was prepared in weekly batches and was fumigated with air containing 900-1300 mg/m^3 ethylene oxide. After fumigation, the ethylene oxide residues in the diet were found to be between 500-1400 mg/kg on the first day of feeding and between 53-400 mg/kg after six days. The experiment was terminated at two years when 13/50 animals (both sexes) were alive in the control group and 16/50 animals (both sexes) in the test group. No increase in tumour incidence was observed in animals fed the ethylene oxide-fumigated diet that died spontaneously or were killed at termination of the study (Bär & Griepentrog, 1969). [The Working Group noted that the ethylene oxide might have been converted to ethylene glycol and ethylene chlorohydrin during contact with the diet for six days and that no data were given on tumour incidence.]

Groups of 50 female Sprague-Dawley rats, about 100 days old, were administered 7.5 or 30 mg/kg bw ethylene oxide (purity, 99.7%) in a commercially-available salad oil [composition unspecified] by gastric intubation twice weekly for 107 weeks (average total doses, 1186 and 5112 mg/kg bw, respectively). Control groups consisted of 50 untreated rats and 50 rats treated with salad oil alone. The survival rate of rats in the low-dose group was comparable to those of the control groups; rats treated with the higher dose died earlier as a result of tumours. Treatment with ethylene oxide resulted in a dose-dependent increase in the incidence of local tumours, mainly squamous-cell carcinomas of the forestomach. The tumour incidences were 0/50 and 0/50 in the untreated controls and the vehicle controls, respectively. The first tumour was observed in the 79th week in the high-dose group. In total, 31/50 animals in this group developed malignant tumours in the stomach - 29 squamous-cell carcinomas in the forestomach and two fibrosarcomas, one of which was located in the glandular stomach. In addition, 4/50 had carcinomas *in situ* and 11/50, papillomas, hyperplasia or hyperkeratosis of the squamous epithelium of the forestomach. In the low-dose group, 8/50 animals developed squamous-cell carcinomas, four had carcinomas *in situ* and nine had papillomas, hyperplasia or hyperkeratosis in the forestomach. Of the 37 squamous-cell carcinomas found in the two dose groups, 10 metastasized and grew invasively into neighbouring organs. There was no increase in the incidence of tumours at other sites in the treated animals as compared to controls. A positive-control group of 50 rats that received 30 mg/kg bw β-propiolactone in salad oil administered by gastric intubation twice weekly for lifespan developed 46/50 stomach tumours [type not specified] (Dunkelberg, 1982).

(b) *Skin application*

Mouse: Thirty female ICR/Ha Swiss mice, eight weeks of age at the start of the treatment, received thrice-weekly applications of a 10% solution of ethylene oxide in acetone (approximately 100 mg of solution per application) on the clipped dorsal skin for life. The median survival time was 493 days; no skin tumour was observed (Van Duuren *et al.*, 1965).

---

[1]The Working Group was aware of a study in progress in mice by inhalation exposure (IARC, 1982b).

(c) Inhalation

Rat: Groups of 120 male and 120 female Fischer 344 rats, eight weeks of age, were exposed to 10, 33 or 100 ppm (18, 59 or 180 mg/m^3) ethylene oxide (purity, 99.9%) vapour, for six hours per day, on five days per week for two years. Two control groups of 120 male and 120 female rats were exposed in inhalation chambers to room air. Post-mortem examinations were made of all animals that died or were killed when moribund and of those killed at scheduled intervals of six, 12, 18 and 24 months. During month 15 of exposure, mortality increased in both treated and control groups due to a viral sialodacryoadenitis. Mortality was higher in the groups inhaling 33 and 100 ppm as compared to other groups and was more frequent in females than in males near the 15th month. With up to 18 months of exposure, no statistically significant increase in tumour incidence was observed in either group. In those killed after 18 months, the incidence of tumours in the brain classified as 'gliomas, malignant reticulosis and granular-cell tumours' was increased for both sexes. Among those killed at 24 months, tumours were found in the brain in 1/48 (control I), 0/49 (control II), 0/51 (10 ppm), 1/39 (33 ppm) and 3/30 (100 ppm) males and 0/60 (control I), 0/56 (control II), 0/54 (10 ppm), 2/48 (33 ppm) and 2/26 (100 ppm) females. Statistical evaluation indicated a treatment-related response, particularly in males at the two highest dose levels. At the end of the experiment, the incidence of mononuclear-cell leukaemia in animals of both sexes and peritoneal mesothelioma in males was also greater in animals exposed to ethylene oxide. In females, mononuclear-cell leukaemia was found in 5/60 (control I), 6/56 (control II), 11/54 (10 ppm), 14/48 (33 ppm) and 15/26 (100 ppm) animals; the increased incidence of leukaemia was statistically significant in the 100-ppm group ($p < 0.001$). A mortality-adjusted trend analysis resulted in a highly significant positive trend ($p < 0.005$). In males, mononuclear-cell leukaemia was found in 5/48 (control I), 8/49 (control II), 9/51 (10 ppm), 12/39 (33 ppm) and 9/30 (100 ppm) animals. The mortality-adjusted trend analysis showed a significant positive trend ($p < 0.05$). In males, peritoneal mesothelioma was found in 1/48 (control I), 1/49 (control II), 2/51 (10 ppm), 4/39 (33 ppm) and 4/30 (100 ppm) animals. Results of trend analysis indicated a highly significant relationship ($p < 0.005$) between exposure to ethylene oxide and development of peritoneal mesothelioma. Although the incidence of pituitary adenomas was not significantly increased in animals of either sex at any single dose level, exposure to ethylene oxide accelerated the appearance of pituitary adenomas in males (trend analysis, $p < 0.001$) (Snellings *et al.*, 1984). [The Working Group noted that the combining of three different histological types of tumours in the brain precludes a proper evaluation of the effects of ethylene oxide on that organ.]

Groups of 80 male weanling Fischer 344 rats were exposed to 0 (control: filtered air), 50 and 100 ppm (92 and 180 mg/m^3) ethylene oxide (purity, 99.7%) vapour for approximately seven hours per day, five days per week for two years. An increase in mortality was observed in the two treated groups as compared to controls. Rats exposed to 50 and 100 ppm ethylene oxide had a higher incidence of inflammatory lesions of the respiratory system, of bronchiectasis and of bronchial epithelial hyperplasia. Treatment with ethylene oxide resulted in an increased incidence of mononuclear-cell leukaemia, peritoneal mesotheliomas and gliomas of the brain. Mononuclear-cell leukaemia was observed in 24/77, 38/79 and 30/76 rats exposed to 0, 50 and 100 ppm ethylene oxide, respectively (interim death and terminal killing). The overall increase in mononuclear-cell leukaemia was statistically significant ($p = 0.03$) in the low-dose group, but not in the high-dose group, in which excessive mortality occurred (survival was 19% compared to 49% in controls). However, comparison of the incidence of mononuclear-cell leukaemia in the terminally killed rats in the high-dose group to that in controls revealed an exposure-related increased incidence of mononuclear-cell leukaemia ($p < 0.01$). Peritoneal mesotheliomas developed in 3/78 controls, 9/79 rats of the 50-ppm group and in 21/79 of the 100-ppm group. The increase in this type of tumour was significant ($p = 0.002$) for the high-dose group only. Results of the Armitage test for trends suggested a proportional increase in the incidence of mesotheliomas with increased exposures. Gliomas (mixed cells resembling astrocytes and oligodendroglia cells) were found in 0/76 controls, 2/77 of the 50-ppm group and 5/79 of the 100-ppm group ($p < 0.05$). Assuming that gliomas occurred in a fatal context, trend analysis indicated a significant increase in gliomas with increased exposure to ethylene oxide. In addition to the gliomas, two additional rats exposed to 50 ppm and four additional rats exposed to 100 ppm ethylene oxide had an increased number of glial cells, termed 'gliosis'. The incidences of other neoplasms were generally comparable among the control and treated groups and bore no relationship to ethylene oxide exposure. A high incidence of proliferative lesions, including nodules that compressed the surrounding tissue, was observed in the adrenal cortex of animals exposed to ethylene oxide. These lesions were never found in the control group, but they were classified as non-neoplastic changes ('multifocal cortical hyperplasia' and 'cortical nodular hyperplasia') (Lynch et al., 1984a).

(d) Subcutaneous and/or intramuscular administration

Mouse: Groups of 100 female NMRI mice, six to eight weeks old, received subcutaneous injections of 0.1, 0.3 or 1.0 mg/mouse ethylene oxide (purity, 99.7%) in tricaprylin, once a week for 95 weeks [mean total doses, 7.3, 22.7 and 64.4 mg/mouse, respectively]. Groups of 200 untreated and 200 tricaprylin-treated mice served as controls. The survival rate of animals treated with the highest dose of ethylene oxide was reduced as compared to that of animals treated with the two lower doses and with controls. Ethylene oxide induced a dose-dependent increase in local tumours, mostly fibrosarcomas. The first tumour appeared in the 50th week of treatment. The incidences of subcutaneous sarcomas (fibrosarcomas, pleomorphic sarcomas and one haemangiosarcoma) were: 0/200 in untreated controls, 4/200 in animals treated with tricaprylin alone, and 5/100 (0.1 mg), 8/100 (0.3 mg) and 11/100 (1 mg) in the ethylene oxide-treated animals [$p < 0.001$, Cochran Armitage test for trend]. The authors analysed the experiment by estimating the adjusted tumour incidence rates at 600 days and established a dose-response relationship for these rates. Tumours other than subcutaneous sarcomas could not be related to treatment with ethylene oxide. In a positive-control group of 100 NMRI mice that received a subcutaneous injection of 2.5 µg benzo[a]pyrene once a week for 95 weeks, 81 mice developed local sarcomas (Dunkelberg, 1981).

3.2 Other relevant biological data

(a) Experimental systems

Toxic effects

The toxicity of ethylene oxide to humans and to experimental animals has been reviewed (Glaser, 1979).

The intragastric LD_{50}s of an aqueous solution of ethylene oxide were 330 and 270 mg/kg bw in rats and guinea-pigs, respectively (Smyth et al., 1941). All of five rats intubated with 200 mg/kg bw ethylene oxide in olive oil died (Hollingsworth et al., 1956). LC_{50}s were 2630 mg/m^3 (1460 ppm) in rats, 1504 mg/m^3 (835 ppm) in mice and 1730 mg/m^3 (960 ppm) in dogs, following four-hour exposures (Jacobson et al., 1956). All of six rats exposed to 15 000 mg/m^3 (8000 ppm) for four hours died (Weil et al., 1963).

In rats, intragastric administration of 15 doses of 100 mg/kg bw ethylene oxide in olive oil during 21 days caused marked loss of body weight, gastric irritation and slight liver damage; no injury resulted from 22 doses of 30 mg/kg bw over 30 days (Hollingsworth et al., 1956). [Given the volatility of ethylene oxide, the doses reported in the intragastric studies may be inaccurate.]

No adverse effect was reported in animals subjected to repeated seven-hour exposures to ethylene oxide vapour on five days per week for six or seven months at 200 mg/m^3 (113 ppm) in guinea-pigs, rabbits and monkeys and 90 mg/m^3 (49 ppm) in mice and rats (Hollingsworth et al., 1956). In a similar experiment, no adverse effect was seen in dogs, mice or rats treated with 183 mg/m^3 (100 ppm) for six months (Jacobson et al., 1956).

Dose-related eye irritation was observed in rabbits exposed to concentrations of over 1800 mg/m^3 (1000 ppm) (McDonald et al., 1973).

Effects on reproduction and prenatal toxicity

A recent review of the literature is available (Kimmel et al., 1984), and only important papers are highlighted in this section.

The reproductive toxicity of ethylene oxide has been determined in mice, rats and rabbits following oral, intravenous and inhalational routes of exposures. In mice, intravenous administration of 0, 75 or 150 mg/kg bw ethylene oxide in 5% dextrose solution on days 4-6, 6-8, 8-10 or 10-12 of gestation significantly increased the incidence of craniofacial defects and fusions of vertebrae in high-dose animals exposed on days 6-8 (19.3%) and 10-12 (9.5%). The incidence ranged from 0-2.3% in the four control groups. The high-dose level resulted in maternal mortality after treatment on days 4-6, 8-10 and 10-12 (LaBorde & Kimmel, 1980). In rabbits, inhalational exposure to 150 ppm (275 mg/m^3) ethylene oxide (purity, 99.7%) vapour for seven hours per day on days 7-19 or 1-19 of gestation resulted in no evidence of maternal toxicity, embryotoxicity or teratogenicity (Hackett et al., 1982).

Male and female Fischer 344 rats were exposed by inhalation to 100 ppm (180 mg/m^3) ethylene oxide vapour for six hours per day, on five days per week for 12 weeks and then mated; exposure continued on seven days per week, and exposure of the females continued through to day 19 of gestation. Fewer implantation sites per female, a smaller ratio of foetuses born to the number of implants, a decreased number of pups born per litter and a tendency for longer lengths of gestation were observed (Snellings et al., 1982). No treatment-related effect was found in two lower-dosage (10 and 33 ppm; 18 and 60 mg/m^3) groups. It was not determined whether this effect was due to exposure of the male, of the female or of both.

Female Sprague-Dawley rats were exposed by inhalation for seven hours per day, on five days per week on days 7-16 of gestation, on days 1-16 of gestation or for three weeks prior to mating and then daily until day 16 of gestation to 150 ppm (measured concentration was within 10% of target concentration) ethylene oxide (purity, 99.7%) vapour. An increased incidence of resorptions (13.6% compared to 5.4% in controls) was reported in the third group. Pregestational exposure appears to be an important factor, as similar effects were not found in females that had gestational exposure only. Foetal growth indices were reduced regardless of whether exposure included the pregestational period or not. The incidence of litters with foetuses with hydroureter was increased only when the mothers were exposed on days 7-16 (42% compared to 22% in controls). A moderate degree of maternal toxicity (reduced weight gain) accompanied these findings (Hackett et al., 1982).

Absorption, distribution, excretion and metabolism

Ethylene oxide is reported to be produced from ethylene in experimental animals (Ehrenberg et al., 1977). Ethylene and ethylene oxide are produced endogenously in experimental animals (Filser & Bolt, 1983).

In mice exposed to [1,2-^{3}H]-ethylene oxide vapour in air for 60-75 min, a mean of 78% of the absorbed radioactivity was excreted in urine within 48 hours. Up to several hours, the highest concentrations of residual radioactivity were found in protein fractions of the spleen, liver and kidney (Ehrenberg et al., 1974). In beagle dogs, the total body clearance for an aqueous solution of ethylene oxide administered intravenously was 20 ml/kg per min. Ethylene glycol was one of the urinary excretion products identified (Martis et al., 1982).

In mice exposed by inhalation to [1,2-^{3}H]-ethylene oxide vapour, kidney DNA was alkylated; the adduct was identified as 7-hydroxyethylguanine. This adduct was also detected in urine, representing 0.007% of the excreted radioactivity over 48 hours (Ehrenberg et al., 1974). The N-7-alkylguanine derivative was also detected in the DNA of liver and testes of rats receiving ethylene oxide by intraperitoneal injection (Osterman-Golkar et al., 1983).

In mice given the compound by inhalation, ethylene oxide has been found to bind covalently *in vivo* to several amino acids in haemoglobin, such as N-1- and N-3-histidine, N-valine and S-cysteine (Segerbäck, 1983).

Ethylene oxide has been shown to react *in vitro* with nucleosides to form 7-alkylguanosine (Brookes & Lawley, 1961) and 1-alkyladenosine (Windmueller & Kaplan, 1961). Esterification of phosphate groups in DNA has also been suggested (Walles & Ehrenberg, 1968).

Mutagenicity and other short-term tests (see also 'Appendix: Activity Profiles for Short-term Tests', p. 338)

Ethylene oxide produced differential killing in DNA-repair-proficient compared to -repair-deficient strains of *Bacillus subtilis* (Tanooka, 1979).

It did not induce mutations in T_2 bacteriophage (Cookson et al., 1971). It was mutagenic to *Salmonella typhimurium* TA1535 and TA100, but not to TA1537, TA1538 or TA98 (Rannug et al., 1976; Pfeiffer & Dunkelberg, 1980; De Flora, 1981). Ethylene oxide was mutagenic in *B. subtilis* (Tanooka, 1979) and *Neurospora crassa* (Kölmark & Westergaard, 1953; Kilbey & Kølmark, 1968; Kølmark & Kilbey, 1968; de Serres, 1983). It induced forward mutations in *Schizosaccharomyces pombe* both in the presence and absence of an exogenous metabolic system (S9) (Migliore et al., 1982).

Ethylene oxide induced mutations at the HGPRT locus of Chinese hamster ovary cells; this activity was unaffected by the presence of S9 from the livers of Aroclor-induced rats (Tan et al., 1981). In an abstract, Hatch et al. (1982) reported that ethylene oxide induced mutations to ouabain and 6-thioguanine resistance in Chinese hamster V79 cells.

Ethylene oxide induced a variety of mutations in barley and rice (Ehrenberg & Gustafsson, 1957; Šulovská et al., 1969; Jana & Roy, 1975; Kucera et al., 1975). Chromosomal aberrations were induced in root tips of barley (Moutschen-Dahmen et al., 1968)

Ethylene oxide induced sex-linked recessive lethal mutations, heritable translocations (Bird, 1952; Nakao & Auerbach, 1961; Watson, 1966) and minute mutations (Fahmy & Fahmy, 1970) in *Drosophila melanogaster*.

Treatment with ethylene oxide induced unscheduled DNA synthesis in human lymphocytes *in vitro* (Pero *et al.*, 1981), chromosomal aberrations in a human FL-cell line (Poirier & Papadopoulo, 1982) and sister chromatid exchanges in human lymphocytes (Garry *et al.*, 1982, 1984). In an abstract, Hatch *et al.* (1982) reported virus-enhanced transformation (simian adenovirus 7) of primary Syrian hamster embryo cells by ethylene oxide.

Dominant lethal mutations were induced in Long-Evans rats treated by inhalation with 1000 ppm (1800 mg/m^3) ethylene oxide vapour for four hours (Embree *et al.*, 1977) and in male (101xC3H)F$_1$ mice inhaling 255 ppm (460 mg/m^3) ethylene oxide vapour for six hours per day for two or 11 weeks (Generoso *et al.*, 1983). Dominant lethal mutations also occurred in random-bred T-stock male mice given an intraperitoneal dose of 150 mg/kg bw (maximum tolerated dose); 25 intraperitoneal injections of 60 mg/kg bw over a period of five weeks induced heritable translocations (Generoso *et al.*, 1980).

Fomenko and Strekalova (1973) and Strekalova *et al.* (1975) reported an increased incidence of chromosomal aberrations in bone-marrow cells of rats exposed by inhalation to concentrations of ethylene oxide vapour ranging from 1-112 mg/m^3 (0.6-63 ppm). Increased incidences of chromosomal abnormalities in bone-marrow cells have been reported in rats exposed to oral doses of 9 mg/kg bw ethylene oxide in aqueous solution (Strekalova, 1971).

Micronuclei were observed in bone-marrow cells of male mice treated intraperitoneally with two doses of 10-200 mg/kg bw ethylene oxide (Conan *et al.*, 1979) and in mice and rats given two intravenous injections of 100, 150 or 200 mg/kg bw and 100 mg/kg bw, respectively (Appelgren *et al.*, 1978).

Ethylene oxide vapour induced sister chromatid exchanges in peripheral lymphocytes of rabbits exposed by inhalation to 50 or 250 ppm (90 or 450 mg/m^3) on five days per week for 12 weeks (Yager & Benz, 1982). Dose-related increases in the incidence of sister chromatid exchanges occurred in peripheral blood lymphocytes of male Fischer 344 rats exposed by inhalation to 50, 150 or 450 ppm (90, 275 or 825 mg/m^3) for six hours per day for one or three days (Kligerman *et al.*, 1983, 1984). In an abstract, Embree and Hine (1975) reported that chromosomal aberrations were induced in bone-marrow cells of male Long-Evans rats exposed to 450 mg/m^3 (250 ppm) ethylene oxide vapour for seven hours per day for three days. In monkeys, inhalation of 50 and 100 ppm (90 and 180 mg/m^3) ethylene oxide vapour led to an increase in the frequency of sister chromatid exchanges and structural chromosomal aberrations in peripheral lymphocytes (Lynch *et al.*, 1984b).

(b) Humans

Toxic effects

Systemic poisoning due to exposure to ethylene oxide is rare, but three cases have been reported in which headache, vomiting, dyspnoea, diarrhoea and lymphocytosis occurred (Hine & Rowe, 1963) and one case with nausea and stomach spasms (Salinas *et al.*, 1981). Skin burns were observed in workers in prolonged contact with a 1% solution of ethylene oxide in water (Sexton & Henson, 1949); one case of corneal burns was reported (McLaughlin, 1946). Workers have developed severe skin irritation, including redness, oedema, blisters and ulceration, after wearing rubber gloves that had absorbed ethylene oxide (Royce & Moore, 1955). Skin burns and dermatitis have been observed in a number of medical personnel after contact with material treated with ethylene oxide (Joyner, 1964; Taylor, 1977) and in hospitalized people in contact with reused surgical gowns, drapes or apparatus sterilized with ethylene oxide (Marx et al., 1969; LaDage, 1970; Hanifin, 1971; Fisher, 1973a,b; Biro et al., 1974; Lebrec et al., 1977).

Five cases of peripheral neuropathy and one case of encephalopathy have been described after employment in chemical sterilization, where transient concentrations of ethylene oxide were thought to have exceeded the odour threshold (700 ppm; 1280 mg/m^3) (Gross et al., 1979; Kuzuhara et al., 1983). Three men developed cataracts after exposure to ethylene oxide sterilizers (Jay et al., 1982).

Effects on reproduction and prenatal toxicity

Hemminki et al. (1982) reported a retrospective study among female sterilizing staff in hospitals in Finland on 1443 pregnancy outcomes that occurred between the early 1950s and 1981. Nursing supervisors from approximately 80 hospitals identified the study participants and the exposure status of each with regard to specific sterilizing agents, which included ethylene oxide, glutaraldehyde and formaldehyde. The pregnancies of sterilizing staff were categorized as 'exposed' or 'unexposed' on the basis of the work history at the beginning of each pregnancy, and comparisons were made between these two groups. In addition, a control group was established consisting of 1179 pregnancies among female nursing auxiliaries who had had no exposure to sterilizing agents, anaesthetic gases or X-rays. Information was collected from each study participant by a self-administered questionnaire. For spontaneous abortions, the most significant increase in rates (adjusted for age, parity, decade of pregnancy, coffee consumption, alcohol consumption and smoking habits) was observed for exposure to ethylene oxide alone: 16.1% for exposed pregnancies, 7.8% for unexposed pregnancies ($p < 0.01$) and 10.5% for controls. In a subsequent analysis (Hemminki et al., 1983), the authors applied a stricter age adjustment and restricted pregnancies among controls to those that began during hospital employment; they found similar results for ethylene oxide exposure.

Pregnancy outcome data for sterilizing staff and controls identified from hospital discharge registries in Finland from 1973 to 1979 were compared (Hemminki et al., 1982). A significant increase in age-adjusted spontaneous abortion rates was observed for ethylene oxide-exposed pregnancies among the sterilizing staff (22.6%) compared to the age-adjusted rate among control pregnancies (9.2%), confirming the finding made on the basis of data obtained from the questionnaires.

Absorption, distribution, excretion and metabolism

2-Hydroxyethyl residues bound to *N*-3-histidine have been detected in the haemoglobin of workers in ethylene oxide sterilization plants (Calleman et al., 1978). No significant difference was detected between the levels of modified haemoglobin in workers exposed to low levels of ethylene oxide and controls; even the control population had measurable levels of modified haemoglobin (van Sittert et al., 1984).

Mutagenicity and chromosomal effects (see also 'Appendix: Activity Profiles for Short-term Tests', p. 338)

Exposure to ethylene oxide in the workplace has been associated with increases in sister chromatid exchange (SCE) frequencies in peripheral blood lymphocytes of exposed workers. Recently reported studies on large groups of individuals suggest an exposure-related response (Table 8).

Table 8. Sister chromatid exchange (SCE) induction in peripheral blood lymphocytes of workers exposed to ethylene oxide

Reference	No. exposed	Mean and range of SCE frequencies	Exposure level[a]	No. of controls	Mean and range of SCE frequencies	Remarks
Garry et al. (1979)	12	8.7	36 ppm	8	6.4 4.90-7.05	4 Exposed with respiratory and neurological symptoms had the highest number of SCEs
Husgafvel-Pursiainen et al. (1980)	4	8.5-13.0	Not given	21	7.0-13.2	Includes smokers and non-smokers in both groups
Hedner et al. (1982)	22	6.8-24.9	Not given	35	7.1-12.2	No significant difference between exposed and controls
Hogstedt et al. (1983)	28	10.2 and 11.4 at factories I and II, respectively	TWA-level <1 ppm	20	10.0 and 8.5 factories I and II, respectively	No difference between exposed and controls
Laurent et al. (1983)	25	9.61-17.57	Not given	10	7.04-8.52	Includes smokers and non-smokers in both groups
Yager et al. (1983)	Low exposure: 9	7.76 6.16-9.62	Cumulative dose <100 mg (6 mo.)	13	7.56 6.80-9.38	Includes smokers and non-smokers in both groups
	High exposure: 5	10.69 8.00-12.98	>100 mg (6 mo.)			
Hansen et al. (1984)	14	7.64 6.2-9.9	<5.0 ppm 8-h TWA	14	7.83 5.8-9.7	Includes smokers and non-smokers in both groups No difference between exposed and controls
Laurent et al. (1984)	Low exposure: 7	12.14 9.61-16.02	Cumulative dose range (2 yrs) 532-714 mg	15 Non-smokers	7.52 5.74-8.54	Includes smokers and non-smokers in the exposed group
	High exposure: 11	13.67 10.52-17.57	1185-5802 mg (2 yrs)	7 Smokers	8.24 7.20-9.86	
Sarto et al. (1984)	Low exposure: 22	11.0	8-h TWA 0.2-0.5 ppm	41	10.0	Tendency to a dose-effect relationship
	High exposure: 19	13.0	8-h TWA 3.7-20 ppm			
Stolley et al. (1984)	61	[b]	8-h TWA 0.5-200 ppm	53		Tendency to a dose-effect relationship

[a]TWA, time-weighted average
[b]See text

In a follow-up period up to 24 months, Stolley et al. (1984) analysed SCE frequencies in peripheral lymphocytes of workers exposed to different concentrations of ethylene oxide (eight-hour time-weighted averages in ppm [mg/m^3]): 0.5 [0.9] (worksite I), 5-10 [9-18] (worksite II), 5-20 [9-36] to 50-200 [90-360] (worksite III)). The SCE frequencies were significantly elevated in the lymphocytes of workers at worksites II and III. In the highest-exposure group (worksite III, two workers), the SCE rate was very high (average, 32.3 SCEs per cell) and remained high even after 24 months (average, 21.1 SCEs per cell), indicating that DNA lesions leading to SCEs were persistent (see also Laurent et al., 1983).

Exposure in vivo to ethylene oxide has been found to lead to structural chromosomal aberrations in human peripheral lymphocytes (Pero et al., 1981; Thiess et al., 1981a; Högstedt et al., 1983). [In all these studies, a culture time of 72 hours was used, which is longer than that currently recommended (Evans, 1984).]

van Sittert et al. (1984) found no significant increase in the frequencies of structural chromosomal aberrations in peripheral lymphocytes of workers exposed to ethylene oxide. These authors used culture times of 48 hours and 72 hours. Metaphases were scored from 48-hour cultures, and from 72-hour cultures in cases where not enough metaphases were found in preparations from 48-hour cultures.

An elevated frequency of micronucleated cells was observed in bone-marrow smears from persons exposed to ethylene oxide (Högstedt et al., 1983).

3.3 Case reports and epidemiological studies of carcinogenicity to humans

Hogstedt et al. (1979b) reported three cases of leukaemia, brought to the attention of the authors by the safety committee of the work force, that occurred between 1972 and 1977 in 70 workers ever employed during that period in a factory in Sweden. From 1968-1977, 50% ethylene oxide and 50% methyl formate had been used for sterilizing hospital equipment. Two cases of leukaemia occurred among 68 women who were not involved in sterilization but were exposed to vapours from sterilized boxes stored for weekly periods in a factory storage hall, where about 30 persons were exposed at any one time. One women with chronic myeloid leukaemia and another with acute myelogenous leukaemia had been exposed for eight hours per day during four and eight years, respectively. The third case was the local male manager, who developed primary macroglobulinaemia (morbus Waldenström) nine years after installation of the sterilizing equipment; he was estimated to have been exposed to ethylene oxide in the storage hall for three hours per week. This man had previously 'had some occasional contact with benzene in laboratory work'. Exposure measurements to ethylene oxide were made in 1977 during the course of the investigation by continuous recording with an infrared spectrophotometer and by gas chromatography of samples collected in impinger bottles. Representative samples taken from different areas in the storage hall at various times showed values ranging from 2-70 ppm (3.6-128 mg/m^3), decreasing during the week of storage of sterilized boxes. The eight-hour, time-weighted average concentration in the breathing zone of the women was calculated by industrial hygienists to have been 20 $\pm$ 10 (SD) ppm (36 $\pm$ 18 mg/m^3) during the period 1968-1977. No exposure measurement was made of methyl formate, but the authors stated that there would have been less exposure to this compound owing to its lower volatility. In addition to the 70 workers described above, another 153 had had occasional exposure and seven sterilizing operators had had peak exposures at some time during the period 1968-1977. The expected number of cases of leukaemia during that period was calculated by multiplying the person-years of observation by the national sex- and age-specific leukaemia incidence rates for

1972. Expected numbers were 0.2 cases for all 230 employees. In a follow-up of this study, Hogstedt et al. (1984) reported an expected number of 0.07 for those exposed in the storage hall.

In an epidemiological follow-up study of cancer incidence and mortality at the same plant (Hogstedt et al., 1984), the 203 workers who had been employed for more than one year were followed up to 1982. A further death from leukaemia - acute blastic leukaemia - was reported in a woman aged 56 who had been employed in various parts of the plant from 1969 to 1972, including brief periods in the storage hall. The case of morbus Waldenström in the male manager had been reclassified as a non-Hodgkin's lymphoma. Altogether, therefore, four deaths from malignancies of the lymphatohaematopoietic system occurred among the 203 workers, compared to 0.3 expected; no further incident case of leukaemia had been reported to the cancer registry up to 1981.

[The Working Group noted that no statistical evaluation was possible because the studies are based on a cluster of case reports.]

A cohort study on male workers employed in a Swedish ethylene oxide-producing plant was undertaken by Hogstedt et al. (1979a), following initial case reports described above. This study consisted of a follow-up of a group of men who had been examined medically in 1960. Mortality and cancer incidence from 1961-1977 were examined and expected numbers were calculated from five-year age-, gender- and calendar year-specific national statistics. In all, 23 deaths were observed compared to 13.5 expected ($p < 0.05$). The excess mortality was due partly to increased mortality from cancer (nine cases observed, 3.4 expected; $p < 0.01$), but also to diseases of the circulatory system. The excess cancer mortality resulted from stomach cancer (three cases observed, 0.4 expected; $p < 0.01$) and leukaemia (two cases observed, 0.14 expected, $p < 0.01$), consisting of one chronic lymphatic leukaemia and one acute myeloid leukaemia. With a requirement of 10 years of exposure/employment and 20 years since first exposure, 13 deaths were observed compared to 4.6 expected; in these cases, five tumours were observed compared to 1.1 expected ($p < 0.01$). No increase in mortality was observed among 86 maintenance workers exposed intermittently to ethylene oxide, or among 66 unexposed men employed in the same factory. Exposure to ethylene oxide during the period 1941-1947 was estimated by a company-affiliated industrial hygiene consultant to have been below 25 mg/m^3, although occasional exposures above the odour threshold (about 1300 mg/m^3) occurred. During the 1950s and through to 1963 an average concentration of 10-50 mg/m^3 ethylene oxide was estimated, but peaks above the odour threshold still occurred. Since the ethylene oxide was produced by the chlorohydrin process, there might have been significant exposure to other chemicals, including large amounts of ethylene dichloride (1,2-dichloroethane, see IARC, 1979d), ethylene (see IARC, 1979a) and ethylene chlorohydrin and small amounts of bis(2-chloroethyl)ether (see IARC, 1975) for one hour per shift. Ethylene oxide production ceased in 1963, and propylene oxide (see this volume, p. 227) was produced in the same plant from 1964 onwards; it is not known whether the same workers were exposed. The authors concluded that the excess mortality and cancer incidence could not be attributed to any particular chemical in the production process, 'but ethylene oxide and ethylene dichloride are the prime suspects'.

In a further follow-up to 1982 (Hogstedt et al., 1984), seven more deaths had occurred among ethylene oxide operators compared to 6.6 expected according to national statistics. Another three cancer deaths had occurred among full-time exposed workers (1.6 expected); two were stomach cancers (0.2 expected) and one an oesophageal cancer (0.04 expected). In the total period 1961-1982, six deaths due to oesophageal cancer and to stomach cancer were observed among full-time exposed ethylene operators (0.7 expected). Three out of five stomach cancer cases had occurred among operators with less than five years of employment. Alimentary-tract cancer was observed in two maintenance workers (0.8 expected) and in one unexposed worker (0.8 expected). Cancer incidence during the period 1961-1981 was significantly in excess among the operators, in whom 17 cases were notified to the cancer registry (7.9 expected according to national statistics). One new case of chronic myeloid leukaemia had been reported to the cancer registry during the follow-up period, compared to 0.06 expected for all leukaemias.

Morgan et al. (1981) reported a retrospective cohort study on 767 men employed between 1955 and 1977 for at least five years and 'potentially exposed to ethylene oxide' in an outdoor reaction system in a chemical plant in eastern Texas, USA. In an industrial hygiene survey in 1977, all readings made in the ethylene oxide-production area were less than 10 ppm (18 mg/m^3) ethylene oxide. In the total cohort, 46 deaths occurred compared to 80 expected on the basis of US vital statistics; 11 malignant neoplasms were observed, with 15.2 expected. Excesses were found for pancreatic cancer (3/0.8), 'brain and CNS cancer' [unspecified] (2/0.7) and Hodgkin's disease (2/0.4), but no death from leukaemia was found. According to the authors, only a 10-fold or greater increase in the risk of leukaemia deaths is likely to have been detected in this study. [The Working Group noted that the criteria for whether a worker was 'potentially exposed' were not described. The marked deficit of observed deaths suggests that important selective factors were present for this particular group of workers.]

Thiess et al. (1981b) reported a mortality study of 602 male active and former employees who had worked for six months or more in an area of alkylene oxide production in the Federal Republic of Germany, who had been exposed to ethylene oxide and propylene oxide, as well as benzene and ethylene chlorohydrin. Industrial hygiene measurements in 1978 showed that average exposure concentrations of ethylene oxide were <4 ppm (7.3 mg/m^3), but no measurement of past levels was available; records of treated cases of intoxication suggest that higher concentrations may occasionally have been encountered in the past. The authors stated that the workers were regularly in brief contact with ethylene oxide, propylene oxide and other substances during sampling operations and when filters were changed. The first worker was employed in 1928, and the period of follow-up was from that year until 30 June 1980. Follow-up of German former employees was 97.6% successful, but 30/66 non-German ex-employees included in the cohort were lost to follow-up. The expected numbers of deaths for the exposed group were calculated for each five-year age group on the person-year principle, using age-specific mortality rates for the populations of Ludwigshafen and Rhinehessia-Palatinate from 1970-1975 and of Germany from 1971-1974. An internal comparison group of 1662 persons employed in a styrene production facility on the same site was used. Of the 602 persons in the cohort, 56 had died, whereas expected numbers were 71.5 (Ludwigshafen), 73.4 (Rhinehessia-Palatinate), 76.6 (Germany) and 57.9 deaths (styrene cohort). There were 14 deaths in this cohort due to cancer, compared with 16.6 expected from national statistics. Comparison with the styrene cohort revealed an increased, but non-significant relative risk of death from cancer in the alkylene oxide conort (14/9.4). There was one case of myeloid leukaemia (0.15 expected) and one case of lymphatic sarcoma with less than 10 years of 'minimum observation time'. Four stomach cancers (2.7 expected) and one brain tumour (0.07 expected) were also observed. [The Working Group questioned the expected numbers used, since they are not calendar period-specific over the whole observation period and it is not clear whether they were computed on the basis of the 92% of identified workers or of the full cohort.]

4. Summary of Data Reported and Evaluation

4.1 Exposure data

Ethylene oxide has been produced commercially since the First World War and is produced in large quantities in many countries. A major source of occupational exposure is its use as a sterilant; another significant source is fumigation of food. Additional sources of exposure are its production and its use as a chemical intermediate.

4.2 Experimental data

Ethylene oxide was tested by intragastric intubation in rats of one strain and induced local tumours, mainly squamous-cell carcinomas of the forestomach, in a dose-dependent manner. When rats were fed diets fumigated with ethylene oxide, no increased incidence of tumours was observed. In two experiments by inhalation exposure in rats of one strain, ethylene oxide increased the incidence of mononuclear-cell leukaemia in animals of both sexes and of peritoneal mesotheliomas in males. In one of these experiments in male rats, gliomas of the brain were induced; a high incidence of proliferative lesions of the adrenal cortex was also found. Ethylene oxide was tested by subcutaneous injection in mice of one strain and produced local tumours, mainly fibrosarcomas, in a dose-dependent manner.

Ethylene oxide has been tested for teratogenicity in mice, rats and rabbits. Teratogenic effects were observed in the offspring of mice given intravenous injections of maternally toxic doses. Fewer implantation sites and increased resorptions were observed in rats following inhalation exposure to ethylene oxide. No effect was seen in rabbits.

Overall assessment of data from short-term tests: ethylene oxide[a]

	Genetic activity			Cell transformation
	DNA damage	Mutation	Chromosomal effects	
Prokaryotes	+	+		
Fungi/green plants		+	+	
Insects		+	+	
Mammalian cells (*in vitro*)	+	+	+	
Mammals (*in vivo*)	+		+	
Humans (*in vivo*)			+	
Degree of evidence in short-term tests for genetic activity: *Sufficient*				Cell transformation: No data

[a]The groups into which the table is divided and the symbol + are defined on pp. 17-18 of the Preamble; the degrees of evidence are defined on p. 18.

Ethylene oxide caused DNA damage in bacteria and was mutagenic to bacteria, plants, fungi and insects. It caused chromosomal aberrations in plants and heritable translocations in insects. In cultured mammalian cells, it was mutagenic and induced DNA damage, chromosomal aberrations and sister chromatid exchanges. Ethylene oxide alkylated DNA in mice and rats *in vivo*. It induced sister chromatid exchanges, chromosomal aberrations, micronuclei, dominant lethal mutations and heritable translocations in mammals *in vivo*. In monkeys, ethylene oxide induced chromosomal aberrations and sister chromatid exchanges in peripheral lymphocytes.

4.3 Human data

From the workforce of a small Swedish factory among whom exposure primarily to ethylene oxide had occurred, case reports of two myeloid leukaemias and one morbus Waldenström, later reclassified as non-Hodgkin's lymphoma, were initially published. In a subsequent five-year follow-up a further death from leukaemia (acute 'blastic') was reported.

A cohort study of Swedish ethylene oxide-production workers found a statistically significant excess of leukaemia based on two deaths (one myeloid and one lymphatic). Again, in a subsequent five-year follow-up a further leukaemia case (myeloid) was registered. There was also a statistically significant excess of stomach cancer. However, these production workers were also exposed to other chemicals. A cohort study of ethylene oxide-production workers in the USA found no case of leukaemia. However, there was only low potential exposure to ethylene oxide among the workforce and an unusually large deficit in the total deaths reported compared to the number expected. A cohort study of factory workers in Germany exposed to ethylene oxide and a mixture of other chemicals reported one death from leukaemia (myeloid) with less than one expected.

A causal relationship between exposure to ethylene oxide and leukaemia may be credible, but the four small epidemiological studies so far available suffer from various disadvantages which make their interpretation difficult. Further epidemiological studies among persons exposed to ethylene oxide alone are desirable.

In a study of hospital sterilizing staff, a statistically significant excess of spontaneous abortions was reported among women exposed to ethylene oxide during pregnancy.

Significant increases in the frequencies of sister chromatid exchanges in peripheral lymphocytes have been associated with occupational exposure to ethylene oxide. There are also indications of increases in the frequencies of chromosomal aberrations in peripheral lymphocytes.

4.4 Evaluation[1]

There is *sufficient evidence* for the carcinogenicity of ethylene oxide to experimental animals.

There is *limited evidence* for the carcinogenicity to humans of exposure to ethylene oxide in combination with other chemicals; there is *inadequate evidence* for the carcinogenicity to humans of exposure to ethylene oxide alone.

Taken together, the data indicate that ethylene oxide is probably carcinogenic to humans.

---

[1]For definitions of the italicized terms, see the Preamble, pp. 15-16.

5. References

American Conference of Governmental Industrial Hygienists (1984) *TLVs^R Threshold Limit Values for Chemical Substances in the Work Environment Adopted by ACGIH for 1984-85*, Cincinatti, OH, p. 19

Anon. (1982a) Producers seek simple limit revision; Users, unions disagree over EPA coverage. *Occup. Saf. Health Rep.*, May, 1028-1030

Anon. (1982b) Johnson & Johnson preliminary results show exposure-linked chromosome effects. *Occup. Saf. Health Rep.*, April, 915, 916

Anon. (1984) EPA moves to reduce ethylene oxide exposure. *Chem. Eng. News*, 30 April, 13

Apol, A. (1978) *Health Hazard Evaluation Determination Report No. 78-42-498, Swedish Hospital, Seattle, Washington (PB81-149700)*, Springfield, VA, National Technical Information Service

Appelgren, L.-E., Eneroth, G., Grant, C., Landström, L.-E. & Tenghagen, K. (1978) Testing of ethylene oxide for mutagenicity using the micronucleus test in mice and rats. *Acta pharmacol. toxicol.*, *43*, 69-71

Bär, F. & Griepentrog, F. (1969) Long-term diet study in rats with feed fumigated with ethylene oxide (Ger.). *Bundesgesundheitsblatt*, *11*, 106-112

Bellenger, P., Pradier, F., Sinegre, M. & Pradeau, D. (1983) Determination of residual ethylene oxide in nonreusable medical devices by head-space analysis (Fr.). *Sci. Tech. Pharm.*, *12*, 37-39 [*Chem. Abstr.*, *98*, 166988b]

Bicchi, C. & Frattini, C. (1981) Determination of ethylene oxide residues in vegetable substrates. *Fitoterapia*, *52*, 261-266 [*Chem. Abstr.*, *97*, 194445v]

Binder, H. & Linder, W. (1972) Determination of ethylene oxide in the smoke of treated and untreated cigarettes (Ger.). *Fachliche Mitt. Oesterr. Tabakregie*, *13*, 215-220 [*Chem. Abstr.*, *78*, 55479a]

Bird, M.J. (1952) Chemical production of mutations in *Drosophila*: Comparison of techniques. *J. Genet.*, *50*, 480-485

Biro, L., Fisher, A.A. & Price, E. (1974) Ethylene oxide burns. A hospital outbreak involving 19 women. *Arch. Dermatol.*, *110*, 924-925

Blome, H. (1982) Determination of ethylene oxide in workplace atmospheres (Ger.). *Staub-Reinhalt. Luft*, *42*, 280-282 [*Chem. Abstr.*, *97*, 114430y]

Bogyo, D.A., Lande, S.S., Meylan, W.M., Howard, P.H. & Santodonato, J. (1980) *Investigation of Selected Potential Environmental Contaminants: Epoxides (PB80-183197)*, Prepared for US Environmental Protection Agency, Springfield, VA, National Technical Information Service, pp. 61, 63, 67-68, 97-104, 162

Bond, E.J. & Dumas, T. (1982) A portable gas chromatograph for macro- and microdetermination of fumigants in the field. *J. agric. Food Chem.*, *30*, 986-988

Brookes, P. & Lawley, P.D. (1961) The alkylation of guanosine and guanylic acid. *J. chem. Soc., Part III*, 3923-3928

Brown, D.J. (1970) Determination of ethylene oxide and ethylene chlorohydrin in plastic and rubber surgical equipment sterilized with ethylene oxide. *J. Assoc. off. anal. Chem.*, 53, 263-267

Burroughs, G.E. (1981) *Health Hazard Evaluation Report No. HETA-81-350-932, Ellis Hospital, Schenectady, New York (PB83-126391)*, Springfield, VA, National Technical Information Service

Calleman, C.J., Ehrenberg, L., Jansson, B., Osterman-Golkar, S., Segerbäck, D., Svensson, K. & Wachtmeister, C.A. (1978) Monitoring and risk assessment by means of alkyl groups in hemoglobin in persons occupationally exposed to ethylene oxide. *J. environ. Pathol. Toxicol.*, 2, 427-442

Cawse, J.N., Henry, J.P., Swartzlander, M.W. & Wadia, P.H. (1980) Ethylene oxide. In: Grayson, M., ed., *Kirk-Othmer Encyclopedia of Chemical Technology*, 3rd ed., Vol. 9, New York, John Wiley & Sons, pp. 432-471

Collins, M. & Barker, N.J. (1983) Direct monitoring of ambient air for ethylene oxide and ethylene dibromide. *Am. Lab.*, 15, 72, 74-76, 78-81

Conan, L., Foucault, B., Siou, G., Chaigneau, M. & Le Moan, G. (1979) On the mutagenic activity of residues of ethylene oxide, ethylene glycol and 2-chloroethanol in plastic materials sterilized with ethylene oxide (Fr.). *Ann. Fals. Exp. Chim.*, 72, 141-151

Cookson, M.J., Sims, P. & Grover, P.L. (1971) Mutagenicity of epoxides of polycyclic hydrocarbons correlates with carcinogenicity of parent hydrocarbons. *Nature-New Biol.*, 234, 186-187

Coretti, K. (1978) Sterilization of spices (Ger.). *Fleischwirtschaft*, 58, 1239-1241, 1255-1256

Coye, M.J. & Belanger, P.L. (1981) *Health Hazard Evaluation Report No. HHE-80-191-829, University of California/Moffitt Hospital, San Francisco, California (PB82-215344)*, Springfield, VA, National Technical Information Service

De Flora, S. (1981) Study of 106 organic and inorganic compounds in the *Salmonella*/microsome test. *Carcinogenesis*, 2, 283-298

Deutsche Forschungsgemeinschaft (1984) *Maximal Work Place Concentrations and Biological Tolerance Values for Compounds in the Work Place* (Ger.), Part XX, Weinheim, Verlag Chemie GmbH, p. 34

Dmitriev, M.T. & Mishchikhin, V.A. (1982) Gas chromatographic determination of ethylene oxide in the atmosphere (Russ.). *Gig. Sanit.*, 4, 65-68 [*Chem. Abstr.*, 96, 222399h]

Dunkelberg, H. (1981) Carcinogenic activity of ethylene oxide and its reaction products 2-chloroethanol, 2-bromoethanol, ethylene glycol and diethylene glycol. I. Carcinogenicity of ethylene oxide in comparison with 1,2-propylene oxide after subcutaneous administration in mice (Ger.). *Zbl. Bakt. Hyg., I. Abt. Orig. B*, 174, 383-404

Dunkelberg, H. (1982) Carcinogenicity of ethylene oxide and 1,2-propylene oxide upon intragastric administration to rats. *Br. J. Cancer*, *46*, 924-933

Ehrenberg, L. & Gustafsson, A. (1957) On the mutagenic action of ethylene oxide and diepoxybutane in barley. *Hereditas*, *43*, 595-602

Ehrenberg, L., Hiesche, K.D., Osterman-Golkar, S. & Wennberg, I. (1974) Evaluation of genetic risks of alkylating agents: Tissue doses in the mouse from air contaminated with ethylene oxide. *Mutat. Res.*, *24*, 83-103

Ehrenberg, L., Osterman-Golkar, S., Segerbäck, D., Svensson, K. & Calleman, C.J. (1977) Evaluation of genetic risks of alkylating agents. III. Alkylation of haemoglobin after metabolic conversion of ethene to ethene oxide *in vivo*. *Mutat. Res.*, *45*, 175-184

Embree, J.W. & Hine, C.H. (1975) Mutagenicity of ethylene oxide (Abstract No. 126). *Toxicol. appl. Pharmacol.*, *33*, 172-173

Embree, J.W., Lyon, J.P. & Hine, C.H. (1977) The mutagenic potential of ethylene oxide using the dominant-lethal assay in rats. *Toxicol. appl. Pharmacol.*, *40*, 261-267

Ethylene Oxide Industry Council (1983) *Hazard Assessment of Ethylene Oxide*, Washington DC, p. 7

Evans, H.J. (1984) *Human peripheral blood lymphocytes for the analysis of chromosome aberrations in mutagen tests*. In: Kilbey, B.J., Legator, M., Nichols, W. & Ramel, C., eds, *Handbook of Mutagenicity Test Procedures*, Amsterdam, Elsevier, pp. 405-427

Fahmy, O.G. & Fahmy, M.J. (1970) Gene elimination in carcinogenesis: Reinterpretation of the somatic mutation theory. *Cancer Res.*, *30*, 192-205

Falcao, R.F. (1981) HMBT colorimetric analysis of ethylene oxide residues (Port.). *Rev. Port. Farm.*, *31*, 245-256 [*Chem. Abstr.*, *97*, 115379a]

Filser, J.G. & Bolt, H.M. (1983) Exhalation of ethylene oxide by rats on exposure to ethylene. *Mutat. Res.*, *120*, 57-60

Fisher, A.A. (1973a) Ethylene oxide dermatitis. *Contact Derm. Newslett.*, *14*, 393-394

Fisher, A.A. (1973b) Post-operative ethylene oxide dermatitis. *Cutis*, *12*, 177-178

Flavor and Extract Manufacturers' Association of the United States (1978) *Scientific Literature Review of Epoxides in Flavor Usage*, Vol. I, *Introduction and Summary Tables of Data, Bibliography (PB-291 103)*, Prepared for the US Food and Drug Administration, Springfield, VA, National Technical Information Service, pp. 4, 13, 59-62, 92-93

Flores, G.H. (1983) Controlling exposure to alkylene oxides. *Chem. Eng. Prog.*, *79*, 39-43

Fomenko, V.N. & Strekalova, E.Y. (1973) The mutagenic effect of some industrial toxins as a function of concentration and exposure time (Russ.). *Toksikol. Nov. Prom. Khim. Veshchestv.*, *7*, 51-57

Garry, V.F., Hozier, J., Jacobs, D., Wade, R.L. & Gray, D.G. (1979) Ethylene oxide: Evidence of human chromosomal effects. *Environ. Mutagenesis*, *1*, 375-382

Garry, V.F., Opp, C.W., Wiencke, J.K. & Lakatua, D. (1982) Ethylene oxide induced sister chromatid exchange in human lymphocytes using a membrane dosimetry system. *Pharmacology*, *25*, 214-221

Gary, V.F., Wiencke, J.K. & Nelson, R.L. (1984) *Ethylene oxide and some factors affecting the mutagen sensitivity of sister chromatid exchange in humans.* In: Tice, R.R. & Hollaender, A., eds, *Sister Chromatid Exchanges: Twenty-five Years of Experimental Research,* Book B, *Genetic Toxicology and Human Studies,* New York, Plenum Publishing Corp., pp. 975-985

Generoso, W.M., Cain, K.T., Krishna, M., Sheu, C.W. & Gryder, R.M. (1980) Heritable translocation and dominant-lethal mutation induction with ethylene oxide in mice. *Mutat. Res.*, *73*, 133-142

Generoso, W.M., Cumming, R.B., Bandy, J.A. & Cain, K.T. (1983) Increased dominant-lethal effects due to prolonged exposure of mice to inhaled ethylene oxide. *Mutat. Res.*, *119*, 377-379

Glaser, Z.R. (1977) *Use of Ethylene as a Sterilant in Medical Facilities (PB-274 795)*, Rockville, MD, National Institute for Occupational Safety and Health

Glaser, Z.R. (1979) Ethylene oxide: Toxicology review and field study results of hospital use. *J. environ. Pathol. Toxicol.*, *2*, 173-208

Goldgraben, R. & Zank, N. (1981) *Mitigation of Worker Exposure to Ethylene Oxide (PB81-233033)*, Prepared for US Environmental Protection Agency, Springfield, VA, National Technical Information Service

Gorman, R. & Horan, J. (1981) *Health Hazard Evaluation Report No. HHE 80-181-909, Ralston Purina Company, Cincinnati, Ohio*, Cincinnati, OH, National Institute for Occupational Safety and Health

Gross, J.A., Haas, M.L. & Swift, R.T. (1979) Ethylene oxide neurotoxicity: Report of four cases and review of the literature. *Neurology*, *29*, 978-983

Hackett, P.L., Brown, M.G., Buschbom, R.L., Clark, M.L., Miller, R.A., Music, R.L., Rowe, S.E., Schirmer, R.E. & Sikov, M.R. (1982) *Teratogenic Study of Ethylene and Propylene Oxide and n-Butyl Acetate (Report for Contract No. 210-80-0013) (PB83-258038)*, Cincinnati, OH, National Institute for Occupational Safety and Health

Hanifin, J.M. (1971) Ethylene oxide dermatitis. *J. Am. med. Assoc.*, *217*, 213

Hansen, J.P., Allen, J., Brock, K., Falconer, J., Helms, M.J., Shaver, G.C. & Strohm, B. (1984) Normal sister chromatid exchange levels in hospital sterilization employees exposed to ethylene oxide. *J. occup. Med.*, *26*, 29-32

Hatch, G.G., Mamay, P.D., Christensen, C.C., Lagenbach, R., Goodheart, C.R. & Nesnow, S. (1982) Enhanced viral transformation of primary Syrian hamster embryo cells and mutagenesis of Chinese hamster lung (V-79) cells exposed to ethylene oxide in sealed treatment chambers (Abstract No. 291). *Am. Assoc. Cancer Res.*, *23*, 74

Hedner, K., Högstedt, B., Kolnig, A.-M., Mark-Vendel, E., Strömbeck, B. & Mitelman, F. (1982) Relationship between sister chromatid exchanges and structural chromosome aberrations in lymphocytes of 100 individuals. *Hereditas, 97*, 237-245

Hemminki, K., Mutanen, P., Saloniemi, I., Niemi, M.-L. & Vainio, H. (1982) Spontaneous abortions in hospital staff engaged in sterilising instruments with chemical agents. *Br. med. J., 285*, 1461-1463

Hemminki, K., Mutanen, P. & Niemi, M.-L. (1983) Spontaneous abortions in hospital sterilising staff. *Br. med. J., 286*, 1976-1977

Hine, C.H. & Rowe, V.K. (1963) Epoxy compounds. In: Patty, F.A., ed., *Industrial Hygiene and Toxicology*, 2nd rev. ed., Vol. 2, New York, Interscience, pp. 1626-1634

Högstedt, B., Gullberg, B., Hedner, K., Kolnig, A.-M., Mitelman, F., Skerfving, S. & Widegren, B. (1983) Chromosome aberrations and micronuclei in bone marrow cells and peripheral blood lymphocytes in humans exposed to ethylene oxide. *Hereditas, 98*, 105-113

Hogstedt, C., Rohlén, O., Berndtsson, B.S., Axelson, O. & Ehrenberg, L. (1979a) A cohort study of mortality and cancer incidence in ethylene oxide production workers. *Br. J. ind. Med., 36*, 276-280

Hogstedt, C., Malmqvist, N. & Wadman, B. (1979b) Leukemia in workers exposed to ethylene oxide. *J. Am. med. Assoc., 241*, 1132-1133

Hogstedt, C., Aringer, L. & Gustavsson, A. (1984) Ethylene oxide and cancer - Review of the literature and follow-up of two studies (Swed.). *Arbete Hälsa, 49*, 1-32

Hollingsworth, R.L., Rowe, V.K., Oyen, F., McCollister, D.D. & Spencer, H.C. (1956) Toxicity of ethylene oxide determined on experimental animals. *Arch. ind. Health, 13*, 217-227

Husgafvel-Pursiainen, K., Mäki-Paakkanen, J., Norppa, H. & Sorsa, M. (1980) Smoking and sister chromatid exchange. *Hereditas, 92*, 247-250

IARC (1975) *IARC Monographs on the Evaluation of Carcinogenic Risk of Chemicals to Man, Vol. 9, Some Aziridines, N-, S- and O-Mustards and Selenium*, Lyon, pp. 117-123

IARC (1976) *IARC Monographs on the Evaluation of Carcinogenic Risk of Chemicals to Man, Vol. 11, Cadmium, Nickel, Some Epoxides, Miscellaneous Industrial Chemicals and General Considerations on Volatile Anaesthetics*, Lyon, pp. 157-167

IARC (1979a) *IARC Monographs on the Evaluation of the Carcinogenic Risk of Chemicals to Humans, Vol. 19, Some Monomers, Plastics and Synthetic Elastomers, and Acrolein*, Lyon, pp. 157-186

IARC (1979b) *IARC Monographs on the Evaluation of the Carcinogenic Risk of Chemicals to Humans, Vol. 19, Some Monomers, Plastics and Synthetic Elastomers, and Acrolein*, Lyon, pp. 303-340

IARC (1979c) *IARC Monographs on the Evaluation of the Carcinogenic Risk of Chemicals to Humans, Vol. 19, Some Monomers, Plastics and Synthetic Elastomers, and Acrolein*, Lyon, pp. 73-113

IARC (1979d) *IARC Monographs on the Evaluation of the Carcinogenic Risk of Chemicals to Humans*, Vol. 20, *Some Halogenated Hydrocarbons*, Lyon, pp. 429-440

IARC (1982a) *IARC Monographs on the Evaluation of the Carcinogenic Risk of Chemicals to Humans*, Suppl. 4, *Chemicals, Industrial Processes and Industries Associated with Cancer in Humans (IARC Monographs, Volumes 1 to 29)*, pp. 126-128

IARC (1982b) *Information Bulletin on the Survey of Chemicals Being Tested for Carcinogenicity*, No. 10, Lyon, p. 210

International Labour Office (1980) *Occupational Exposure Limits for Airborne Toxic Substances*, 2nd (rev.) ed. (*Occupational Safety and Health Series No. 37*), Geneva, pp. 114-115

Jacobson, K.H., Hackley, E.B. & Feinsilver, L. (1956) The toxicity of inhaled ethylene oxide and propylene oxide vapors. *Arch. ind. Health*, 13, 237-244

Jana, M.K. & Roy, K. (1975) Effectiveness and efficiency of ethyl methanesulphonate and ethylene oxide for the induction of mutations in rice. *Mutat. Res.*, 28, 211-215

Jay, W.M., Swift, T.R. & Hull, D.S. (1982) Possible relationship of ethylene oxide exposure to cataract formation. *Am. J. Ophthalmol.*, 93, 727-732

Johnson, P.L., Evans, W. & Verst, G. (1979) *Health Evaluation and Technical Assistance Report No. TA78-56, St Mary's Hospital, Rhinelander, Wisconsin (PB82-182601)*, Springfield, VA, National Technical Information Service

Jordy, A. (1981) Residues in meat products made with ethylene oxide-gassed spices (Ger.). *Fleischwirtschaft*, 61, 1667-1668, 1673-1674, 1712

Joyner, R.E. (1964) Chronic toxicity of ethylene oxide. *Arch. environ. Health*, 8, 700-710

Kilbey, B.J. & Kølmark, H.G. (1968) A mutagenic after-effect associated with ethylene oxide in *Neurospora crassa*. *Molec. gen. Genet.*, 101, 185-188

Kimmel, C.A., LaBorde, J.B. & Hardin, B.D. (1984) *Reproductive and developmental toxicology of selected epoxides*. In: Kacew, S. & Reasor, M., eds, *Toxicology and the Newborn*, Amsterdam, Elsevier, pp. 1-32

Kiss, T. & Kovacs, H.K., (1982) Determination of the residual ethylene oxide contamination in penicillins sterilized with ethylene oxide (Hung.). *Acta pharm. hung.*, 52, 280-285 [*Chem. Abstr.*, 98, 78230t]

Kligerman, A.D., Erexson, G.L., Phelps, M.E. & Wilmer, J.L. (1983) Sister-chromatid exchange induction in peripheral blood lymphocytes of rats exposed to ethylene oxide by inhalation. *Mutat. Res.*, 120, 33-47

Kligerman, A.D., Erexson, G.L. & Wilmer, J.L. (1984) *Development of rodent peripheral blood lymphocyte culture systems to detect cytogenetic damage* in vivo. In: Tice, R.R. & Hollaender, A., eds, *Sister Chromatid Exchange: 25 Years of Experimental Research*, New York, Plenum Publishing Corp., pp. 569-584

Koketsu, M. & Alli, B.O. (1977) *Jefferson Chemical Company, Port Neches, Texas, Task II, Ethylene Oxide Survey Report of the Plant Contact July 11-12, 1977 (PB81-229874)*, Springfield, VA, National Technical Information Service, pp. 2, 11-12

Kolb, B. (1982) Multiple headspace extraction - A procedure for eliminating the influence of the sample matrix in quantitative headspace gas chromatography. *Chromatographia, 15*, 587-594

Kølmark, H.G. & Kilbey, B.J. (1968) Kinetic studies of mutation induction by epoxides in *Neurospora crassa*. *Molec. gen. Genet., 101*, 89-98

Kölmark, G. & Westergaard, M. (1953) Further studies on chemically induced reversions at the adenine locus of *Neurospora*. *Hereditas, 39*, 209-224

Kucera, J., Lundqvist, U. & Gustafsson, A. (1975) Induction of breviaristatum mutants in barley. *Hereditas, 80*, 263-278

Kuzuhara, S., Kanazawa, I., Nakanishi, T. & Egashira, T. (1983) Ethylene oxide polyneuropathy. *Neurology, 33*, 377-380

LaBorde, J.B. & Kimmel, C.A. (1980) The teratogenicity of ethylene oxide administered intravenously to mice. *Toxicol. appl. Pharmacol., 56*, 16-22

LaDage, L.H. (1970) Facial 'irritation' from ethylene oxide sterilization of anesthesia mask? *Plastic reconst. Surg., 45*, 179

Laurent, C., Frederic, J. & Maréchal, F. (1983) Increased sister chromatid exchange frequency in workers exposed to ethylene oxide (Fr.). *Ann. Génét., 26*, 138-142

Laurent, C., Frederic, J. & Léonard, A.Y. (1984) Sister chromatid exchange frequency in workers exposed to high levels of ethylene oxide, in a hospital sterilization service. *Int. Arch. occup. environ. Health* (in press)

Lebrec, D., Masquet, C. & Rueff, B. (1977) Collapsus after endovascular exploration using ethylene oxide sterilized catheters (Fr.). *Nouv. Presse méd., 6*, 2991

Lovegren, B.C. & Koketsu, M. (1977a) *BASF-Wyandotte Corporation, Geismar, Louisiana, Task II, Ethylene Oxide Survey Report of the Plant Contact June 27-28, 1977 (PB81-229775)*, Springfield, VA, National Technical Information Service, pp. 2, 12

Lovegren, B.C. & Koketsu, M. (1977b) *Union Carbide Corporation, Institute, West Virginia, Task II, Ethylene Oxide Survey Report of the Plant Contact, July 15-16, 1977 (PB82-106709)*, Springfield, VA, National Technical Information Service, p. 12

Lovegren, B.C. & Koketsu, M. (1977c) *Union Carbide Corporation, Texas City, Texas, Task II, Ethylene Oxide Survey Report of the Plant Contact, June 8-9, 1977 (PB82-108218)*, Springfield, VA, National Technical Information Service

Lynch, D.W., Lewis, T.R., Moorman, W.J., Burg, J.R., Groth, D.H., Khan, A., Ackerman, L.J. & Cockrell, B.Y. (1984a) Carcinogenic and toxicological effects of inhaled ethylene oxide and propylene oxide. *Toxicol. appl. Pharmacol., 76*, 69-84

Lynch, D.W., Lewis, T.R., Moorman, W.J., Burg, J.R., Gulati, D.K., Kaur, P. & Sabharwal, P.S. (1984b) Sister-chromatid exchanges and chromosome aberrations in lymphocytes from monkeys exposed to ethylene oxide and propylene oxide by inhalation. *Toxicol. appl. Pharmacol.*, 76, 85-95

Manoff, T.C., Billings, C.E., Krusé, C.W., Olivieri, V.P. & Reed, J.C. (1982) *Ethylene Oxide Sterilizer Exposures (prepared under NIOSH Grant DTMD-07090)*, Baltimore, MD, School of Hygiene and Public Health

Martis, L., Kroes, R., Darby, T.D. & Woods, E.F. (1982) Disposition kinetics of ethylene oxide, ethylene glycol, and 2-chlorethanol in the dog. *J. Toxicol. environ. Health*, 10, 847-856

Marx, G.F., Steen, S.N., Schapira, M., Erlanger, H.L., Arkins, R.E., Jadwat, C.M. & Kepes, E. (1969) Hazards associated with ethylene oxide sterilization. *N.Y. State J. Med.*, 69, 1319-1320

McDonald, T.O., Kasten, K., Hervey, R., Gregg, S., Borgmann, A.R. & Murchison, T. (1973) Acute ocular toxicity of ethylene oxide, ethylene glycol, and ethylene chlorohydrin. *Bull. parent. Drug Assoc.*, 27, 153-164

McLaughlin, R.S. (1946) Chemical burns of the human cornea. *Am. J. Ophthalmol.*, 29, 1355-1362

Migliore, L., Rossi, A.M. & Loprieno, N. (1982) Mutagenic action of structurally related alkene oxides on *Schizosaccharomyces pombe*: The influence, '*in vitro*', of mouse-liver metabolizing system. *Mutat. Res.*, 102, 425-437

Morgan, R.W., Claxton, K.W., Divine, B.J., Kaplan, S.D. & Harris, V.B. (1981) Mortality among ethylene oxide workers. *J. occup. Med.*, 23, 767-770

Moseley, C. (1979) *Hazard Evaluation and Technical Assistance Report No. TA 79-16, Johnson Memorial Hospital, Stafford Springs, Connecticut (PB80-148133)*, Springfield, VA, National Technical Information Service

Mouilleseaux, A., Laurent, A.M., Fabre, M., Jouan, M. & Festy, B. (1983) Atmospheric concentration of ethylene oxide in the occupational environment of disinfection and sterilization facilities (Fr.). *Arch. Mal. prof. méd. Trav. Sécur. soc.*, 44, 1-14

Moutschen-Dahmen, J., Moutschen-Dahmen, M. & Ehrenberg, L. (1968) Note on the chromosome breaking activity of ethylene oxide and ethyleneimine. *Hereditas*, 60, 267-269

Nakao, Y. & Auerbach, C. (1961) Test of a possible correlation between cross-linking and chromosome breaking abilities of chemical mutagens. *Z. Vererb.*, 92, 457-461

National Finnish Board of Occupational Safety and Health (1981) *Airborne Contaminants in the Workplace (Safety Bulletin No. 3)*, Helsinki, p. 12

National Institute for Occupational Safety and Health (1977) *NIOSH Manual of Analytical Methods. Part II Standards Completion Program Validated Methods*, 2nd Ed., Vol. 3 *(DHEW (NIOSH) Publ. No. 77-157-C)*, Washington DC, US Government Printing Office, pp. S286-1-S286-9

National Institute for Occupational Safety and Health (1980) *Projected Number of Occupational Exposures to Chemical and Physical Hazards*, Cincinnati, OH, p. 63

National Institute for Occupational Safety and Health (1981a) *National Occupational Hazard Survey* (microfiche), Cincinnati, OH, p. 4100 (generics excluded)

National Institute for Occupational Safety and Health (1981b) *Ethylene Oxide (Current Intelligence Bulletin 35; DHHS (NIOSH) Publ. No. 81-130)*, Washington DC, US Government Printing Office, pp. 3, 4

National Institute for Occupational Safety and Health (1983) NIOSH recommendations for occupational health standards. *Morbidity and Mortality Weekly Report*, *32*, Suppl., 135

National Research Council/National Academy of Sciences (1979) *The 1977 Survey of Industry on the Use of Food Additives*, Part 2 (*PB80-113418*), Springfield, VA, National Technical Information Service, pp. 497, 795

National Swedish Board of Occupational Safety and Health (1981) *Hygiene Standards (AFS 1981-8)*, Stockholm, p. 13

Oser, J.L., Crandall, M., Phillips, R. & Marlow, D. (1978a) *Indepth Industrial Hygiene Report of Ethylene Oxide Exposure at Union Carbide Corporation, Institute, West Virginia* (*PB82-114786*), Springfield, VA, National Technical Information Service, pp. 3, 19-21

Oser, J.L., Crandall, M. & Rinsky, R. (1978b) *Industrial Hygiene Survey of Dow Chemical Company, Plaquemine, Louisiana* (*PB81-229924*), Springfield, VA, National Technical Information Service, pp. 2, 8-9

Oser, J.L., Young, M., Boyle, T. & Marlow, D. (1979) *Indepth Industrial Hygiene Report of Ethylene Oxide Exposure at Union Carbide Corporation, South Charleston, West Virginia* (*PB82-110024*), Springfield, VA, National Technical Information Service, pp. 2, 10-12

Osterman-Golkar, S., Farmer, P.B., Segerbäck, D., Bailey, E., Calleman, C.J., Svensson, K. & Ehrenberg, L. (1983) Dosimetry of ethylene oxide in the rat by quantitation of alkylated histidine in hemoglobin. *Teratog. Carcinog. Mutagenesis*, *3*, 395-405

Pero, R.W., Widegren, B., Hogstedt, B. & Mitelman, F. (1981) In vivo and in vitro ethylene oxide exposure of human lymphocytes assessed by chemical stimulation of unscheduled DNA synthesis. *Mutat. Res.*, *83*, 271-289

Pfeiffer, E.H. & Dunkelberg, H. (1980) Mutagenicity of ethylene oxide and propylene oxide and of the glycols and halohydrins formed from them during the fumigation of foodstuffs. *Food Cosmet. Toxicol.*, *18*, 115-118

Poirier, V. & Papadopoulo, D. (1982) Chromosomal aberrations induced by ethylene oxide in a human amniotic cell line in vitro. *Mutat. Res.*, *104*, 255-260

Pritts, I.M., McConnaughey, P.W., Roberts, C.C. & McKee, E.S. (1982) *Ethylene oxide detection and protection*. In: *Toxic Materials in the Atmosphere (ASTM STP 786)*, Philadelphia, PA, American Society for Testing and Materials, pp. 14-25

Qazi, A.H. & Ketcham, N.H. (1977) A new method for monitoring personal exposure to ethylene oxide in the occupational environment. *Am. ind. Hyg. Assoc. J.*, *38*, 635-647

Rannug, U., Göthe, R. & Wachmeister, C.A. (1976) The mutagenicity of chloroethylene oxide, chloroacetaldehyde, 2-chloroethanol and chloroacetic acid, conceivable metabolites of vinyl chloride. *Chem.-biol. Interactions*, *12*, 251-263

Ricottilli, F., Branca, P. & Martire, N. (1981) Gas-chromatographic determination of ethylene oxide residues in food products in heat-sealed packages (Ital.). *Boll. Chim. Unione Ital. Lab. Prov., Parte Sci.*, *32*, 133-140 [*Chem. Abstr.*, *96*, 4991d]

Royce, A. & Moore, W.K.S. (1955) Occupational dermatitis caused by ethylene oxide. *Br. J. ind. Med.*, *12*, 169-171

Ruhe, R.L. (1977) *Hazard Evaluation and Technical Assistance Report No. TA 76-92, Harpers Ferry Center Museum Laboratory, Harpers Ferry, West Virginia*, Cincinnati, OH, National Institute for Occupational Safety and Health

Ruhe, R.L. (1978) *Health Hazard Evaluation Determination Report No. HE 78-70-528, Hospital Medical Corporation, Littleton, Colorado* (*PB81-143695*), Springfield, VA, National Technical Information Service

Salinas, E., Sasich, L., Hall, D.H., Kennedy, R.M. & Morriss, H. (1981) Acute ethylene oxide intoxication. *Drug. Intell. clin. Pharm.*, *15*, 384-386

Sarto, F., Cominato, I., Pinton, A.M., Brovedani, P.G., Faccioli, C.M., Bianchi, V. & Levis, A.G. (1984) Cytogenetic damage in workers exposed to ethylene oxide. *Mutat. Res.*, *138*, 185-195

Sawicki, E. (1976) *Analysis of atmospheric carcinogens and their cofactors.* In: Rosenfeld, C. & Davis, W., eds, *Environmental Pollution and Carcinogenic Risks* (*INSERM Symposium Series Vol. 52/IARC Scientific Publications No. 13*), Paris, Institut National de la Santé et de la Recherche Médicale/Lyon, International Agency for Research on Cancer, pp. 315-316

Schutte, N.P. (1977) *Hazard Evaluation and Technical Assistance Report No. TA 77-11, Xomed Company, Cincinnati, Ohio* (*PB 278 834*), Springfield, VA, National Technical Information Service

Segerbäck, D. (1983) Alkylation of DNA and hemoglobin in the mouse following exposure to ethene and ethene oxide. *Chem.-biol. Interactions*, *45*, 139-151

de Serres, F.J. (1983) The use of *Neurospora* in the evaluation of the mutagenic activity of environmental chemicals. *Environ. Mutagenesis*, *5*, 341-351

Sexton, R.J. & Henson, E.V. (1949) Determatological injuries by ethylene oxide. *J. ind. Hyg. Toxicol.*, *31*, 297-300

van Sittert, N.J., de Jong, G., Clare, M.G., Davies, R., Dean, B.J., Wren, L.J. & Wright, A.S. (1984) Cytogenetic, immunological and haematological effects in workers in an ethylene oxide manufacturing plant. *Br. J. ind. Med.* (in press)

Smyth, H.F., Jr, Seaton, J. & Fischer, L. (1941) The single dose toxicity of some glycols and derivatives. *J. ind. Hyg. Toxicol.*, *23*, 259-268

Snellings, W.M., Zelenak, J.P. & Weil, C.S. (1982) Effects on reproduction in Fischer 344 rats exposed to ethylene oxide by inhalation for one generation. *Toxicol. appl. Pharmacol.*, *63*, 382-388

Snellings, W.M., Weil, C.S. & Maronpot, R.R. (1984) A two-year inhalation study of the carcinogenic potential of ethylene oxide in Fischer 344 rats. *Toxicol. appl. Pharmacol.*, *75*, 105-117

Stephenson, R.L., Albrecht, W.N. & Drotman, D.P. (1980) *Technical Assistance Report No. TA-80-42 at Hardin Memorial Hospital, Elizabethtown, Kentucky (PB82-150897)*, Springfield, VA, National Technical Information Service

Stolley, P.D., Soper, K.A., Galloway, S.M., Nichols, W.W., Norman, S.A. & Wolman, S.R. (1984) Sister chromatid exchanges in association with occupational exposure to ethylene oxide. *Mutat. Res.*, *129*, 89-102

Strekalova, E.Y. (1971) The mutagenic effect of ethylene oxide on mammals (Russ.). *Toksikol. Nov. Prom. Khim. Veshchestv.*, *12*, 72-78

Strekalova, E.Y., Chirkova, Y.M. & Golubovich, Y.Y. (1975) Mutagenic action of ethylene oxide on sex and somatic cells in male white rats (Russ.). *Toksikol. Nov. Prom. Khim. Veshchestv.*, *6*, 11-16

Šulovská, K., Lindgren, D., Eriksson, G. & Ehrenberg, L. (1969) The mutagenic effect of low concentrations of ethylene oxide in air. *Hereditas*, *62*, 264-265

Swan, J.D. (1954) Determination of epoxides with sodium sulfite. *Anal. Chem.*, *26*, 878-880

Systems Applications, Inc. (1981) *Human Exposure to Atmospheric Concentrations of Selected Chemicals*, Vol. II, Appendix A-14, Research Triangle Park, NC, US Environmental Protection Agency

Tan, E.-L., Cumming, R.B. & Hsie, A.W. (1981) Mutagenicity and cytotoxicity of ethylene oxide in the CHO/HGPRT system. *Environ. Mutagenesis*, *3*, 683-686

Tanooka, H. (1979) Application of *Bacillus subtilis* spores in the detection of gas mutagens: A case of ethylene oxide. *Mutat. Res.*, *64*, 433-435

Taylor, J.S. (1977) Dermatologic hazards from ethylene oxide. *Cutis*, *19*, 189-192

Tharr, D.G. & Donohue, M. (1980) *Health Hazard Evaluation Report No. HE-79-80, 81-746, at Cobe Laboratories, Inc., Lakewood and Arvada, Colorado (PB82-107210)*, Springfield, VA, National Technical Information Service

Thiess, A.M., Schwegler, H., Fleig, I. & Stocker, W.G. (1981a) Mutagenicity study of workers exposed to alkylene oxides (ethylene oxide/propylene oxide) and derivatives. *J. occup. Med.*, *23*, 343-347

Thiess, A.M., Frentzel-Beyme, R., Link, R. & Stocker, W.G. (1981b) *Mortality study on employees exposed to alkylene oxides (ethylene oxide/propylene oxide) and their derivatives.* In: *Prevention of Occupational Cancer - International Symposium* (*Occupational Safety and Health Series No. 46*), Geneva, International Labour Office, pp. 249-259

Tsuge, M. & Senba, T. (1981) Gas chromatographic determination of ethylene oxide in materials for medical devices (Jpn.). *Bunseki Kagaku, 30,* 814-816 [*Chem. Abstr., 96,* 57846s]

US Department of Commerce (1983) *US Imports for Consumption and General Imports* (*FT246/Annual 1982*), Bureau of the Census, Washington DC, US Government Printing Office, p. 1-291

US Department of Commerce (1984) *US Exports, Schedule E Commodity by Country* (*FT410/December 1983*), Bureau of the Census, Washington DC, US Government Printing Office, p. 2-78

US Department of Labor (1982) Occupational exposure to ethylene oxide. *US Code Fed. Regul., Title 29,* Part 1910; *Fed. Regist., 47* (No. 17), pp. 3566-3571

US Department of Labor (1983) Occupational exposure to ethylene oxide. *US Code Fed. Regul., Title 29,* Part 1910; *Fed. Regist., 48* (No. 78), pp. 17284, 17290-17291, 17297, 17299-17300, 17316-17319

US Department of Labor (1984) Occupational exposure to ethylene oxide; final standard. *US Code Fed. Regul., Title 29,* Part 1910; *Fed. Regist., 49* (No. 122), p. 25734

US Department of Transportation (1982) Performance-oriented packagings standards. *US Code Fed. Regul., Title 49,* Parts 171, 172, 173, 178; *Fed. Regist., 47* (No. 73), pp. 16268, 16273, 16276

US Environmental Protection Agency (1978) Notice of rebuttable presumption against registration and continued registration of pesticide products containing ethylene oxide. *Fed. Regist., 43* (No. 19), pp. 3801-3815

US Environmental Protection Agency (1980) Hazardous waste management system: Identification and listing of hazardous wastes. *US Code Fed. Regul., Title 40,* Part 261; *Fed. Regist., 45* (No. 98), pp. 33084, 33122-33127, 33131-33133

US Environmental Protection Agency (1982) Protection of environment. *US Code Fed. Regul., Title 40,* Part 180.151

US Environmental Protection Agency (1983) Notification requirements; Reportable quantity adjustments. *US Code Fed. Regul., Title 40,* Part 302; *Fed. Regist., 48* (No. 102), pp. 23552, 23584

US Environmental Protection Agency (1984) Ethylene oxide; Response to the Interagency Testing Committee. *Fed. Regist., 49* (No. 1), 200-205

US Food and Drug Administration (1978) Ethylene oxide, ethylene chlorohydrin, and ethylene glycol. *US Code Fed. Regul., Title 21,* Parts 211, 821; *Fed. Regist., 43* (No. 122), pp. 27471-27483

US Food and Drug Administration (1980) Food and drugs. *US Code Fed. Regul., Title 21*, Parts 176.180, 178.3520, 193.200

US International Trade Commission (1983) *Synthetic Organic Chemicals, US Production and Sales, 1982* (*USITC Publication 1422*), Washington DC, US Government Printing Office, pp. 262, 296

US Tariff Commission (1922) *Census of Dyes and Other Synthetic Chemicals, 1921* (*Tariff Information Series No. 26*), Washington DC, US Government Printing Office, p. 150

Van Duuren, B.L., Orris, L. & Nelson, N. (1965) Carcinogenicity of epoxides, lactones, and peroxy compounds. Part II. *J. natl Cancer Inst., 35*, 707-717

Vanell, L.D. (1982) On-site monitoring of ethylene oxide sterilizers. *Int. Lab., 12*, 82, 84-85

Verschueren, K. (1977) *Handbook of Environmental Data on Organic Chemicals*, New York, Van Nostrand Reinhold, p. 329

Walles, S. & Ehrenberg, L. (1968) Effects of β-hydroxyethylation and β-methoxyethylation on DNA *in vitro. Acta chem. scand., 22*, 2727-2729

Watson, W.A.F. (1966) Further evidence of an essential difference between the genetical effects of mono- and bifunctional alkylating agents. *Mutat. Res., 3*, 455-457

Weil, C.S., Condra, N., Haun, C. & Striegel, J.A. (1963) Experimental carcinogenicity and acute toxicity of representative epoxides. *Am. ind. Hyg. Assoc. J., 24*, 305-325

Windmueller, H.G. & Kaplan, N.O. (1961) The preparation and properties of N-hydroxyethyl derivatives of adenosine, adenosine triphosphate, and nicotinamide adenine dinucleotide. *J. biol. Chem., 236*, 2716-2726

Yager, J.W. & Benz, R.D. (1982) Sister chromatid exchanges induced in rabbit lymphocytes by ethylene oxide after inhalation exposure. *Environ. Mutagenesis, 4*, 121-134

Yager, J.W., Hines, C.J. & Spear, R.C. (1983) Exposure to ethylene oxide at work increases sister chromatid exchanges in human peripheral lymphocytes. *Science, 219*, 1221-1223

PROPYLENE OXIDE

This substance was considered by a previous working group, in February 1976 (IARC, 1976). Since that time, new data have become available, and these have been incorporated into the monograph and taken into account in the present evaluation.

1. Chemical and Physical Data

1.1 Synonyms and trade names

Chem. Abstr. Services Reg. No.: 75-56-9

Chem. Abstr. Name: Oxirane, methyl-

IUPAC Systematic Name: Propylene oxide

Synonyms: Epoxy propane; epoxypropane; 1,2-epoxypropane; 2,3-epoxypropane; methyl ethylene oxide; methyl oxirane; methyloxirane; NCI-C50099; PO; propene oxide; propylene epoxide; 1,2-propylene oxide

1.2 Structural and molecular formulae and molecular weight

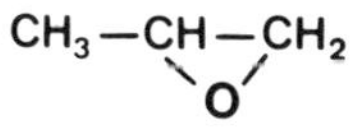

$$CH_3-CH-CH_2$$
$$\diagdown O \diagup$$

C_3H_6O Mol. wt: 58.1

1.3 Chemical and physical properties of the pure substance

From Kirk and Dempsey (1982), unless otherwise specified

(a) *Description:* Colourless liquid (a racemic mixture of two optical isomers)

(b) *Boiling-point*: 34.2°C

(c) *Freezing-point*: -112°C

(d) *Density*: 0.830 at 20°C

(e) *Refractive index*: n_D^{25} 1.36322

(f) *Optical rotation*: $[\alpha]_D^{18}$, +12.72° (*R*-isomer) and -8.26° (*S*-isomer)

(g) *Spectroscopy data:* Infrared and nuclear magnetic resonance spectral data have been reported (National Toxicology Program, 1984).

(h) *Solubility*: Solubility in water (20°C), 40.5 wt % (Windholz, 1983); miscible with most organic solvents

(i) *Viscosity*: 0.28 cP at 25°C (Bogyo *et al.*, 1980)

(j) *Volatility*: Vapour pressure, 439 mm Hg at 20°C

(k) *Stability*: Flash-point (closed-cup), -35°C; inflammable (Windholz, 1983)

(l) *Reactivity*: Very reactive (e.g., polymerizes, reacts with compounds having labile hydrogen)

(m) *Conversion factor*: 1 ppm = 2.37 mg/m^3 at 760 mm Hg and 25°C (Verschueren, 1977)

1.4 Technical products and impurities

Propylene oxide is available in the USA with the following typical specifications: water, 500 mg/kg max; total aldehydes, 100 mg/kg max; chlorides (as chlorine), 40 mg/kg max; acidity (as acetic acid), 20 mg/kg max; and density (25/25°C), 0.829-0.831 (Kirk & Dempsey, 1982).

Propylene oxide is available in western Europe with the following specifications: purity, 99.9% min; water, 200 mg/kg max; aldehydes (as propionaldehyde), 100 mg/kg max; and chlorine, 50 mg/kg max.

It has been reported that the chlorohydrin process (see section 2.1(*a*) below) can produce monochloroacetone, 1,2-dichloro-3-propanol, and propylene dichloride in small amounts. Acetaldehyde (see monograph, p. 101) and propionaldehyde are produced in small amounts as by-products of the peroxidation processes (see section 2.1(*a*) below) (Kirk & Dempsey, 1982).

2. Production, Use, Occurrence and Analysis

2.1 Production and use

(a) *Production*

Propylene oxide was first prepared in 1860 by Oser (Kirk & Dempsey, 1982) by the reaction of propylene chlorohydrin with potassium hydroxide (Windholz, 1983). Until 1969, essentially all the propylene oxide produced in the USA was made by the so-called chlorohydrin process in which propylene is treated with hypochlorous acid (chlorine and water) to produce propylene chlorohydrin; this is converted to propylene oxide using calcium hydroxide of sodium hydroxide. A plant using a version of the peroxidation process was started in the USA in 1969. Peroxidation processes use an oxidant such as an organic hydroperoxide (*tert*-butyl hydroperoxide or ethylbenzene hydroperoxide) or peracetic acid to convert propylene to propylene oxide. Currently, about one-half of US propylene oxide-production capacity is based on the chlorohydrin process, and the other half is based on the peroxidation process using an organic hydroperoxide (Kirk & Dempsey, 1982).

Propylene oxide was first produced commercially in the USA in 1925 (US Tariff Commission, 1926). US production reached a peak in 1979 when 1020 thousand tonnes were produced by six companies (US International Trade Commission, 1980). Total production by the four plants of the two US producing companies (one uses the chlorohydrin process and the other the peroxidation process) is estimated to have been 762 thousand tonnes in 1982. US imports of propylene oxide (mostly from Canada and Brazil) amounted to 23.1 thousand tonnes in 1982 (US Department of Commerce, 1983). US exports (mostly to Mexico, Australia and Japan) were 75.4 thousand tonnes in 1983 (US Department of Commerce, 1984).

Propylene oxide is produced by one company in Canada and one in Brazil.

It is produced by four companies in the Federal Republic of Germany, two in the Netherlands and one each in France, Italy and Spain. Six of these companies use the chlorohydrin process and three use the peroxidation process. Total western European production peaked at about 850 thousand tonnes in 1979 and amounted to about 810 thousand tonnes in 1982. At least one-half of the total was produced in Germany. Propylene oxide is also produced at two plants in Romania and at one plant each in Bulgaria, the German Democratic Republic, Poland, the USSR and Yugoslavia.

Commercial production of propylene oxide started in Japan in 1959. Five Japanese companies currently manufacture it at six plants (only one uses the peroxidation process), and production in 1982 is estimated to have been 186 thousand tonnes. Japanese imports amounted to 19 thousand tonnes in 1982 and exports totalled 10 thousand tonnes.

Propylene oxide is produced by one company in Taiwan and at one plant in India.

(b) Use

Almost all propylene oxide is used as a chemical intermediate, mainly in the production of polyols used in polyurethane manufacture and as a chemical intermediate in the manufacture of propylene glycol. The use pattern for the estimated 730 thousand tonnes of propylene oxide used in the USA in 1982 was as follows: polyurethane polyols, 64%; propylene glycol, 19%; dipropylene glycol, 2%; glycol ethers, 2%; and other uses (including minor quantities used in nonintermediate applications), 13%.

The polyols made from propylene oxide by reaction with glycols or other hydroxy compounds are used in the manufacture of polyurethane polymers. About half of the total is used in polyurethane foams (predominantly flexible foams) (see IARC, 1979), and the other half in non-foam applications (e.g., surface coatings, elastomers, etc.).

Propylene glycol, made by the reaction of propylene oxide with water, is used mainly for production of unsaturated polyester resins; a variety of smaller applications include use as a solvent for foods, drugs, cosmetics and surface coatings, and as a tobacco humectant.

Other uses for propylene oxide include use as a chemical intermediate in the manufacture of: dipropylene glycol (largely used for unsaturated polyester resin production); glycol ethers (solvents); nonurethane polyether polyols (mostly surfactants and lubricants); isopropanolamines; various surfactants (e.g., ethoxylated and propoxylated higher alcohols); propylene carbonate; and hydroxypropyl cellulose and similar products. It was also used in the USA until 1982 to make glycerol through allyl alcohol.

Propylene oxide itself is used as a fumigant, principally for sterilizing packaged food products in fumigation chambers (Berg, 1984). It has also been reported to have found use as a stabilizer in dichloromethane (US Environmental Protection Agency, 1984).

The use pattern for the estimated 830 thousand tonnes of propylene oxide used in western Europe in 1982 was as follows: polyols, 72%; propylene glycol, 23%; and other uses, 5%.

An estimated 190 thousand tonnes of propylene oxide were used in Japan in 1982, with the following use pattern: polypropylene glycol, 77%; and other uses (mainly propylene glycol), 23%.

Occupational exposure to propylene oxide has been limited by regulation or recommended guidelines in at least 12 countries. The standards are listed in Table 1.

The US Food and Drug Administration (1980) has approved use of propylene oxide as a direct and indirect food additive for the following purposes: (1) as an etherifying agent in the production of modified food starch (at use levels of 25% max or less); and (2) as a package fumigant for certain fruit products and as a fumigant for bulk quantities of several food pro-

Table 1. National occupational exposure limits for propylene oxide[a]

Country	Year	Concentration		Interpretation[b]	Status
		mg/m^3	ppm		
Australia	1978	240	100	TWA	Guideline
Belgium	1978	240	100	TWA	Regulation
Finland	1981	240	100	TWA[c]	Regulation
Germany, Federal Republic of	1984	120	50	TWA	Guideline
Italy	1978	240	100	TWA	Guideline
Netherlands	1978	240	100	TWA	Guideline
Romania	1975	100	-	TWA	Regulation
		200	-	Maximum	
Sweden	1981	12	5	TWA	Guideline
		25	10	STEL	
Switzerland	1978	240	100	TWA	Regulation
USA[d]					
OSHA	1978	240	100	TWA	Regulation
ACGIH	1984/85	50	20	TWA	Guideline
USSR	1977	1	-	Maximum[c]	Regulation
Yugoslavia	1971	240	100	Ceiling	Regulation

[a]From International Labour Office (1980); National Finnish Board of Occupational Safety and Health (1981); National Swedish Board of Occupational Safety and Health (1981); American Conference of Governmental Industrial Hygienists (1984); Deutsche Forschungsgemeinschaft (1984)
[b]TWA, time-weighted average; STEL, short-term exposure limit
[c]Skin irritant notation added
[d]OSHA, Occupational Safety and Health Administration; ACGIH, American Conference of Governmental Industrial Hygienists

ducts, provided residues of propylene oxide or propylene glycol do not exceed specified limits.

The US Environmental Protection Agency (1982) exempted propylene oxide from a tolerance on its residues on agricultural products when used as a stabilizer in accordance with good agricultural practice as an inert (or occasionally active) ingredient in pesticide formulations applied to growing crops, raw agricultural commodities after harvest, or animals. That Agency has also identified propylene oxide as a toxic waste and requires that persons who generate, transport, treat, store or dispose of it comply with the regulations of a federal hazardous waste management programme (US Environmental Protection Agency, 1980). The US Environmental Protection Agency (1983) also requires that notification be given whenever discharges containing 45.4 kg or more of propylene oxide are made into waterways.

As part of the Hazardous Materials Regulations of the US Department of Transportation (1982), shipments of propylene oxide are subject to a variety of labelling, packaging, quantity and shipping restrictions consistent with its designation as a hazardous material.

2.2 Occurrence

(a) Natural occurrence

Propylene oxide is not known to occur as a natural product.

(b) Occupational exposure

In its 1974 National Occupational Hazard Survey, the National Institute for Occupational Safety and Health (1981) found that US workers in 78 industries were exposed to propylene oxide.

Occupational exposure to propylene oxide was evaluated for two units of a large chemical manufacturing facility producing more than 200 chemical products, including derivatives of propylene oxide. Propylene oxide was not detectable (<0.1 mg/m^3) in all but one sample, which contained 3.6 mg/m^3 and was obtained for an inside operator at a flexible polyol unit (Oser et al., 1978). In a similar study, occupational exposure to propylene oxide was evaluated for three units of another large chemical manufacturing facility which produced derivatives of propylene oxide among its many chemical products. Levels of worker exposure were reported to be 0.5-5.9 mg/m^3 in the polymer polyol unit, not detectable (<0.6 mg/m^3)-1.2 mg/m^3 in the oxide adducts unit, and not detectable (<0.5 mg/m^3) in the flexible polyol unit (Oser et al., 1979). In a study by one US manufacturer in 1979, typical average daily exposures of workers to propylene oxide were 0.5-5 mg/m^3; worst-case peak exposures were 59-9000 mg/m^3, the highest exposure being that of maintenance workers cleaning pumps (Flores, 1983).

(c) Air

Propylene oxide has been tentatively identified in atmospheric air samples in the USA (Sawicki, 1976). It has also been observed as a product of the combustion of hydrocarbon fuels and in automobile exhausts (Bogyo et al., 1980).

Estimated US emissions of propylene oxide from its production and use in 1978 are shown in Table 2.

Table 2. Estimated emissions of propylene oxide in the USA in 1978[a]

Source	Estimated emission (kg)
Production	526 940
Urethane polyols	67 120
Propylene glycol	6 320
Surfactant polyols	7 190
Di-/tripropylene glycols	1 320
Glycol ethers	530
Miscellaneous	1 740
Total	611 160

[a]From Systems Applications, Inc. (1981)

(d) *Water*

Little, if any, propylene oxide is present in the waste-water from propylene oxide manufacture *via* the chlorohydrin process. It was detected in the effluent from one chemical plant in Bandenburg, KY, USA (Bogyo *et al.*, 1980).

(e) *Other*

Propylene oxide has been found as a trace level impurity in poly(propylene oxide) (Mokeyeva *et al.*, 1983).

2.3 Analysis

Methods used for the analysis of propylene oxide in a variety of matrices are listed in Table 3.

Table 3. Methods for the analysis of propylene oxide

Sample matrix	Sample preparation	Assay procedure[a]	Limit of detection	Reference
Air (workplace)	Adsorb (charcoal); desorb (carbon disulphide)	GC/FID	25 mg/m^3 (5-litre sample)	National Institute for Occupational Safety and Health (1977)
	Concentrate (permeable silicone membrane)	Microwave absorption spectrometry	5-16 mg/m^3	Breuer (1981)
Air (ambient)	Adsorb (Tenax GC); desorb thermally	GC/MS	60 ng/m^3	Krost *et al.* (1982)
Air	Adsorb (Porapak N porous polymer); desorb thermally	GC/FID	0.002 mg/m^3	Russell (1975)
Aqueous solution	React with sodium sulphide	Titration	Not given	Swan (1954)
	React with periodate; react with cadmium iodide-starch	Spectrophotometry	Not given	Mishmash & Meloan (1972)
Acetone solution	React with 4-(*p*-nitrobenzyl)-pyridine/triethylamine	Spectrophotometry	<0.6 µg	Agarwal *et al.* (1979)
Poly(propylene oxide)	Inject into headspace	GC	Not given	Mokeyeva *et al.* (1983)

[a]Abbreviations: GC/FID, gas chromatography/flame ionization detection; GC/MS, gas chromatography/mass spectrometry; GC, gas chromatography

3. Biological Data Relevant to the Evaluation of Carcinogenic Risk to Humans

3.1 Carcinogenicity studies in animals

(a) *Oral administration*

Rat: Groups of 50 female Sprague-Dawley rats, about 100 days old, were administered 15 or 60 mg/kg bw propylene oxide (purity, 99%) in a commercially available salad oil [composition unspecified] by gastric intubation twice weekly for 109.5 weeks (average total doses, 2714 and 10 798 mg/kg bw, respectively). Control groups consisted of 50 untreated rats and 50 rats treated with salad oil alone. The survival rates in rats treated with propylene oxide were comparable to those in the controls. Treatment with propylene oxide resulted in a dose-dependent increase in the incidence of local tumours, mainly squamous-cell carcinomas of the forestomach. No squamous-cell carcinoma was found in untreated or vehicle controls. The first tumour was observed in the 79th week in the high-dose group. The incidences of malignant stomach tumours (squamous-cell carcinomas of the forestomach) were 2/50 in the low-dose group; one animal in the high-dose group had an adenocarcinoma of the stomach and one had a carcinoma *in situ* of the forestomach. In addition, 7/50 low-dose and 17/50 high-dose animals developed papillomas, hyperplasia or hyperkeratosis of the forestomach. The incidences of tumours at various other sites in treated animals were not increased as compared to controls. A positive control group of 50 rats receiving 30 mg/kg bw β-propiolactone in salad oil by gastric intubation twice weekly for lifespan had a higher incidence of stomach tumours (46/50) [type unspecified] than animals treated with propylene oxide (Dunkelberg, 1982).

(b) *Inhalation*

Mouse: Groups of 50 male and 50 female B6C3F$_1$ mice, seven to nine weeks old, were exposed to 200 or 400 ppm (474 or 948 mg/m^3) propylene oxide (purity, >99.9%) vapour for six hours per day on five days per week for 103 weeks. Control groups of 50 male and 50 female mice were exposed in inhalation chambers to room air. The number of animals surviving to the end of the experiment was lower in the treated groups than in controls (males: control, 42/50, low-dose, 34/50, high-dose, 29/50; females: control, 38/50; low-dose, 29/50; high-dose, 10/50). Propylene oxide caused an increase in the incidence and severity of inflammation of the respiratory epithelium of the nasal turbinates; squamous metaplasia was observed in one low-dose male and in two high-dose female mice. One squamous-cell carcinoma and one papilloma occurred in the nasal cavity of two different high-dose male mice, and two high-dose female mice had adenocarcinomas of the nasal cavity. Three high-dose males and three high-dose females had a saccular dilatation of submucosal turbinate vessels (classified as angiectasis). In the high-dose group, haemangiomas developed in the nasal cavities of 5/50 male and 3/50 female mice, and haemangiosarcomas developed in the nasal cavities of 5/50 males and 2/50 females. The increased incidences of haemangiomas in males and females and of haemangiosarcomas in males were statistically significant. Vascular tumours were not observed in the nasal turbinates of low-dose or control mice (National Toxicology Program, 1984).

Rat: Groups of 50 male and 50 female Fischer 344/N rats, seven to eight weeks old, were exposed to 200 or 400 ppm (474 or 948 mg/m^3) propylene oxide (purity, >99.9%) vapour for six hours per day on five days per week for 103 weeks. Control groups of 50 male and 50 female rats were exposed in inhalation chambers to room air. Survival of rats exposed to propylene oxide was comparable with that of the controls. Between 29 and 35 animals per group survived to sacrifice; terminal body weights were lower in high-dose males and high-dose females than in controls. Increases in the incidences of suppurative inflammation, epithelial hyperplasia and squamous metaplasia of the respiratory epithelium of the nasal turbinates occurred in treated animals; papillary adenomas, involving the respiratory epithelium and underlying submucosal glands of the nasal turbinates, were observed in 3/50 female and 2/50 male rats exposed to 400 ppm propylene oxide. The incidence of adenomas in females was significantly different from that in controls as determined by trend tests. In high-dose female rats, the incidences of C-cell adenomas and C-cell carcinomas of the thyroid were increased, but only the combined incidence of these tumours was significant (control, 2/45; low-dose, 2/35; high-dose, 7/37) (National Toxicology Program, 1984).

Groups of 80 male weanling Fischer 344 rats were exposed to 0 (control; filtered air), 100 and 300 ppm (237 and 711 mg/m^3) propylene oxide (purity, 98%) vapour for about seven hours per day on five days per week for 104 weeks. Increased mortality was observed in the two groups of rats exposed to propylene oxide compared to the controls. Rats exposed to 100 or 300 ppm propylene oxide had an increased incidence of inflammatory lesions of the respiratory system and of a 'complex epithelial hyperplasia' in the nasal cavity, which was dose-dependent (controls, 0/76; low-dose 2/77; high-dose 11/78). In addition, two rats in the high-dose group developed nasal-cavity adenomas, which were not seen in controls. The proliferative lesions in the nasal mucosa appeared to be treatment-related, but the degree to which their development was influenced by the intercurrent inflammatory disease could not be ascertained. Adrenal phaeochromocytomas developed in 8/78 controls compared with 25/78 rats of the low-dose group and 22/80 rats of the high-dose group. The increased incidence of this tumour was statistically significant ($p < 0.05$) in the exposed animals. An increase in the incidence of peritoneal mesotheliomas was also found in the groups exposed to propylene oxide (controls, 3/78; low-dose, 8/78; high-dose, 9/80). The incidence of other neoplasms in the exposed animals was comparable to that of the controls (Lynch et al., 1984).

(c) *Subcutaneous and/or intramuscular administration*

Mouse: Groups of 100 female NMRI mice, six to eight weeks old, received subcutaneous injections of 0.1, 0.3, 1.0 or 2.5 mg/mouse propylene oxide (purity, 99%) in tricaprylin once a week for 95 weeks (mean total doses: 6.8, 21.7, 72.8 or 165.4 mg/mouse, respectively). Groups of 200 untreated and 200 tricaprylin-treated mice served as controls. Survival rates in the animals treated with propylene oxide were comparable to those in controls. Propylene oxide induced local tumours, mostly fibrosarcomas. The incidences of local sarcomas (fibrosarcomas, pleomorphic sarcomas) were 0/200 in untreated controls, 4/200 in tricaprylin controls and 3/100 (0.1 mg), 2/100 (0.3 mg), 12/100 (1.0 mg) and 15/100 (2.5 mg) in propylene oxide-treated animals [$p < 0.001$ by the Cochran Armitage test for trend]. The authors analysed the experiment by estimating the adjusted tumour incidence rates at 600 days and established a dose-response relationship for these rates. Tumours other than local sarcomas could not be related to propylene oxide treatment. In a positive-control group of 100 NMRI mice receiving weekly subcutaneous injections of 2.5 µg benzo[a]pyrene for 95 weeks, 81 animals developed local sarcomas (Dunkelberg, 1981).

3.2 Other relevant biological data

(a) *Experimental systems*

Toxic effects

The intragastric LD_{50}s of propylene oxide in rats and guinea-pigs were 1.14 and 0.69 g/kg bw, respectively (Smyth et al., 1941); the percutaneous LD_{50} in rabbits was 1.3 g/kg bw (Weil et al., 1963). The four-hour inhalational LC_{50}s of propylene oxide vapour were 4126 mg/m³ (1740 ppm) in mice and 9486 mg/m³ (4000 ppm) in rats (Jacobson et al., 1956).

No organ injury was seen in rats exposed by inhalation to calculated concentrations of 9480 mg/m³ (4000 ppm) propylene oxide vapour for 0.5 hour, to 4740 mg/m³ (2000 ppm) for two hours or to 2370 mg/m³ (1000 ppm) for seven hours; 4/10 rats died after four hours' exposure to the highest concentration (Rowe et al., 1956).

Exposure of rats and guinea-pigs to propylene oxide vapour produced eye irritation, nasal irritation, difficulty in breathing, drowsiness and weakness. While no toxic effect was seen in monkeys, rabbits, rats or male guinea-pigs exposed to 462 mg/m³ (195 ppm) propylene oxide vapour on five days per week for six or seven months, an increase in the average weight of lungs was observed in female guinea-pigs; no adverse effect was noted in the four species when exposed to 242 mg/m³ (102 ppm). With similar exposures to 1083 mg/m³ (457 ppm), no adverse effect was seen in rabbits and monkeys; irritation of the eye and respiratory passages was noted in rats and guinea-pigs (Rowe et al., 1956).

Repeated intragastric doses of 0.2 g/kg bw administered as a 10% solution in olive oil on five days per week for 24 days produced no toxic effect in rats (Rowe et al., 1956).

In rabbits, hyperaemia and oedema of the shaved skin resulted from contact with undiluted or aqueous solutions of propylene oxide (Rowe et al., 1956).

No compound-related gross or microscopic pathological effect was observed in mice or rats exposed to up to 500 ppm (1185 mg/m³) propylene oxide vapour for six hours per day on five days per week for 13 weeks. Suppurative inflammation and squamous metaplasia of the respiratory epithelium were observed in mice and rats exposed to 200 and 400 ppm (474 and 948 mg/m³) propylene oxide vapour at a similar exposure regimen for 104 weeks. The measured concentrations of gas agreed with the calculated ones (National Toxicology Program, 1984).

Effects on reproduction and prenatal toxicity

New Zealand white rabbits were exposed by inhalation to 0 (n = 17) or 500 ppm (1185 mg/m³) propylene oxide (purity, 99%) vapour for seven hours per day on either gestation days 7-19 (n = 11) or 1-19 (n = 19). Foetuses were examined on day 30. Food consumption, but not maternal body weight, was generally depressed in the treated groups during the periods of exposure. The overall resorption rate was not increased in the treated groups (0.71 resorptions per litter in the controls, 0.73 in the group exposed on days 7-19 and 1.58 in the group exposed on days 1-19). However, in females that had been exposed throughout gestation, the resorption rate was increased (4.3%) as compared to either controls (1.5%) or females exposed from days 7-19 only (1.3%) (Hackett et al., 1982).

Sprague-Dawley rats were exposed by inhalation to 0 (n = 46) or 500 ppm propylene oxide (purity, 99%) vapour for seven hours per day either from three weeks prior to gestation to day 16 of gestation (n = 43), on days 1-16 of gestation (n = 41) or on days 7-16 of gestation (n = 44). Foetuses were examined on day 21. Food consumption was reduced in females that received the pregestational exposure, and maternal weight gain tended to be lower in all treated groups during exposure. The numbers of corpora lutea (13.8 compared to 15.4 in controls) and implantation sites per dam (12.3 compared to 13.9 in controls) and, consequently, the number of live foetuses per litter (11.7 compared to 13.0 in controls) were lower in the rats receiving pregestational exposure. Foetal growth was reduced in rats receiving only the gestational exposures. No treatment-related major malformation was seen, but the incidence of rib dysmorphology (primarily wavy ribs) was increased in all propylene oxide-treated groups (Hackett et al., 1982).

Absorption, distribution, excretion and metabolism

Propylene oxide reacts with DNA at neutral pH to yield two principal products, *N*-7-(2-hydroxypropyl)guanine and *N*-3-(2-hydroxypropyl)adenine (Lawley & Jarman, 1972). In-vitro reaction products of propylene oxide have been reported with deoxyadenosine and deoxyguanosine (Hemminki *et al.*, 1980), deoxycytidine (Djuric & Sinsheimer, 1984a) and thymidine (Djuric & Sinsheimer, 1984b).

Mutagenicity and other short-term tests (see also 'Appendix: Activity Profiles for Short-term Tests', p. 340)

Propylene oxide did not induce mutations in T_2 bacteriophage (Cookson *et al.*, 1971). It was mutagenic to *Salmonella typhimurium* TA1535 and TA100 both in the presence and absence of an exogenous metabolic system; no activity was observed in strains TA1537 or TA98. It was also mutagenic to *Escherichia coli* WP2 and CM891 both in the presence and absence of an Aroclor-induced rat-liver metabolic system (Wade *et al.*, 1978; Bootman *et al.*, 1979; Pfeiffer & Dunkelberg, 1980; Yamaguchi, 1982).

Propylene oxide induced mutations in *Klebsiella pneumoniae* (Voogd *et al.*, 1981) and in *Neurospora crassa* (Kølmark & Giles, 1955). In *Schizosaccharomyces pombe*, the induction of mutations by propylene oxide was slightly reduced by a metabolic system from phenobarbital-induced mouse liver (Migliore *et al.*, 1982).

Sex-linked recessive lethal mutations were observed in *Drosophila melanogaster* exposed to 1530 mg/m^3 (645 ppm) propylene oxide vapour for 24 hours (Hardin *et al.*, 1983).

Propylene oxide induced DNA single-strand breaks in rat hepatocytes *in vitro*, as evaluated by alkaline elution (Sina *et al.*, 1983).

Propylene oxide induced chromosomal aberrations in an epithelial-type cell line (RL_1) derived from rat liver (Dean & Hodson-Walker, 1979) and in human lymphocytes treated *in vitro* (Bootman *et al.*, 1979).

An increase in the incidence of micronucleated cells in the bone marrow was observed in male CD-1 mice given two intraperitoneal injections of 300 mg/kg bw propylene oxide 30 and six hours prior to killing; no such increase was seen in mice receiving intraperitoneal injections of 75 or 150 mg/kg bw, nor in mice given two doses (30 and six hours prior to killing) of 100, 250 or 500 mg/kg bw propylene oxide by gavage (Bootman *et al.*, 1979).

No dominant lethal mutation was observed in the offspring of male CD-1 mice treated orally with 50 or 250 mg/kg bw propylene oxide every 24 hours for 14 days and mated with virgin females every week for six weeks (Bootman *et al.*, 1979), nor in Sprague-Dawley rats treated by inhalation with concentrations of 305 ppm (290-323 ppm) (723/687-765 mg/m^3) propylene oxide (purity, 98%) vapour for seven hours per day for five days and mated with virgin females every five days for 30 days; propylene oxide did not lead to sperm abnormalities in mice (Hardin *et al.*, 1983).

(b) Humans

Toxic effects

Three cases of corneal burns from exposure to propylene oxide have been described (McLaughlin, 1946).

Effects on reproduction and prenatal toxicity

No data were available to the Working Group.

Absorption, distribution, excretion and metabolism

N^T-(2-Hydroxypropyl)histidine was determined in haemoglobin of workers producing hydroxypropylated starch, as a measure of exposure. In those exposed to propylene oxide, the level of alkylhistidine was 4.5-13 nmol/g haemoglobin, as compared to 0.1-0.38 nmol/g haemoglobin in workers exposed to ethylene oxide and <0.1 nmol/g haemoglobin in non-exposed workers (Osterman-Golkar *et al.*, 1984).

Mutagenicity and chromosomal effects (see also 'Appendix: Activity Profiles for Short-term Tests', p. 340)

Lymphocytes obtained from workers exposed to a mixture of alkylene oxides including ethylene oxide, propylene oxide, butylene oxide, dioxane, epichlorohydrin, dichloropropane, ethylene chlorohydrin and propylene chlorohydrin, showed an increased incidence of chromosomal aberrations (Thiess *et al.*, 1981a). [The Working Group noted the unquantified mixed exposure and that a culture time of 70-72 hours was used, which is longer than that currently recommended (Evans, 1984).]

3.3 Case reports and epidemiological studies of carcinogenicity to humans

Thiess *et al.* (1981b) reported a mortality study of 602 active employees and former employees who had worked for six months or more in an area of alkylene oxide production in Germany. The men had been exposed to ethylene oxide and, later, to propylene oxide, as well as to benzene and ethylene chlorohydrin. This study is reported in detail in the monograph on ethylene oxide (p. 189). It included workers from 1928, although production of propylene oxide began only in 1959. No industrial hygiene measurement of this compound was reported. The authors stated that workers were regularly in brief contact with propylene oxide, ethylene oxide and other substances during sampling operations and when filters were changed. The observed and expected numbers of cancer deaths did not differ significantly. [The Working Group questioned the calculation of expected numbers (see monograph on ethylene oxide). Since propylene oxide production started 30 years after the beginning of the observation period, only a portion of the study population would have been exposed to propylene oxide.

4. Summary of Data Reported and Evaluation

4.1 Exposure data

Propylene oxide has been produced commercially since 1925 and is produced in large quantities in several countries. It is used primarily as an intermediate and also finds limited use as a fumigant, principally for sterilizing packaged food products.

4.2 Experimental data

Propylene oxide was tested by oral gavage in rats of one strain and produced local tumours, mainly squamous-cell carcinomas and papillomas of the forestomach, in a dose-dependent manner. It was tested by inhalation in mice of one strain and in two experiments in rats of one strain. It produced haemangiomas and haemangiosarcomas of the

Overall assessment of data from short-term tests: propylene oxide[a]

	Genetic activity			Cell transformation
	DNA damage	Mutation	Chromosomal effects	
Prokaryotes		+		
Fungi/green plants		+		
Insects		+		
Mammalian cells (*in vitro*)	+		+	
Mammals (*in vivo*)			+	
Humans (*in vivo*)				
Degree of evidence in short-term tests for genetic activity: *Sufficient*				Cell transformation: No data

[a]The groups into which the table is divided and the symbol + are defined on pp. 17-18 of the Preamble; the degrees of evidence are defined on p. 18.

nasal turbinates in mice; an increased incidence of papillary adenomas of the nasal turbinates was observed in rats in both studies. In one experiment in male rats, an increased incidence of adrenal phaeochromocytomas and of peritoneal mesotheliomas was observed. Propylene oxide was also tested by subcutaneous administration in mice of one strain and induced local sarcomas, mainly fibrosarcomas, in a dose-dependent manner.

Exposure of rats by inhalation to propylene oxide at maternally toxic doses decreased the number of corpora lutea, implantation sites and live foetuses. No teratogenic effect was observed in rats or rabbits.

Propylene oxide was mutagenic to bacteria, fungi and insects. It induced DNA damage and chromosomal aberrations in cultured mammalian cells, and micronuclei in bone-marrow cells of mice *in vivo*. Dominant lethal mutations were not induced in mice or rats.

4.3 Human data

A study of workers, some of whom were exposed to propylene oxide but also to other chemicals, was inconclusive in relation to the carcinogenicity of propylene oxide.

No adequate data were available to the Working Group on mutagenic or chromosomal effects of propylene oxide in humans.

4.4 Evaluation[1]

There is *sufficient evidence*[2] for the carcinogenicity of propylene oxide to experimental animals.

There is *inadequate evidence* for the carcinogenicity of propylene oxide to humans.

---

[1]For definitions of the italicized terms, see the Preamble, pp. 15-16.
[2]In the absence of adequate data on humans, it is reasonable, for practical purposes, to regard chemicals for which there is *sufficient evidence* of carcinogenicity in experimental animals as if they represented a carcinogenic risk to humans.

5. References

Agarwal, S.C., Van Duuren, B.L. & Kneip, T.J. (1979) Detection of epoxides with 4-(*p*-nitrobenzyl)pyridine. *Bull. environ. Contam. Toxicol.*, 23, 825-829

American Conference of Governmental Industrial Hygienists (1984) *TLVs Threshold Limit Values for Chemical Substances in the Work Environment Adopted by ACGIH for 1984-85*, Cincinnati, OH, p. 28

Berg, G.L., ed. (1984) *Farm Chemicals Handbook, 1984*, Willoughby, OH, Meister Publishing Co., p. C190

Bogyo, D.A., Lande, S.S., Meylan, W.M., Howard, P.H. & Santodonato, J. (1980) *Investigation of Selected Potential Environmental Contaminants: Epoxides (EPA-560/11-80-005)*, Washington DC, US Environmental Protection Agency, pp. 4, 66-67, 96, 104, 162

Bootman, J., Lodge, D.C. & Whalley, H.E. (1979) Mutagenic activity of propylene oxide in bacterial and mammalian systems. *Mutat. Res.*, 67, 101-112

Breuer, G.M. (1981) *Evaluation of a Prototype Portable Microwave Multi-Gas Analyzer*, Cincinnati, OH, National Institute for Occupational Safety and Health

Cookson, M.J., Sims, P. & Grover, P.L. (1971) Mutagenicity of epoxides of polycyclic hydrocarbons correlates with carcinogenicity of the parent hydrocarbons. *Nature-New Biol.*, 234, 186-187

Dean, B.J. & Hodson-Walker, G. (1979) An in vitro chromosome assay using cultured rat-liver cells. *Mutat. Res.*, 64, 329-337

Deutsche Forschungsgemeinschaft (1984) *Maximal Work Place Concentrations and Biological Tolerance Values for Compounds in the Work Place* (Ger.), Part XX, Weinheim, Verlag Chemie GmbH, p. 48

Djuric, Z. & Sinsheimer, J.E. (1984a) Reactivity of propylene oxides towards deoxycytidine and identification of reaction products. *Chem.-biol. Interactions*, 50, 219-231

Djuric, Z. & Sinsheimer, J.E. (1984b) Characterization and quantitation of 3-alkylthymidines from reactions of mutagenic propylene oxides with thymidine. *Nucleic Acids Res.* (in press)

Dunkelberg, H. (1981) Carcinogenic activity of ethylene oxide and its reaction products 2-chloroethanol, 2-bromoethanol, ethylene glycol and diethylene glycol. I. Carcinogenicity of ethylene oxide in comparison with 1,2-propylene oxide after subcutaneous administration in mice (Ger.). *Zbl. Bakt. Hyg., I. Abt. Orig. B*, 174, 383-404

Dunkelberg, H. (1982) Carcinogenicity of ethylene oxide and 1,2-propylene oxide upon intragastric administration to rats. *Br. J. Cancer*, 46, 924-933

Evans, H.J. (1984) *Human peripheral blood lymphocytes for the analysis of chromosome aberrations in mutagen tests*. In: Kilbey, B.J., Legator, M., Nichols, W. & Ramel, C., eds, *Handbook of Mutagenicity Test Procedures*, Amsterdam, Elserier, pp. 405-427

Flores, G.H. (1983) Controlling exposure to alkylene oxides. *Chem. Eng. Prog.*, *79*, 39-43

Hackett, P.L., Brown, M.G., Buschbom, R.L., Clark, M.L., Miller, R.A., Music, R.L., Rowe, S.E., Schirmer, R.E. & Sikov, M.R. (1982) *Teratogenic Study of Ethylene and Propylene Oxide and n-Butyl Acetate (NIOSH Contract No. 210-80-0013)*, Cincinnati, OH, National Institute for Occupational Safety and Health

Hardin, B.D., Schuler, R.L., McGinnis, P.M., Niemeier, R.W. & Smith, R.J. (1983) Evaluation of propylene oxide for mutagenic activity in 3 in vivo test systems. *Mutat. Res.*, *117*, 337-344

Hemminki, K., Paasivirta, J., Kurkirinne, T. & Virkki, L. (1980) Alkylation products of DNA bases by simple epoxides. *Chem.-biol. Interactions*, *30*, 259-270

IARC (1976) *IARC Monographs on the Evaluation of Carcinogenic Risk of Chemicals to Man*, Vol. 11, *Cadmium, Nickel, Some Epoxides, Miscellaneous Industrial Chemicals and General Considerations on Volatile Anaesthetics*, Lyon, pp. 191-199

IARC (1979) *IARC Monographs on the Evaluation of the Carcinogenic Risk of Chemicals to Humans*, Vol. 19, *Some Monomers, Plastics and Synthetic Elastomers, and Acrolein*, Lyon, pp. 303-340

International Labour Office (1980) *Occupational Exposure Limits for Airborne Toxic Substances*, 2nd (rev.) ed. (*Occupational Safety and Health Series No. 37*), Geneva, pp. 182, 183

Jacobson, K.H., Hackley, E.B. & Feinsilver, L. (1956) The toxicity of inhaled ethylene oxide and propylene oxide vapors. *Arch. ind. Health*, *13*, 237-244

Kirk, R.O. & Dempsey, T.J. (1982) *Propylene oxide*. In: Grayson, M., ed., *Kirk-Othmer Encyclopedia of Chemical Technology*, 3rd ed., Vol. 19, New York, John Wiley & Sons, pp. 246-274

Kølmark, G. & Giles, N.H. (1955) Comparative studies of monoepoxides as inducers of reverse mutations in *Neurospora*. *Genetics*, *40*, 890-902

Krost, K.J., Pellizzari, E.D., Walburn, S.G. & Hubbard, S.A. (1982) Collection and analysis of hazardous organic emissions. *Anal. Chem.*, *54*, 810-817

Lawley, P.D. & Jarman, M. (1972) Alkylation by propylene oxide of deoxyribonucleic acid, adenine, guanosine and deoxyguanylic acid. *Biochem. J.*, *126*, 893-900

Lynch, D.W., Lewis, T.R., Moorman, W.J., Burg, J.A., Groth, D.H., Khan, A., Ackerman, L.J. & Cockrell, B.Y. (1984) Carcinogenic and toxicologic effects of inhaled ethylene oxide and propylene oxide in F344 rats. *Toxicol. appl. Pharmacol.*, *76*, 69-84

McLaughlin, R.S. (1946) Chemical burns of the human cornea. *Am. J. Ophthalmol.*, *29*, 1355-1362

Migliore, L., Rossi, A.M. & Loprieno, N. (1982) Mutagenic action of structurally related alkene oxides on *Schizosaccharomyces pombe*: The inflence, '*in vitro*', of mouse-liver metabolizing system. *Mutat. Res.*, *102*, 425-437

Mishmash, H.E. & Meloan, C.E. (1972) Indirect spectrophotometric determination of nanomole quantities of oxiranes. *Anal. Chem.*, *44*, 835, 836

Mokeyeva, R.N., Tsarfin, Y.A. & Ernst, W. (1983) Determination of volatile microimpurities in alkaline polymerizates and neutral propylene oxide polyethers by gas chromatographic headspace analysis. *J. Chromatogr.*, *264*, 272-278

National Finnish Board of Occupational Safety and Health (1981) *Airborne Contaminants in the Work Places* (*Safety Bull. No. 3*), Helsinki, pp. 11, 31

National Institute for Occupational Safety and Health (1977) *NIOSH Manual of Analytical Methods. Part II. Standards Completion Program Validated Methods*, 2nd ed., Vol. 2 (*DHEW (NIOSH) Publication No. 77-157-B*), Cincinnati, OH, pp. S75-1 - S75-8

National Institute for Occupational Safety and Health (1981) *National Occupational Hazard Survey* (microfiche), Cincinnati, OH, pp. 7760-7761 (generics excluded)

National Swedish Board of Occupational Safety and Health (1981) *Hygienic Limit Values* (*AFS 1981:8*), Stockholm, p. 20

National Toxicology Program (1984) *Toxicology and Carcinogenesis Studies of Propylene Oxide (CAS No. 75-56-9) in F344/N Rats and B6C3F$_1$ Mice (Inhalation Study)* (*NTP-83-020; NIH Publ. No. 84-2523*), Research Triangle Park, NC, US Department of Health and Human Services

Oser, J., Crandall, M., Phillips, R. & Marlow, D. (1978) *Indepth Industrial Hygiene Report of Ethylene Oxide Exposure at Union Carbide Corporation Institute, West Virginia* (*PB82-114786*), Springfield, VA, National Technical Information Service, pp. 14-15, 22

Oser, J., Young, M., Boyle, T. & Marlow, D. (1979) *Indepth Industrial Hygiene Report of Ethylene Oxide Exposure at Union Carbide Corporation, South Charleston, West Virginia* (*PB82-110024*), Springfield, VA, National Technical Information Service, pp. 6, 13

Osterman-Golkar, S., Bailey, E., Farmer, P.B., Gorf, S.M. & Lamb, J.H. (1984) Monitoring exposure to propylene oxide through the determination of hemoglobin alkylation. *Scand. J. Work Environ. Health*, *10*, 99-102

Pfeiffer, E.H. & Dunkelberg, H. (1980) Mutagenicity of ethylene oxide and propylene oxide and of the glycols and halohydrins formed from them during the fumigation of foodstuffs. *Food Cosmet. Toxicol.*, *18*, 115-118

Rowe, V.K., Hollingsworth, R.L., Oyen, F., McCollister, D.D. & Spencer, H.C. (1956) Toxicity of propylene oxide determined on experimental animals. *Arch. ind. Health*, *13*, 228-236

Russell, J.W. (1975) Analysis of air pollutants using sampling tubes and gas chromatography. *Environ. Sci. Technol.*, *9*, 1175-1178

Sawicki, E. (1976) Analysis of atmospheric carcinogens and their cofactors. In: Rosenfeld, C. & Davis, W., eds, *Environmental Pollution and Carcinogenic Risks* (*INSERM Symposium Series Vol. 52/IARC Scientific Publications No. 13*), Paris, Institut National de la Santé et de la Recherche Médicale/Lyon, International Agency for Research on Cancer, pp. 315, 316

Sina, J.F., Bean, C.L., Dysart, G.R., Taylor, V.I. & Bradley, M.O. (1983) Evaluation of the alkaline elution/rat hepatocyte assay as a predictor of carcinogenic/mutagenic potential. *Mutat. Res.*, *113*, 357-391

Smyth, H.F., Jr, Seaton, J. & Fischer, L. (1941) The single dose toxicity of some glycols and derivatives. *J. ind. Hyg. Toxicol.*, *23*, 259-268

Swan, J.D. (1954) Determination of epoxides with sodium sulfite. *Anal. Chem.*, *26*, 878-880

Systems Applications, Inc. (1981) *Human Exposure to Atmospheric Concentrations of Selected Chemicals*, Vol. II, Appendix A-26, Research Triangle Park, NC, US Environmental Protection Agency

Thiess, A.M., Schwegler, H., Fleig, I. & Stocker, W.G. (1981a) Mutagenicity study of workers exposed to alkylene oxides (ethylene oxide/propylene oxide) and derivatives. *J. occup. Med.*, *23*, 343-347

Thiess, A.M., Frentzel-Beyme, R., Link, R. & Stocker, W.G. (1981b) *Mortality study on employees exposed to alkylene oxides (ethylene oxide/propylene oxide) and their derivatives*. In: *Prevention of Occupational Cancer - International Symposium* (*Occupational Safety and Health Series No. 46*), Geneva, International Labour Office, pp. 249-259

US Department of Commerce (1983) *US Imports for Consumption and General Imports* (*FT246/Annual 1982*), Bureau of the Census, Washington DC, US Government Printing Office, p. 1-291

US Department of Commerce (1984) *US Exports, Schedule E Commodity by Country* (*FT410/December 1983*), Bureau of the Census, Washington DC, US Government Printing Office, p. 2-78

US Department of Transportation (1982) Performance-oriented packagings standards. *US Code Fed. Regul., Title 49*, Parts 171, 172, 173, 178; *Fed. Regist.*, *47* (No. 73), pp. 16268, 16273, 16279

US Environmental Protection Agency (1980) Hazardous waste management system: Identification and listing of hazardous wastes. *US Code Fed. Regul., Title 40*, Part 261; *Fed. Regist.*, *45* (No. 98), pp. 33084, 33122-33127, 33131-33133

US Environmental Protection Agency (1982) Protection of environment. *US Code Fed. Regul., Title 40*, Parts 116.4, 180.1001

US Environmental Protection Agency (1983) Notification requirements; Reportable quantity adjustments. *US Code Fed. Regul., Title 40*, Part 302; *Fed. Regist.*, *48* (No. 102), pp. 23552, 23592

US Environmental Protection Agency (1984) Propylene oxide; Proposed test rule. *US Code Fed. Regul., Title 40*, Part 799; *Fed. Regist., 49* (No. 2), pp. 430-438

US Food and Drug Administration (1980) Food and drugs. *US Code Fed. Regul., Title 21*, Parts 172.892, 193.380

US International Trade Commission (1980) *Synthetic Organic Chemicals, US Production and Sales, 1979* (*USITC Publication 1099*), Washington DC, US Government Printing Office, pp. 269, 299

US Tariff Commission (1926) *Census of Dyes and Other Synthetic Organic Chemicals, 1925* (*Tariff Information Series No. 26*), Washington DC, US Government Printing Office, p. 147

Verschueren, K. (1977) *Handbook of Environmental Data on Organic Compounds*, New York, Van Nostrand Reinhold Co., p. 544

Voogd, C.E., van der Stell, J.J. & Jacobs, J.J.J.A.A. (1981) The mutagenic action of aliphatic epoxides. *Mutat. Res., 89*, 269-282

Wade, D.R., Airy, S.C. & Sinsheimer, J.E. (1978) Mutagenicity of aliphatic epoxides. *Mutat. Res., 58*, 217-223

Weil, C.S., Condra, N., Haun, C. & Striegel, J.A. (1963) Experimental carcinogenicity and acute toxicity of representative epoxides. *Am. ind. Hyg. Assoc. J., 24*, 305-325

Windholz, M., ed. (1983) *The Merck Index*, 10th ed., Rahway, NJ, Merck & Co., p. 1131

Yamaguchi, T. (1982) Mutagenicity of trioses and methyl glyoxal on *Salmonella typhimurium*. *Agric. biol. Chem., 46*, 849-851

STYRENE OXIDE

This substance was considered by previous working groups, in February 1976 (IARC, 1976), February 1978 (IARC, 1979a) and February 1982 (IARC, 1982a). Since that time, new data have become available, and these have been incorporated into the monograph and taken into account in the present evaluation.

1. Chemical and Physical Data

1.1 Synonyms and trade names

Chem. Abstr. Services Reg. No.: 96-09-3

Chem. Abstr. Name: Oxirane, phenyl-

IUPAC Systematic Name: (Epoxyethyl)benzene

Synonyms: Epoxyethylbenzene; 1,2-epoxyethylbenzene; (1,2-epoxyethyl)benzene; epoxystyrene; α,β-epoxystyrene; NCI-C54977; phenethylene oxide; 1-phenyl-1,2-epoxyethane; phenylethylene oxide; phenyl oxirane; phenyloxirane; 1-phenyloxirane; 2-phenyloxirane; styrene epoxide; styrene 7,8-oxide; styrene-7,8-oxide; styryl oxide

1.2 Structural and molecular formulae and molecular weight

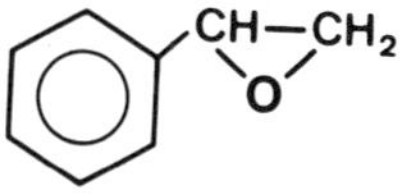

C_8H_8O Mol. wt: 120.2

1.3 Chemical and physical properties of the pure substance

From National Research Council (1981), unless otherwise specified

(a) *Description*: Colourless to pale, straw-coloured liquid (Hawley, 1981)

(b) *Boiling-point*: 194.1°C

(c) *Freezing-point*: -36.8°C

(d) *Density*: Specific gravity (20/20°C), 1.0540

(e) *Refractive index*: n_D^{20} 1.5339

(f) *Spectroscopy data*: Ultraviolet, infrared, nuclear magnetic resonance and mass spectral data have been reported

(g) *Solubility*: Slightly soluble in water (0.3%); completely soluble in acetone, benzene, carbon tetrachloride, diethyl ether, heptane and methanol

(h) *Viscosity*: 1.99 cP at 20°C

(i) *Volatility*: Vapour pressure, 0.3 mm Hg at 20°C

(j) *Stability*: Flash-point, 80°C

(k) *Reactivity*: Reacts with compounds having labile hydrogen

(l) *Conversion factor*: 1 ppm = 4.91 mg/m^3 at 760 mm Hg and 25°C

1.4 Technical products and impurities

Typical properties of the styrene oxide available from the sole US producing company are: purity, 97.5%; apparent specific gravity (20/20°C), 1.0540; boiling-point, 194.0°C; vapour pressure, <1 mm Hg at 20°C; freezing-point, -36.8°C; and solubility in water (20°C), 0.3% (Union Carbide Corporation, 1979).

2. Production, Use, Occurrence and Analysis

2.1 Production and use

(a) *Production*

Styrene oxide was first made in 1905 by the reaction of styrene iodohydrin with potassium hydroxide (Fourneau & Tiffeneau, 1905). It is believed to be produced commercially either by treating styrene chlorohydrin with alkali or by epoxidizing styrene (see IARC, 1979b) with peracetic acid (Lapkin, 1967).

Commercial production of styrene oxide in the USA was first reported in 1974 (US International Trade Commission, 1976). Currently, the sole producer is estimated to produce 450-900 thousand kg per year. Imports of styrene oxide through the principal US customs districts have been reported in only two years. In 1979, they amounted to 24 thousand kg (US International Trade Commission, 1980), and in 1982, they totalled 276 thousand kg (US International Trade Commission, 1983). Separate data on US exports of styrene oxide are not published.

Commercial production in Japan of styrene oxide started in 1964. Two Japanese companies currently manufacture styrene oxide, and their combined 1983 production is estimated to have been 3-4 million kg, about one million kg of which was exported.

(b) Use

Styrene oxide is used as a reactive plasticizer or diluent in epoxy resins to lengthen the polymer segments between cross-links and to produce a slight softening and flexibility with improved impact strength (Sears & Touchette, 1982).

Styrene oxide is also used as a chemical intermediate in one of the commercial processes for making β-phenethyl alcohol, a fragrance material. Ringk and Theimer (1978) reported that 20% of all β-phenethyl alcohol is produced by this process, and they estimated that total US production of β-phenethyl alcohol amounted to 1.2 million kg in 1976 (this would mean that 240 thousand kg were made from styrene oxide).

Styrene oxide has also been used to make a polymer with linoleic acid dimer, ethylene diamine and 2-ethoxyethyl acetate. It may also have been used as a chemical intermediate to make special polyols.

In Japan, approximately half of the styrene oxide is used to make β-phenethyl alcohol, and the remainder is used in epoxy resins and other applications.

Styrene oxide has been approved by the US Food and Drug Administration (1980) for use as a cross-linking agent for epoxy resins in coatings for containers with a volume of 1000 gallons (3785 l) or more intended for repeated use in contact with alcoholic beverages containing up to 8% of ethanol by volume.

2.2 Occurrence

(a) Occupational exposure

In its 1974 National Occupational Hazard Survey, the National Institute for Occupational Safety and Health (1981) reported that US workers in two industries were exposed to styrene oxide. The principal industry in which exposure was found was fabricated rubber products; some exposure was also noted in the paints and allied products industry.

Exposure to styrene oxide can occur when styrene is used in the reinforced plastics industry: styrene oxide was found in the air of a plant in Finland. The average breathing-zone concentration during lamination processes was 0.2 mg/m^3 during hand application (the simultaneous concentration of styrene was 560 mg/m^3) and 0.6 mg/m^3 during spray application (the simultaneous concentration of styrene was 550 mg/m^3) (Pfäffli *et al.*, 1979). Styrene oxide concentrations ranging from <0.02-0.6 mg/m^3 were observed in the air during the manufacture of reinforced polyester plastics in Norway (the simultaneous concentrations of styrene ranged from 70-1200 mg/m^3) (Fjeldstad *et al.*, 1979).

(b) Air

Styrene oxide has been tentatively identified in air samples collected in the Los Angeles Basin, USA, using gas chromatography and mass spectrometry analysis (Pellizzari *et al.*, 1976). Sawicki (1976) has also reported the tentative identification of styrene oxide in US atmospheric air samples using similar analytical methods.

(c) Water

Styrene oxide has been detected in effluent-water from latex manufacturing plants in Louisville, KY, and from chemical manufacturing plants in Louisville and in Memphis, TN (Shackelford & Keith, 1976).

(d) Tobacco and tobacco smoke

Styrene oxide has been detected as a volatile component of a Burley tobacco concentrate (Demole & Berthet, 1972).

(e) Biological fluids

Low concentrations of styrene oxide (0.05 μmol/l) have been detected in the venous blood of four workers exposed to styrene of unspecified purity (Wigaeus et al., 1983).

(f) Other

Styrene oxide has been detected as an impurity in commercial samples of styrene chlorohydrin (Dolgopolov & Lishcheta, 1971).

2.3 Analysis

Reported methods for the analysis of styrene oxide in a variety of matrices are listed in Table 1.

Table 1. Methods for the analysis of styrene oxide

Sample matrix	Sample preparation	Assay procedure[a]	Limit of detection	Reference
Ambient air	Collect on sorbent; desorb thermally	GC/MS	2 ng/m^3	Pellizzari et al. (1976); Krost et al. (1982)
Workplace air	Collect on sorbent; extract (ethyl acetate)	GC/FID	0.2 ng in extract (0.1 μg/sample)	Stampfer & Hermes (1981)
	Collect on charcoal; extract (dichloromethane)	GC/FID; GC/MS	Not given	Pfäffli et al. (1979)
Drinking water	Concentrate; extract (ethanol); react with 4-nitrothiophenol	HPLC/UV	Not given	Cheh & Carlson (1981)
Biological media	Form picrate	GC/FID or TLC	Not given	Leibman & Ortiz (1970)
Mouse blood	Extract (dichloromethane); use para-methylanisole as an internal standard	GC/FID or GC/MS	10 ng/ml	Bidoli et al. (1980)
Rat-liver homogenate	React with nicotinamide; incubate	Fluorimetry	24-60 ng	Nelis & Sinsheimer (1981)
Commercial styrene chlorohydrin		TLC/Spectrophotometry	1.5 μg	Dolgopolov & Lishcheta (1971)
Aqueous solution	React with periodate; react with cadmium iodide-starch	Spectrophotometry	Not given	Mishmash & Meloan (1972)
	React with sodium sulphite	Titration	Not given	Swan (1954)
Acetone solution	React with 4-(p-nitrobenzyl)-pyridine/triethylamine	Spectrophotometry	12 μg max	Agarwal et al. (1979)

[a]Abbreviations: GC/MS, gas chromatography/mass spectrometry; GC/FID, gas chromatography/flame ionization detection; HPLC/UV, high-performance liquid chromatography/ultraviolet absorbance detection; TLC, thin-layer chromatography

3. Biological Data Relevant to the Evaluation of Carcinogenic Risk to Humans

3.1 Carcinogenicity studies in animals[1]

(a) Oral administration

Rat: Groups of 40 male and 40 female Sprague-Dawley rats, 13 weeks old, received 50 or 250 mg/kg bw styrene oxide [purity unspecified] in olive oil by intragastric intubation once daily, on four to five days per week for 52 weeks. Control groups of 40 male and 40 female rats received olive oil alone. The preliminary results reported refer to an observation period of 135 weeks from the start of the bioassay and are concerned solely with oncological lesions of the stomach. Treatment with styrene oxide induced a dose-dependent increase in the incidence of squamous-cell carcinomas and papillomas of the forestomach. The first epithelial tumour of the forestomach appeared after 51 weeks. The incidences (referring to corrected numbers of animals alive at 51 weeks) of squamous-cell carcinomas (invasive plus *in situ*) were 0/37 in control, 6/31 in low-dose and 12/28 in high-dose males, respectively; and 0/28, 6/31 and 15/30, respectively, in females. The incidences of papillomas of the forestomach were 0/37, 0/31 and 3/28, respectively, in males; and 0/28, 2/31 and 6/30, respectively, in females (Maltoni *et al.*, 1979).

(b) Skin application

Mouse: A group of 40 13-week-old C3H mice received skin applications by brush of a 5% solution of styrene oxide in acetone on the clipped dorsal skin thrice weekly for life. No skin tumour was observed at 24 months; 37 mice survived 12 months, 33 mice survived 17 months and 17 mice survived 24 months. Another group of C3H mice was similarly treated with a 10% solution of styrene oxide in acetone; 18 mice survived 12 months, only two mice survived to 17 months, and no skin tumour was observed (Weil *et al.*, 1963).

Of 30 male Swiss ICR/Ha mice, eight weeks of age at the start of the treatment, given thrice-weekly applications of 100 mg of a 10% solution of styrene oxide in benzene on the clipped dorsal skin for life, three developed skin tumours; one of these was a squamous-cell carcinoma. The median survival time was 431 days. Of 150 benzene-painted controls, 11 developed skin tumours, one of which was a squamous-cell carcinoma. The study was considered to be negative by life-table analysis (Van Duuren *et al.*, 1963).

(c) Pre- and post-natal administration

Rat: A group of 14 female BDIV inbred rats [age unspecified] were given 200 mg/kg bw styrene oxide (purity, 97%) *orally* by gavage on day 17 of pregnancy. Their offspring (62 females and 43 males) received 96 weekly oral doses of styrene oxide (100-150 mg/kg bw) in olive oil from four weeks until 120 weeks of age, at which time the experiment was terminated (estimated total doses, 2.5 g for females and 5.0 g for males). Groups of control rats received olive oil only. Of the treated progeny, 60 males and 42 females were alive at the time of appearance of the first tumour. A statistically significant increase in the incidence of forestomach tumours was observed in the styrene oxide-treated progeny of both sexes:

---

[1]The Working Group was aware of studies in progress in mice and rats by oral administration (IARC, 1982b).

papillomas in 9/102; carcinoma *in situ* in 10/102; and carcinomas in 26/102. In the progeny of controls, the incidence of forestomach tumours was: papillomas, 2/104; and carcinomas, 1/104. No difference between treated and control groups was observed in the incidence of tumours occurring at other sites (Ponomarkov *et al.*, 1984). [The Working Group could not evaluate the effects of the prenatal treatment, since no appropriate control group was available.]

3.2 Other relevant biological data

(a) *Experimental systems*

Toxic effects

The toxicity of styrene oxide has recently been reviewed (International Programme on Chemical Safety, 1983).

In rats, the oral LD_{50} of styrene oxide was reported to be 4290 mg/kg bw (Smyth *et al.*, 1954) or 3000 mg/kg bw (Weil *et al.*, 1963); the intraperitoneal LD_{50} was 460-610 mg/kg bw (Ohtsuji & Ikeda, 1971). The LD_{50} by skin application in rabbits was reported as 1184 mg/kg bw (Smyth *et al.*, 1954) or 930 mg/kg bw (Weil *et al.*, 1963). Inhalation of 4900 mg/m^3 (1000 ppm) in air killed 2/6 rats within four hours (Weil *et al.*, 1963).

Styrene oxide causes corneal injury in rabbits (Weil *et al.*, 1963); even dilutions as low as 1% cause eye irritation (Hine & Rowe, 1963). Intradermal injections sensitized the skin of guinea-pigs (Weil *et al.*, 1963).

One intraperitoneal dose of 375 mg/kg bw styrene oxide caused a decrease in the activities of rat-liver mixed-function oxidases and in cytochrome P-450 content (Parkki *et al.*, 1976). Styrene oxide decreased rat-liver glutathione content *in vivo* at doses of 50 and 200 mg/kg bw (Marniemi *et al.*, 1977).

Effects on reproduction and prenatal toxicity

Doses of 0, 0.5, 1, 2, 2.5 or 5 µmol/egg styrene oxide (purum grade) dissolved in ethanol were injected into the air space of groups of 10-20 White Leghorn SK 12 chicken eggs on day three of incubation. Embryos were examined on day 14 of incubation. Concentrations above 0.1 µmol [data for this dose were not presented] reduced embryonic viability (LD_{50}, 1.5 µmol/egg), and malformations were observed in 7% of the treated eggs and 0% of control eggs. The lowest effective dose that produced malformations was 0.5 µmol/egg (Vainio *et al.*, 1977). No dose-response relationship was observed.

Doses of 0.8 µmol/egg styrene oxide (purity, 97%) dissolved in vegetable oil were injected into the air space of White-Leghorn 'mittari' and SK 12 chicken eggs on day three of incubation. In additional groups, 0.1 µmol trichloropropylene oxide (TCPO), an inhibitor of epoxide hydrolase, was injected simultaneously with styrene oxide as a check on the effects of metabolism on embryotoxicity. Embryos were examined on day 14 of incubation. Styrene oxide treatment alone resulted in embryolethality and malformations; addition of TCPO to the styrene oxide treatment augmented these effects (Kankaanpää *et al.*, 1979).

Groups of 23-24 New Zealand white rabbits were exposed by inhalation to 0, 15 or 50 ppm (74 or 245 mg/m^3) (measured concentration of 14.6 or 51 ppm) styrene oxide (purity,

99%) vapour for seven hours per day on days 1-24 of gestation. Foetuses were examined on day 30. Exposure to styrene oxide resulted in maternal toxicity (increased mortality, decreased food consumption and weight gain) and increased the frequency of resorptions. Maternal mortality was 1/23, 4/24 and 19/24, and the resorption rates were 0.25, 0.93 and 1.5 per litter in the control, low- and high-dose groups, respectively (Hardin et al., 1981; Sikov et al., 1981, 1984).

Six groups of at least 31 Sprague-Dawley rats were exposed by inhalation to 100 ppm (490 mg/m^3) or 300 ppm (1470 mg/m^3) styrene oxide (purity, 99%) vapour for seven hours per day either during a three-week pregestational period only, during a three-week pregestational period and through days 1-19 of gestation, or on gestation days 1-19 only. A control group was exposed to air during the whole period. Foetuses were examined on day 21. There was extensive mortality in rats that received prolonged exposure to 100 ppm; exposures at 300 ppm were discontinued after one day due to mortality. Maternal weight gain was reduced in all groups receiving 100 ppm. Gestational exposure decreased fecundity by increasing the preimplantation loss of embryos. Foetal weights and lengths were reduced, and the incidences of ossification defects of the sternebrae and occipital bones were increased by gestational exposure (Hardin et al., 1981; Sikov et al., 1981, 1984).

Absorption, distribution, excretion and metabolism

The metabolism and pharmacokinetics of styrene oxide have been reviewed by Leibman (1975) and Vainio et al. (1984a,b) (see Fig. 1).

After mice received an intraperitoneal injection of styrene, the highest concentrations of styrene oxide were detected in the kidneys, subcutaneous adipose tissue and blood from one to five hours after injection (Nordqvist et al., 1983; Löf et al., 1984). Styrene oxide is a metabolite of styrene (Leibman & Ortiz, 1970; Norppa et al., 1980) (see Fig. 1). It is also formed non-enzymatically through catalysis by oxyhaemoglobin (Belvedere et al., 1983).

Styrene oxide is hydrolysed *in vitro* to styrene glycol (phenylethylene glycol) by microsomal epoxide hydrolase from the liver, kidneys, intestine, lungs and skin of several mammalian species (Oesch, 1973).

The biotransformation of styrene oxide to styrene glycol was stimulated by pretreatment of rats with phenobarbital or 3-methylcholanthrene (Oesch et al., 1971); further metabolism of styrene glycol to mandelic acid was not stimulated (Ohtsuji & Ikeda, 1971). Isolated, perfused rat liver rapidly metabolized styrene oxide to styrene glycol, mandelic acid and glutathione conjugates (Ryan & Bend, 1977; Steele et al., 1981). Microsomal conjugation of styrene oxide with glutathione yields about 60% S-(1-phenyl-2-hydroxyethyl)glutathione and 40% S-(2-phenyl-2-hydroxyethyl)glutathione (Pachecka et al., 1979).

The main route of excretion of styrene oxide metabolites in animals is *via* the kidney; in rabbits, about 80% of a single oral dose was excreted in the urine (James & White, 1967). Acidic urinary metabolites of styrene oxide derived from glutathione conjugates are species dependent. In rats, the only products detected are the mercapturic acids. In guinea-pigs, the major bivalent sulphur acids are the corresponding mercaptoacetic acids, together with mercaptolactic and mercaptopyruvic and mercapturic acids. Some reduction of styrene oxide to styrene may occur in rats and guinea-pigs, with subsequent formation of the dihydrodiol, 3,4-dihydroxy-3,4-dihydro-1-vinylbenzene, which has been found as a urinary metabolite of both styrene and styrene oxide (Nakatsu et al., 1983).

Fig. 1. Metabolic pathways of styrene oxide[a]

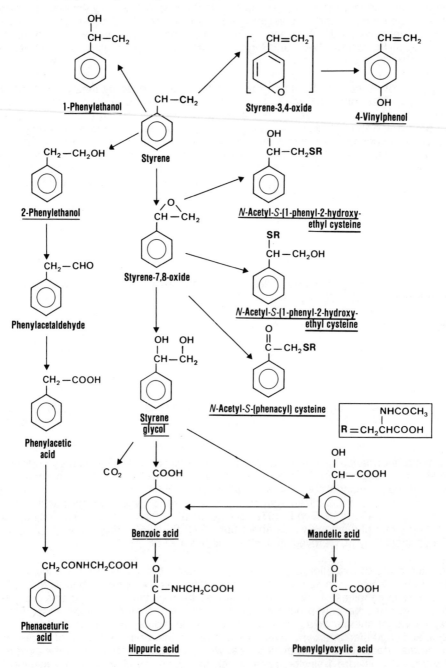

[a]From Vainio et al. (1984a); underlined metabolites are excreted in urine.

[7-^{3}H]-Styrene oxide injected intraperitoneally or incubated *in vitro* binds covalently to microsomes, protein and nucleic-acid fractions of rat liver (Marniemi *et al.*, 1977); in perfused rat liver, binding to RNA and DNA is also detected (Van Anda *et al.*, 1979). When styrene oxide was reacted with polyamino acids *in vitro*, preferential binding to polycysteine was noted (Hemminki, 1983). Reaction products of styrene oxide with nucleosides were identified as 7-alkylguanine and 3-alkylcytosine (Hemminki *et al.*, 1980; Sugiura & Goto, 1981); adducts at N^2- and O^6-guanosine have been described (Hemminki & Hesso, 1984). Esterification of phosphate groups in thymidine monophosphate has been noted (Hemminki & Suni, 1984).

Mutagenicity and other short-term tests (see also 'Appendix: Activity Profiles for Short-term Tests', p. 341)

Styrene oxide was mutagenic to *Salmonella typhimurium* TA1535 and TA100 in the absence of an exogenous metabolic system (S9); it was not mutagenic to strains TA1537, TA1538 or TA98 in the presence or absence of S9 (Milvy & Garro, 1976; Vainio *et al.*, 1976; Stoltz & Withey, 1977; Loprieno *et al.*, 1978; de Meester *et al.*, 1978; Sugiura *et al.*, 1978a; Ueno *et al.*, 1978; Wade *et al.*, 1978; Watabe *et al.*, 1978; Busk, 1979a; Yoshikawa *et al.*, 1980; De Flora, 1981; de Meester *et al.*, 1981; Turchi *et al.*, 1981; Glatt *et al.*, 1983). Activity was reduced by glutathione and S9 (de Meester *et al.*, 1978; Busk, 1979b; Yoshikawa *et al.*, 1980; De Flora, 1981). The (R) enantiomer of styrene oxide is more mutagenic to *S. typhimurium* TA100 than is the (S) enantiomer (Pagano *et al.*, 1982). Mutations were also induced by styrene oxide in *Escherichia coli* WP2 in the absence of an exogenous metabolic system (Sugiura *et al.*, 1978b; Sugiura & Goto, 1981).

Styrene oxide was mutagenic to *Klebsiella pneumoniae* (Voogd *et al.*, 1981). It induced forward mutations in *Schizosaccharomyces pombe in vitro* and in the host-mediated assay in which mice were given oral doses of 100 mg/kg bw. [The authors report the latter as a negative result; however, the data indicate a positive effect.] The compound induced mitotic gene conversion in *Saccharomyces cerevisiae in vitro*, and in the host-mediated assay in which mice were given oral doses of 100 mg/kg bw (Loprieno *et al.*, 1976).

Chromosomal aberrations and micronucleated cells were observed in root-tip cells of *Allium cepa* treated with styrene oxide (Linnainmaa *et al.*, 1978a,b).

Sex-linked recessive lethal mutations were observed in *Drosophila melanogaster* exposed to 200 ppm (980 mg/m^3) styrene oxide vapour for six hours per day for four days, or fed a 200 mg/kg solution in 1% sucrose for 24 hours (Donner *et al.*, 1979).

Treatment of a primary culture of rat hepatocytes with 0.3 mM [36 µg/ml] styrene oxide induced DNA single-strand breaks, as evaluated by alkaline elution (Sina *et al.*, 1983). Unscheduled DNA synthesis was induced in a human heteroploid cell line following exposure to styrene oxide in the absence of an exogenous metabolic system (Loprieno *et al.*, 1978). It was reported in an abstract that styrene oxide induces unscheduled DNA synthesis in human amniotic cells (Audette *et al.*, 1979).

The compound induced forward mutations (HGPRT locus) in Chinese hamster V79 cells (Loprieno *et al.*, 1976; Bonatti *et al.*, 1978; Loprieno *et al.*, 1978; Beije & Jenssen, 1982). Perfusion of styrene oxide through isolated liver abolished its mutagenic effect to V79 cells (Beije & Jenssen, 1982). It was positive in the mouse-lymphoma L5178Y assay (TK$^{+/-}$); activity was reduced by addition of S9 (Amacher & Turner, 1982).

Styrene oxide induces chromosomal aberrations in Chinese hamster V79 cells (Turchi *et al.*, 1981) and in human lymphocytes (Fabry *et al.*, 1978; Linnainmaa *et al.*, 1978a,b; Norppa et al., 1981). It induces micronuclei in Chinese hamster V79 cells (Turchi et al., 1981) and cultured human lymphocytes (Linnainmaa et al., 1978a,b), and sister chromatid exchanges in Chinese hamster ovary cells (de Raat, 1978) and cultured human lymphocytes (Norppa et al., 1981).

Styrene oxide injected intraperitoneally into male mice at 1.8-7.0 mmol/kg bw induced single-strand breaks in the DNA of liver, lung, kidney, testis and brain (Walles & Orsén, 1983). It did not induce chromosomal aberrations or micronuclei in the bone-marrow cells of BALB/c mice treated intraperitoneally with 250 mg/kg bw, or dominant lethal mutations or translocations in meiotic male germ cells of BALB/c mice (Fabry et al., 1978).

Styrene oxide was negative in a micronucleus test with Chinese hamsters given a single intraperitoneal injection of 250 mg/kg bw (Penttilä et al., 1980).

No increase in the incidence of chromosomal aberrations or sister chromatid exchanges was observed in bone-marrow cells of male Chinese hamsters treated by inhalation with 25, 50, 75 or 100 ppm (122, 245, 368 or 4900 mg/m^3) styrene oxide vapour for two, four and 21 (25 ppm only) days (Norppa et al., 1979). In a preliminary report, inhalation exposure of mice to 50 ppm (245 mg/m^3) styrene oxide vapour induced a slight increase in sister chromatid exchanges in regenerating liver cells and alveolar macrophages, but not in bone-marrow cells (Conner et al., 1982).

Styrene oxide induced chromosomal aberrations in the bone-marrow cells of male CD-1 mice treated by gavage with 50, 500 or 1000 mg/kg bw (Loprieno et al., 1978).

(b) Humans

Toxic effects

Acute exposure to styrene oxide causes skin and eye irritation and skin sensitization (Hine et al., 1981).

Effects on reproduction and prenatal toxicity

No data were available to the Working Group.

Absorption, distribution, excretion and metabolism

No data were available on the absorption, distribution, excretion and metabolism after human exposure to styrene oxide alone. However, the main urinary excretion products of styrene include mandelic acid and phenylglyoxylic acid, both of which involve styrene oxide as an intermediate (Leibman, 1975; Vainio et al., 1984b). Low concentrations of styrene oxide (0.05 μmol/l) have been detected in the venous blood of four workers exposed to styrene of unspecified purity (Wigaeus et al., 1983).

Mutagenicity and chromosomal effects

No data were available to the Working Group.

3.3 Case reports and epidemiological studies of carcinogenicity to humans

No data were available to the Working Group.

STYRENE OXIDE

4. Summary of Data Reported and Evaluation

4.1 Exposure data

Commercial production of styrene oxide was first reported in 1974. This compound is used as a reactive plasticizer and diluent for epoxy resins and as a chemical intermediate. It has been found in low concentrations in the air of workplace environments where styrene was used, and has been detected in the blood of workers exposed to styrene.

4.2 Experimental data

Styrene oxide was tested by intragastric intubation in rats of one strain and induced squamous-cell carcinomas and papillomas of the forestomach. Prenatal exposure followed by postnatal oral administration of styrene oxide to rats of another strain also produced squamous-cell carcinomas and increased the incidence of papillomas of the forestomach. No increase in the incidence of skin tumours was observed in mice of two strains following topical application of styrene oxide.

Exposure of rats and rabbits by inhalation to styrene oxide vapour at maternally toxic doses did not result in malformations; however, there was an increase in preimplantation losses and ossification defects in rats and an increased resorption frequency in rabbits.

Styrene oxide was mutagenic to bacteria, yeast and insects; it induced chromosomal aberrations and micronuclei in plants. The compound was mutagenic to mammalian cells *in vitro*; it induced DNA damage in mammalian cells both *in vivo* and *in vitro*, chromosomal

Overall assessment of data from short-term tests: styrene oxide[a]

	Genetic activity			Cell transformation
	DNA damage	Mutation	Chromosomal effects	
Prokaryotes		+		
Fungi/green plants		+	+	
Insects		+		
Mammalian cells (*in vitro*)	+	+	+	
Mammals (*in vivo*)	+		?	
Humans (*in vivo*)				
Degree of evidence in short-term tests for genetic activity: *Sufficient*				Cell transformation: No data

[a]The groups into which the table is divided and the symbols + and ? are defined on pp. 17-18 of the Preamble; the degrees of evidence are defined on p. 18.

aberrations and sister chromatid exchanges *in vitro*. In several studies in mice and hamsters *in vivo*, no dominant lethal mutations, chromosomal aberrations, micronuclei or sister chromatid exchanges were induced; however, in one study in mice, styrene oxide induced chromosomal aberrations.

4.3 Human data

No case report or epidemiological study of the carcinogenicity of styrene oxide to humans was available to the Working Group.

4.4 Evaluation[1]

There is *sufficient evidence*[2] for the carcinogenicity of styrene oxide to experimental animals.

No data on the carcinogenicity of styrene oxide to humans were available to the Working Group.

5. References

Agarwal, S.C., Van Duuren, B.L. & Kneip, T.J. (1979) Detection of epoxides with 4-(*p*-nitrobenzyl)pyridine. *Bull. environ. Contam. Toxicol.*, *23*, 825-829

Amacher, D. & Turner, G.N. (1982) Mutagenic evaluation of carcinogens and non-carcinogens in the L5178Y/TK assay utilizing postmitochondrial fractions (S9) from normal rat liver. *Mutat. Res.*, *97*, 49-65

Audette, M., Pagé, M. & Chouinard, L. (1979) Cytotoxicity and DNA repair induced by styrene oxide *in vitro* (Abstract No. ME1605). *J. Cell Biol.*, *83*, 301a

Beije, B. & Jenssen, D. (1982) Investigation of styrene in the liver perfusion/cell culture system. No indication of styrene-7,8-oxide as the principal mutagenic metabolite produced by the intact rat liver. *Chem.-biol. Interactions*, *39*, 57-76

Belvedere, G., Tursi, F. & Vainio, H. (1983) *Non-microsomal activation of styrene to styrene oxide*. In: Rydström, I., Montelius, J. & Bengtsson, M., eds, *Extrahepatic Drug Metabolism and Chemical Carcinogenesis*, Amsterdam, Elsevier Science Publishers, pp. 193-200

Bidoli, F., Airoldi, L. & Pantarotto, C. (1980) Quantitative determination of styrene-7,8-oxide in blood by combined gas chromatography-multiple ion detection mass fragmentography. *J. Chromatogr.*, *196*, 314-318

Bonatti, S., Abbondandolo, A., Corti, G., Fiori, R. & Mazzaccaro, A. (1978) The expression curve of mutants induced by styrene oxide at the HGPRT locus in V79 cells. *Mutat. Res.*, *52*, 295-300

---

[1] For definitions of the italicized terms, see the Preamble, pp. 15-16.
[2] In the absence of adequate data on humans, it is reasonable, for practical purposes, to regard chemicals for which there is *sufficient evidence* of carcinogenicity in animals as if they presented a carcinogenic risk to humans.

Busk, L. (1979a) Mutagenic effects of styrene and styrene oxide. *Mutat. Res.*, *67*, 201-208

Busk, L. (1979b) Mutagenic effects of styrene oxide in the Ames test (Abstract No. 44). *Mutat. Res.*, *64*, 129

Cheh, A.M. & Carlson, R.E. (1981) Determination of potentially mutagenic and carcinogenic electrophiles in environmental samples. *Anal. Chem.*, *53*, 1001-1006

Conner, M.K., Alarie, Y. & Dombroske, R.L. (1982) Multiple tissue comparisons of sister chromatid exchanges induced by inhaled styrene. *Environ. Sci. Res.*, *25*, 433-441

De Flora, S. (1981) Study of 106 organic and inorganic compounds in the *Salmonella*/microsome test. *Carcinogenesis*, *2*, 283-298

Demole, E. & Berthet, D. (1972) A chemical study of Burley tobacco flavour (*Nicotiana tabacum* L.). I. Volatile to medium-volatile constituents (b.p. < 84°/0.001 Torr). *Helv. chim. Acta*, *55*, 1866-1882

Dolgopolov, V.D. & Lishcheta, L.I. (1971) Qualitative determination of by-products in commercial samples of styrene chlorohydrin (Russ.). *Khim.-Farm. Zh.*, *5*, 55-56 [*Chem. Abstr.*, *76*, 27967b]

Donner, M., Sorsa, M. & Vainio, H. (1979) Recessive lethals induced by styrene and styrene oxide in *Drosophila melanogaster*. *Mutat. Res.*, *67*, 373-376

Fabry, L., Léonard, A. & Roberfroid, M. (1978) Mutagenicity tests with styrene oxide in mammals. *Mutat. Res.*, *51*, 377-381

Fjeldstad, P.E., Thorud, S. & Wannag, A. (1979) Styrene oxide in the manufacture of reinforced polyester plastics. *Scand. J. Work Environ. Health*, *5*, 162-163

Fourneau & Tiffeneau (1905) Some monosubstituted aromatic ethylene oxides (Fr.). *C.R. Acad. Sci. (Paris)*, *140*, 1595-1597

Glatt, H., Jung, R. & Oesch, F. (1983) Bacterial mutagenicity investigation of epoxides: Drugs, drug metabolites, steroids and pesticides. *Mutat. Res.*, *11*, 99-118

Hardin, B.D., Bond, G.P., Sikov, M.R., Andrew, F.D., Beliles, R.P. & Niemeier, R.W. (1981) Testing of selected workplace chemicals for teratogenic potential. *Scand. J. Work Environ. Health*, *7*, Suppl. 4, 66-75

Hawley, G.G., ed. (1981) *The Condensed Chemical Dictionary*, 10th ed., New York, Van Nostrand-Reinhold Co., p. 976

Hemminki, K. (1983) Reactions of methylnitrosourea, epichlorohydrin, styrene oxide and acetoxyacetylaminofluorene with polyamino acids. *Carcinogenesis*, *4*, 1-3

Hemminki, K. & Hesso, A. (1984) Reaction products of styrene oxide with guanosine in aqueous media. *Carcinogenesis*, *5*, 601-607

Hemminki, K. & Suni, R. (1984) Formation of phosphodiesters in thymidine monophosphate by styrene oxide. *Toxicol. Lett.*, *21*, 59-63

Hemminki, K., Paasivirta, J., Kurkirinne, T. & Virkki, L. (1980) Alkylation products of DNA bases by simple epoxides. *Chem.-biol. Interactions*, *30*, 259-270

Hine, C.H. & Rowe, V.K. (1963) *Styrene oxide*. In: Patty, F.A., ed., *Industrial Hygiene and Toxicology*, 2nd rev. ed., Vol. 2, New York, Interscience, pp. 1649-1651

Hine, C., Rowe, V.K., White, E.R., Darmer, K.I., Jr & Youngblood, G.T. (1981) *Epoxy compounds*. In: Clayton, G.D. & Clayton, F.E., eds, *Patty's Industrial Hygiene and Toxicology*, 3rd rev. ed., New York, Wiley-Interscience, pp. 2192-2194

IARC (1976) *IARC Monographs on the Evaluation of Carcinogenic Risk of Chemicals to Man*, Vol. 11, *Cadmium, Nickel, Some Epoxides, Miscellaneous Industrial Chemicals and General Considerations on Volatile Anaesthetics*, Lyon, pp. 201-208

IARC (1979a) *IARC Monographs on the Evaluation of the Carcinogenic Risk of Chemicals to Humans*, Vol. 19, *Some Monomers, Plastics and Synthetic Elastomers, and Acrolein*, Lyon, pp. 275-283

IARC (1979b) *IARC Monographs on the Evaluation of the Carcinogenic Risk of Chemicals to Humans*, Vol. 19, *Some Monomers, Plastics and Synthetic Elastomers, and Acrolein*, Lyon, pp. 231-274

IARC (1982a) *IARC Monographs on the Evaluation of the Carcinogenic Risk of Chemicals to Humans*, Suppl. 4, *Chemicals, Industrial Processes and Industries Associated with Cancer in Humans (IARC Monographs, Volumes 1 to 29)*, Lyon, pp. 229-233

IARC (1982b) *Information Bulletin on the Survey of Chemicals Being Tested for Carcinogenicity*, No. 10, Lyon, p. 150

International Programme on Chemical Safety (1983) *Styrene (Environmental Health Criteria No. 26)*, Geneva, World Health Organization

James, S.P. & White, D.A. (1967) The metabolism of phenethyl bromide, styrene and styrene oxide in the rabbit and rat. *Biochem. J.*, *104*, 914-921

Kankaanpää, J.T.J., Hemminki, K. & Vainio, H. (1979) Embryotoxicity and teratogenicity of styrene and styrene oxide on chick embryos enhanced by trichloropropylene oxide. *Acta pharmacol. toxicol.*, *45*, 399-402

Krost, K.J., Pellizzari, E.D., Walburn, S.G., & Hubbard, S.A. (1982) Collection and analysis of hazardous organic emissions. *Anal. Chem.*, *54*, 810-817

Lapkin, M. (1967) *Epoxides*. In: Standen, A., ed., *Kirk-Othmer Encyclopedia of Chemical Technology*, 2nd ed., Vol. 8, New York, Interscience Publishers, p. 289

Leibman, K.C. (1975) Metabolism and toxicity of styrene. *Environ. Health Perspect.*, *11*, 115-119

Leibman, K.C. & Ortiz, E. (1970) Epoxide intermediates in microsomal oxidation of olefins to glycols. *J. Pharmacol. exp. Ther.*, *173*, 242-246

Linnainmaa, K., Meretoja, T., Sorsa, M. & Vainio, H. (1978a) Cytogenetic effects of styrene and styrene oxide on human lymphocytes and *Allium cepa*. *Scand. J. Work Environ. Health*, *4*, Suppl. 2, 156-162

Linnainmaa, K., Meretoja, T., Sorsa, M. & Vainio, H. (1978b) Cytogenetic effects of styrene and styrene oxide. *Mutat. Res.*, *58*, 277-286

Löf, A., Gullstrand, E., Lundgren, E. & Nordqvist, M.B. (1984) Occurrence of styrene-7,8-oxide and styrene glycol in mouse after the administration of styrene. *Scand. J. Work Environ. Health*, *10*, 179-187

Loprieno, N., Abbondandolo, A., Barale, R., Baroncelli, S., Bonatti, S., Bronzetti, G., Cammellini, A., Corsi, C., Corti, G., Frezza, D., Leporini, C., Mazzaccaro, A., Nieri, R., Rosellini, D. & Rossi, A.M. (1976) Mutagenicity of industrial compounds: Styrene and its possible metabolite styrene oxide. *Mutat. Res.*, *40*, 317-324

Loprieno, N., Presciuttini, S., Sbrana, I., Stretti, G., Zaccaro, L., Abbondandolo, A., Bonatti, S., Fiorio, R. & Mazzaccaro, A. (1978) Mutagenicity of industrial compounds. VII. Styrene and styrene oxide: II. Point mutations, chromosome aberrations and DNA repair induction analyses. *Scand. J. Work Environ. Health*, *4*, Suppl. 2, 169-178

Maltoni, C., Failla, G. & Kassapidis, G. (1979) First experimental demonstration of the carcinogenic effects of styrene oxide. Long-term bioassays on Sprague-Dawley rats by oral administration. *Med. Lav.*, *5*, 358-362

Marniemi, J., Suolinna, E.-M., Kaartinen, N. & Vainio, H. (1977) *Covalent binding of styrene oxide to rat liver macromolecules* in vivo *and* in vitro. In: Ullrich, V., Roots, I., Hildebrandt, A., Estabrook, R.W. & Conney, A.H., eds, *Microsomes and Drug Oxidations*, Oxford, Pergamon Press, pp. 698-702

de Meester, C., Poncelet, F., Roberfroid, M., Rondelet, J. & Mercier, M. (1978) Mutagenicity of styrene and styrene oxide. *Mutat. Res.*, *56*, 147-152

de Meester, C., Duverger-van Bogaort, M., Lambotte-Vandepaer, M., Mercier, M. & Poncelet, F. (1981) Mutagenicity of styrene in the *Salmonella typhimurium* test system. *Mutat. Res.*, *90*, 443-450

Milvy, P. & Garro, A.J. (1976) Mutagenic activity of styrene oxide (1,2-epoxyethylbenzene), a presumed styrene metabolite. *Mutat. Res.*, *40*, 15-18

Mishmash, H.E. & Meloan, C.E. (1972) Indirect spectrophotometric determination of nanomole quantities of oxiranes. *Anal. Chem.*, *44*, 835-836

Nakatsu, K., Hugenroth, S., Sheng, L.-S., Horning, E.C. & Horning, M.G. (1983) Metabolism of styrene oxide in the rat and guinea pig. *Drug Metab. Disposition*, *11*, 463-470

National Institute for Occupational Safety and Health (1981) *National Occupational Hazard Survey* (microfiche), Cincinnati, OH, p. 8804 (generics excluded)

National Research Council (1981) *The Alkyl Benzenes (PB82-160334)*, Prepared for the US Environmental Protection Agency, Springfield, VA, National Technical Information Service, pp. 109-123

Nelis, H.J.C.F. & Sinsheimer, J.E. (1981) A sensitive fluorimetric procedure for the determination of aliphatic epoxides under physiological conditions. *Anal. Biochem.*, 115, 151-157

Nordqvist, M., Ljungquist, E. & Löf, A. (1983) *Metabolic conversion of of styrene to styrene glycol in the mouse. Occurrence of the intermediate styrene-7,8-oxide.* In: Rydström, J., Montelius, J. & Bengtsson, M., eds, *Extrahepatic Drug Metabolism and Chemical Carcinogenesis*, Amsterdam, Elsevier Science Publishers, pp. 219-220

Norppa, H., Elovaara, E., Husgafvel-Pursiainen, K., Sorsa, M. & Vainio, H. (1979) Effects of styrene oxide on chromosome aberrations, sister chromatid exchange and hepatic drug biotransformation in Chinese hamsters *in vivo*. *Chem.-biol. Interactions*, 26, 305-315

Norppa, H., Sorsa, M., Pfäffli, P. & Vainio, H. (1980) Styrene and styrene oxide induce SCEs and are metabolised in human lymphocyte cultures. *Carcinogenesis*, 1, 357-361

Norppa, H., Hemminki, K., Sorsa, M. & Vainio, H. (1981) Effect of monosubstituted epoxides on chromosome aberrations and SCE in cultured human lymphocytes. *Mutat. Res.*, 91, 243-250

Norppa, H., Vainio, H. & Sorsa, M. (1983) Metabolic activation of styrene by erythrocytes detected as increased sister chromatid exchanges in cultured human lymphocytes. *Cancer Res.*, 43, 3579-3582

Oesch, F. (1973) Mammalian epoxide hydrases: Inducible enzymes catalysing the inactivation of carcinogenic and cytotoxic metabolites derived from aromatic and olefinic compounds. *Xenobiotica*, 3, 305-340

Oesch, F., Jerina, D.M. & Daly, J. (1971) A radiometric assay for hepatic epoxide hydrase activity with [7-^{3}H] styrene oxide. *Biochim. biophys. Acta*, 227, 685-691

Ohtsuji, H. & Ikeda, M. (1971) The metabolism of styrene in the rat and the stimulatory effect of phenobarbital. *Toxicol. appl. Pharmacol.*, 18, 321-328

Pachecka, J., Gariboldi, P., Cantoni, L., Belvedere, G., Mussini, E. & Salmona, M. (1979) Isolation and structure determination of enzymatically formed styrene oxide glutathione conjugates. *Chem.-biol. Interactions*, 27, 313-321

Pagano, D.A., Yagen, B., Hernandez, O., Bend., J.R. & Zeiger, E. (1982) Mutagenicity of (R) and (S) styrene 7,8-oxide and the intermediary mercapturic acid metabolites formed from styrene 7,8-oxide. *Environ. Mutagenesis*, 4, 575-584

Parkki, M.G., Marniemi, J. & Vainio, H. (1976) Action of styrene and its metabolites styrene oxide and styrene glycol on activities of xenobiotic biotransformation enzymes in rat liver *in vivo*. *Toxicol. appl. Pharmacol.*, 38, 59-70

Pellizzari, E.D., Bunch, J.E., Berkley, R.E. & McRae, J. (1976) Determination of trace hazardous organic vapor pollutants in ambient atmospheres by gas chromatography/mass spectrometry/computer. *Anal. Chem.*, 48, 803-807

Penttilä, M., Sorsa, M. & Vainio, H. (1980) Inability of styrene to induce nondisjunction in *Drosophila* or a positive micronucleus test in the Chinese hamster. *Toxicol. Lett.*, 6, 119-123

Pfäffli, P., Vainio, H. & Hesso, A. (1979) Styrene and styrene oxide concentrations in the air during the lamination process in the reinforced plastics industry. *Scand. J. Work Environ. Health, 5,* 158-161

Ponomarkov, V., Cabral, J.R.P., Wahrendorf, J. & Galendo, D. (1984) A carcinogenicity study of styrene-7,8-oxide in rats. *Cancer Lett., 24,* 95-101

de Raat, W.K. (1978) Induction of sister chromatid exchanges by styrene and its presumed metabolite styrene oxide in the presence of rat liver homogenate. *Chem.-biol. Interactions, 20,* 163-170

Ringk, W. & Theimer, E.T. (1978) *Benzyl alcohol and β-phenethyl alcohol.* In: Grayson, M., ed., *Kirk-Othmer Encyclopedia of Chemical Technology,* 3rd ed., Vol. 3, New York, John Wiley & Sons, pp. 796, 799-800, 802

Ryan, A.J. & Bend, J.R. (1977) The metabolism of styrene oxide in the isolated perfused rat liver. Identification and quantitation of major metabolites. *Drug Metab. Disposition, 5,* 363-367

Sawicki, E. (1976) *Analysis of atmospheric carcinogens and their cofactors.* In: Rosenfeld, C. & Davis W., eds, *Environmental Pollution and Carcinogenic Risks (INSERM Symposium Series Vol. 52/IARC Scientific Publications No. 13),* Paris, Institut National de la Santé et de la Recherche Médicale/Lyon, International Agency for Research on Cancer, pp. 315-316

Sears, J.K. & Touchette, N.W. (1982) *Plasticizers.* In: Grayson, M., ed., *Kirk-Othmer Encyclopedia of Chemical Technology,* 3rd ed., Vol. 18, New York, John Wiley & Sons, pp. 171, 183

Shackelford, W.M. & Keith, L.H. (1976) *Frequency of Organic Compounds Identified in Water (EPA-600/4-76-062),* Athens, GA, US Environmental Protection Agency, p. 214

Sikov, M.R., Cannon, W.C., Carr, D.B., Miller, R.A., Montgomery, L.F. & Phelps, D.W. (1981) *Teratogenic Assessment of Butylene Oxide, Styrene Oxide and Methyl Bromide (DHHS (NIOSH) Publication No. 81-124),* US Department of Health and Human Services, Cincinnati, OH, National Institute for Occupational Safety and Health

Sikov, M.R., Niemeier, R.W., Cannon, W.C., Carr, D.B. & Hardin, B.D. (1984) Reproductive toxicology of inhaled styrene oxide in rats and rabbits. *Fundam. appl. Toxicol.* (in press)

Sina, J.F., Bean, C.L., Dysart, G.R., Taylor, V.I. & Bradley, M.O. (1983) Evaluation of the alkaline elution/rat hepatocyte assay as a predictor of carcinogenic/mutagenic potential. *Mutat. Res., 113,* 357-391

Smyth, H.F., Jr, Carpenter, C.P., Weil, C.S. & Pozzani, U.C. (1954) Range-finding toxicity data. List V. *Arch. ind. Hyg. occup. Med., 10,* 61-68

Stampfer, J.F. & Hermes, R.E. (1981) Development of sampling and analytical method for styrene oxide. *Am. ind. Hyg. Assoc. J., 42,* 699-706

Steele, J.W., Yagen, B., Hernandez, O., Cox, R.H., Smith, B.R. & Bend, J.R. (1981) The metabolism and excretion of styrene oxide-glutathione conjugates in the rat and by isolated perfused liver, lung and kidney preparations. *J. Pharmacol. exp. Ther.*, *219*, 35-41

Stoltz, D.R. & Withey, R.J. (1977) Mutagenicity testing of styrene and styrene epoxide in *Salmonella typhimurium*. *Bull. environ. Contam. Toxicol.*, *17*, 739-742

Sugiura, K. & Goto, M. (1981) Mutagenicities of styrene oxide derivatives on bacterial test systems: Relationship between mutagenic potencies and chemical reactivity. *Chem.-biol. Interactions*, *35*, 71-91

Sugiura, K., Kimura, T. & Goto, M. (1978a) Mutagenicities of styrene oxide derivatives on *Salmonella typhimurium* (TA100). Relationship between mutagenic potencies and chemical reactivity. *Mutat. Res.*, *58*, 159-165

Sugiura, K., Yamanaka, S., Fukasawa, S. & Goto, M. (1978b) The mutagenicity of substituted and unsubstituted styrene oxides in *E. coli*: Relationship between mutagenic potencies and physico-chemical properties. *Chemosphere*, *9*, 737-742

Swan, J.D. (1954) Determination of epoxides with sodium sulfite. *Anal. Chem.*, *26*, 878-880

Turchi, G., Bonatti, S., Citti, L., Gervasi, P.G., Abbondandolo, A. & Presciuttini, S. (1981) Alkylating properties and genetic activity of 4-vinylcyclohexene metabolites and structurally related epoxides. *Mutat. Res.*, *83*, 419-430

Ueno, Y., Kubota, K., Ito, T. & Nakamura, Y. (1978) Mutagenicity of carcinogenic mycotoxins in *Salmonella typhimurium*. *Cancer Res.*, *38*, 536-542

Union Carbide Corporation (1979) *Chemicals and Plastics Physical Properties, 1979-1980*, New York, p. 17

US Food and Drug Administration (1980) Food and drugs. Resinous and polymeric coatings. *US Code Fed. Regul.*, Title 21, Part 175.300

US International Trade Commission (1976) *Synthetic Organic Chemicals, US Production and Sales, 1974* (*USITC Publication 776*), Washington DC, US Government Printing Office, p. 44

US International Trade Commission (1980) *Imports of Benzenoid Chemicals and Products, 1979* (*USITC Publication 1083*), Washington DC, US Government Printing Office, p. 20

US International Trade Commission (1983) *Imports of Benzenoid Chemicals and Products, 1982* (*USITC Publication 1401*), Washington DC, US Government Printing Office, p. 17

Vainio, H., Pääkkönen, R., Rönnholm, K., Raunio, V. & Pelkonen, O. (1976) A study on the mutagenic activity of styrene and styrene oxide. *Scand. J. Work Environ. Health*, *3*, 147-151

Vainio, H., Hemminki, K. & Elovaara, E. (1977) Toxicity of styrene and styrene oxide on chick embryos. *Toxicology*, *8*, 319-325

Vainio, H., Norppa, H. & Belvedere, G. (1984a) *Metabolism and mutagenicity of styrene and styrene oxide.* In: Järvisalo, J., Pfäffli, P. & Vainio, H., eds, *Industrial Hazards of Plastics and Synthetic Elastomers,* New York, Alan R. Liss, pp. 215-225

Vainio, H., Hietanen, E. & Belvedere, G. (1984b) *Pharmacokinetics and metabolism of styrene.* In: Bridges, J.W. & Chasseaud, L.F., eds, *Progress in Drug Metabolism,* Vol. 8, London, Taylor & Francis Ltd, pp. 203-239

Van Anda, J., Smith, B.R., Fouts, J.R. & Bend, J.R. (1979) Concentration-dependent metabolism and toxicity of [^{14}C]styrene oxide in the isolated perfused rat liver. *J. Pharmacol. exp. Ther., 211,* 207-212

Van Duuren, B.L., Nelson, N., Orris, L., Palmes, E.D. & Schmitt, F.L. (1963) Carcinogenicity of epoxides, lactones, and peroxy compounds. *J. natl Cancer Inst., 31,* 41-55

Voogd, C.E., van der Stell, J.J. & Jacobs, J.J.J.A.A. (1981) The mutagenic action of aliphatic epoxides. *Mutat. Res., 89,* 269-282

Wade, D.R., Airy, S.C. & Sinsheimer, J.E. (1978) Mutagenicity of aliphatic epoxides. *Mutat. Res., 58,* 217-223

Walles, S.A.S. & Orsén, I. (1983) Single-strand breaks in DNA of various organs of mice induced by styrene and styrene oxide. *Cancer Lett., 21,* 9-15

Watabe, T., Isobe, M., Sawahata, T., Yoshikawa, K., Yamada, S. & Takabatake, E. (1978) Metabolism and mutagenicity of styrene. *Scand. J. Work Environ. Health, 4,* Suppl. 2, 142-155

Weil, C.S., Condra, N., Haun, C. & Striegel, J.A. (1963) Experimental carcinogenicity and acute toxicity of representative epoxides. *Am. ind. Hyg. Assoc. J., 24,* 305-325

Wigaeus, E., Löf, A., Bjurström, R. & Nordqvist, M.B. (1983) Exposure to styrene. Uptake, distribution, metabolism and elimination in man. *Scand. J. Work Environ. Health, 9,* 479-488

Yoshikawa, K., Isobe, M., Watabe, T. & Takabatake, E. (1980) Studies on metabolism and toxicity of styrene. III. The effect of metabolic inactivation by rat-liver S9 on the mutagenicity of phenyloxirane toward *Salmonella typhimurium. Mutat. Res., 78,* 219-226

PEROXIDES

BENZOYL PEROXIDE

1. Chemical and Physical Data

1.1 Synonyms and trade names

Chem. Abstr. Services Reg. No.: 94-36-0

Chem. Abstr. Name: Peroxide, dibenzoyl

IUPAC Systematic Name: Benzoyl peroxide

Synonyms: Benzoic acid, peroxide; benzoperoxide; benzoyl superoxide; dibenzoyl peroxide; dibenzoylperoxide; diphenylglyoxal peroxide

Trade Names: Acetoxyl; Acnegel; Aztec BPO; Benox A-70, A-80, B-50, B-120, B-125, B-135 and L40; Benoxyl; Benzac; Benzaknen; BZF-60; BZQ 25, 40, 50, 50 Pigmented and 55; BZW 70, 75 and 80; Cadet BCP, BPO-70, BP0-78 and BPO-78FP; Cadox BCP, BFF-50, BFF-60W, BP, BP-55, BS and 40E; Clearasil Benzoyl Peroxide Lotion; Clearasil BP Acne Treatment; Cuticura Acne Cream; Debroxide; Dry and Clear Acne Cream; Dry and Clear Acne Medication; Epiclear; Fostex; G20; Garox BZP, QZA and 55A; Incidol; Loroxide; Lucidol; Lucidol B50, CH50, G20, KL50, 50P, 70, 70S, 78 and 98; Luperco AA, ACP, ACP-50, AFR, AFR-400, AMB, ANS, ANS-50, ANS-50-AT, ANS-60, AST, ATC, BP, BP-55 and WET; Luperox FL; Nayper B and BO; Norox BZP-C-35 and BZP-250; Novadelox; Oxy-5; Oxy-10; Oxylite; Oxy Wash; Panoxyl; Pan Oxyl; Pan Oxyl Bar; Poroxdox; Persadox HP; Quinolor Compound; Quinolor Compound Ointment; Sulfoxyl; Superox 705, 706, 717, 718 and 742; Theraderm; Topex; Vanoxide; Xerac

1.2 Structural and molecular formulae and molecular weight

$C_{14}H_{10}O_4$ Mol. wt: 242.2

1.3 Chemical and physical properties of the pure substance

From Dailey (1978), unless otherwise specified

(a) *Description*: A tasteless, white, granular, crystalline solid with a slight almond-like odour resembling that of benzaldehyde

(b) *Melting-point*: 103-105°C, with explosive decomposition above 105°C

(c) *Density*: Specific gravity (25°C), 1.3340

(d) *Spectroscopy data*: A review on spectroscopy data has been published (Silbert, 1971).

(e) *Solubility*: Very sparingly soluble in water; sparingly soluble in carbon disulphide (1 g in 40 ml), ethanol and olive oil (1 g in 50 ml); soluble in benzene, chloroform and diethyl ether

(f) *Stability*: Inflammable; may decompose explosively if subjected to excessive heat, friction or sudden shock (National Institute for Occupational Safety and Health, 1977)

(g) *Reactivity*: Highly reactive oxidizing material. Can decompose on photolysis (<300 nm) or by catalysis with transition metal ions (Sheppard & Mageli, 1982)

(h) *Conversion factor*: 1 ppm = 10.1 mg/m^3 at 760 mm Hg and 20°C (Verschueren, 1977)

1.4 Technical products and impurities

Benzoyl peroxide is available in the USA as granules containing 98.5 ± 1% benzoyl peroxide, as granules containing 75-80% and 70 ± 2 % benzoyl peroxide with water as a diluent, as pastes containing from 24-27% up to 55% min benzoyl peroxide with proprietary diluents, as a paste containing 50-52% benzoyl peroxide with tricresyl phosphate as a diluent, as a paste containing 50% min benzoyl peroxide with silicone oil as a diluent, as a powder containing 35-37% benzoyl peroxide with dicalcium phosphate as a diluent, and as a powder containing 32-33% benzoyl peroxide with wheat starch as a diluent (Pennwalt Corporation, 1983).

Flour bleach typically contains 32% benzoyl peroxide and 68% corn-starch (National Institute for Occupational Safety and Health, 1983).

Benzoyl peroxide is available in western Europe in the following forms: 98% active water-wet solid; 50% and 55% active pastes in plasticizers; 36% active liquid in a plasticizer; and 50% active paste in silicone oil (Luperox GmbH, 1979).

In the USA, to meet the requirements of the Food Chemicals Codex, benzoyl peroxide must pass an identification test and meet the following requirements: purity, 96% min; heavy metals, 0.004% max; lead, 0.001% max; and arsenic, 0.0003% max (National Research Council, 1981).

Hydrous benzoyl peroxide USP must pass a thin-layer chromatography identification test and meet the following specifications: purity, 65-82%; water, about 26%; and acidity as benzoic acid, 1.5% max (US Pharmacopeial Convention, Inc., 1980).

BENZOYL PEROXIDE

2. Production, Use, Occurrence and Analysis

2.1 Production and use

(a) Production

Benzoyl peroxide was first prepared by Brodie in 1858. It can be synthesized by the reaction of benzoyl chloride (see IARC, 1982a) with sodium hydroxide and hydrogen peroxide (i.e., sodium peroxide), and this is the method used for its commercial production (Dailey, 1978).

US commercial production of benzoyl peroxide was started in 1927, although it was reportedly being used as early as 1921 in Germany (Dailey, 1978). Annual US production reached a peak in 1974 when six companies reported combined production of 4.1 million kg (US International Trade Commission, 1976). In 1982, five US companies reported combined production of 2.3 million kg (US International Trade Commission, 1983a). Six US companies currently produce benzoyl peroxide.

Imports of benzoyl peroxide through the principal US customs districts reached a peak in 1975 when 45 thousand kg were imported (US International Trade Commission, 1977). Imports through the principal US customs districts totalled only 350 kg in 1982 (US International Trade Commission, 1983b).

One company produces benzoyl peroxide in Canada (Anon., 1981).

Benzoyl peroxide is produced by three companies each in Germany and the UK, by two companies in the Netherlands, and by one company in France. Total annual production in western Europe is estimated to be in the range of 1-5 million kg.

The commercial production of benzoyl peroxide in Japan started around 1944. Six Japanese companies currently manufacture it from benzoyl chloride and sodium peroxide, and 1983 production is estimated to have been 300-400 thousand kg.

(b) Use

Benzoyl peroxide is used principally as a source of free radicals in the plastics and rubber industry; it also finds use as a food additive and in non-prescription drugs. Applications in a number of other areas have been reported, but their present commercial status is unknown.

Benzoyl peroxide is widely used as an initiator in free-radical polymerizations, primarily in the curing of unsaturated polyester resins (used in glass-fibre construction, e.g., boats and automobile repairs). Production of polystyrene (see IARC, 1979) and related resins is the next most important use, and approximately 3.1 million kg of benzoyl peroxide were used in polyester resins in the USA in 1978 and approximately 0.8 million kg in styrene polymers (Burke, 1979). Benzoyl peroxide has been recommended for use in the polymerization/curing of vinyl acetate and styrenated alkyd resins (Pennwalt Corporation, 1983). It is also used to cross-link resins and elastomers and may be used in this way in the curing of rubber and silicone elastomers (see IARC, 1982b). Benzoyl peroxide has also been reported to be the most commonly used initiator in dental applications, where it is typically used in the polymerization of methacrylate monomers to produce resin cements and restorative resins (Paffenbarger & Rupp, 1979). A number of paint-on artificial fingernail products were reported, in 1982, to contain benzoyl peroxide as the initiator for polymerizing the alkyl methacrylate monomers in such products (Fuller, 1982).

Benzoyl peroxide was used to bleach edible oils in the early 1900s, but this practice is reported to be rare now (National Institute for Occupational Safety and Health, 1977); however, it continues to be used to bleach foods (e.g., flour, milk used for cheese, and lecithin). Although a 1977 survey on the use of food additives in US industry did not report its use, 668 thousand kg were added to food in 1975 (Federation of American Societies for Experimental Biology, 1980).

Benzoyl peroxide is used in topical non-prescription medications for the treatment of acne. These products are sold as lotions, gels, creams and ointments containing 2.5-10% benzoyl peroxide; similar products have been suggested for use in the treatment of athlete's foot (American Pharmaceutical Association, 1982). It was used in Germany as an antiseptic and local anaesthetic in the treatment of burns and ulcers. It was also, reportedly, formerly taken internally (Dailey, 1978). It is used in the treatment of bed sores and was used formerly in the treatment of dermatitis due to poison ivy (National Institute for Occupational Safety and Health, 1977).

Benzoyl peroxide has also been reported to be used in the following applications: (1) as a fixing agent in light microscopy in Germany (Dailey, 1978); (2) as an initiator in systems used to prepare polymers for use in roof bolting in mines; (3) formerly, as a bleaching agent for textiles and paper (National Institute for Occupational Safety and Health, 1977); (4) as a drying agent for unsaturated oils (Hawley, 1981); (5) as a burn-out agent for cellulose acetate in mixed fabrics with viscose, silk or cotton to produce a lace-like appearance; (6) in printing pastes; (7) in special fast-drying printing inks for printing on plastic surfaces; (8) as an initiator for addition and substitution reactions in organic synthesis (Mackison et al., 1981); and (9) in the embossing of vinyl flooring (Hawley, 1981). The commercial status of these possible uses is unknown.

Canadian usage of benzoyl peroxide in 1980 has been estimated at 265 thousand kg (Anon., 1981).

In Japan, benzoyl peroxide is used principally in the plastics industry; lesser quantities are used as a bleaching agent and in cosmetics and pharmaceuticals.

Occupational exposure to benzoyl peroxide has been limited by regulation or recommended guidelines in at least eight countries. The standards are listed in Table 1.

Benzoyl peroxide has been approved by the US Food and Drug Administration for use as a direct and indirect food additive in the following ways: as a bleaching agent for flour and for milk used in the preparation of certain cheeses and lecithin; as an ingredient in the preparation of hydroxylated lecithin; as a component of adhesives; as a preservative in paper and paperboard; as a catalyst in the formulation of polyester resins; and as an accelerator in the production of rubber articles. In 1982, it was proposed that benzoyl peroxide be affirmed as generally recognized as safe as a direct human food ingredient when used as a bleaching agent following conditions of use of current good manufacturing practice (US Food and Drug Administration, 1982a).

In 1982, an advisory review panel on over-the-counter drugs concluded that benzoyl peroxide is safe and effective for topical use in the treatment of acne when used at concentrations of 2.5-10% in over-the-counter products (US Food and Drug Administration, 1982b).

Table 1. National occupational exposure limits for benzoyl peroxide[a]

Country	Year	Concentration mg/m³	ppm	Interpretation[b]	Status
Australia	1978	5	-	TWA	Guideline
Belgium	1978	5	-	TWA	Regulation
Bulgaria	1971	0.05	-	TWA	Regulation
Finland	1981	5	-	TWA	Guideline
		10	-	STEL	
Germany, Federal Republic of	1984	5	-	TWA	Guideline
Netherlands	1978	5	-	TWA	Guideline
Switzerland	1978	5	-	TWA	Regulation
USA[c]					
OSHA	1978	5	-	TWA	Regulation
ACGIH	1984-85	-	10	Maximum (30 min)	Guideline
		5	-	TWA	

[a]From International Labour Office (1980); National Finnish Board of Occupational Safety and Health (1981); American Conference of Governmental Industrial Hygienists (1984); Deutsche Forschungsgemeinschaft (1984)
[b]TWA, time-weighted average; STEL, short-term exposure limit
[c]OSHA, Occupational Safety and Health Administration; ACGIH, American Conference of Governmental Industrial Hygienists

As part of the Hazardous Materials Regulations of the US Department of Transportation (1980), shipments of benzoyl peroxide are subject to a variety of labelling, packaging, quantity, and shipping restrictions consistent with its designation as a hazardous material.

2.2 Occurrence

(a) Natural occurrence

Benzoyl peroxide is not known to occur as a natural product.

(b) Occupational exposure

On the basis of the 1974 National Occupational Hazard Survey, the National Institute for Occupational Safety and Health (1980, 1981) estimated that about 58 000 US workers in 47 industries were exposed to benzoyl peroxide. The principal industry in which exposure was found was the industrial patterns manufacturing industry.

Benzoyl peroxide levels were below the limit of detection (0.002 mg/sample) of the analytical method used for three personal breathing-zone air samples taken in April 1980 from the production area at a plant in Baton Rouge, LA, where plastic tanks, hoods and ducts were being manufactured from glass fibre during the time of sampling (Markel & Jannerfeldt, 1981).

Exposure to airborne benzoyl peroxide was investigated in December 1979 in the production area of a plant in Little Rock, AR, where large-diameter glass-fibre reinforced plastic pipe was being manufactured. Ten out of 12 personal breathing-zone and general area air samples were below the limit of detection (0.8 µg/sample); of the remaining two samples, one from personal breathing-zone air contained 0.10 mg/m³ and the other from general area air contained 0.01 mg/m³ (Markel & Wilcox, 1981).

Exposures to benzoyl peroxide resulting from its use in various applications have been implied. In 1975, interviewed welders employed at a plant in La Grange, IL, in the manufacture of diesel locomotives reported exposure to a plastic body filler made of a talc-polyester resin and benzoyl peroxide (Bloom, 1975). Polyester processors at a Netherlands aircraft factory in 1957 were possibly exposed to benzoyl peroxide. In 1960, 34 of an unspecified number of workers in Czechoslovakia were assessed as having experienced some exposure to benzoyl peroxide owing to its use as a hardener for epoxy resins. In 1976, telephone-repair workers in the USA were reported to be exposed to a styrene hardener containing 50% benzoyl peroxide when new and replacement telephone cables were installed (National Institute for Occupational Safety and Health, 1977).

(c) *Food, beverages and feeds*

It was concluded in 1953 that the greater part of the benzoyl peroxide that is added to flour (at typical concentrations of about 15 mg/kg) as a bleaching agent decomposes into benzoic acid within a few days, although small amounts of the peroxide may persist for several weeks (Dailey, 1978). It is used at a level of 0.002% to bleach milk in the preparation of certain cheeses. Weighted means of the usual levels of addition of benzoyl peroxide to some US foods in 1970 were as follows: baked goods, baking mixes, 56 mg/kg; grain products, 41 mg/kg; and fats and oils, 100 mg/kg (Federation of American Societies for Experimental Biology, 1980).

(d) *Other*

Benzoyl peroxide is present in dental restorative resins owing to its use as a catalyst. It has been detected in different brands of restoration resins at concentrations ranging from 0.32-2.59% (Asmussen, 1980; Miyazaki *et al.*, 1981).

2.3 Analysis

General methods for the analysis of benzoyl peroxide in various matrices have been reviewed (Mair & Hall, 1971).

The National Institute for Occupational Safety and Health (1977) reported methods for the determination of benzoyl peroxide in pharmaceuticals, fats and oils. Methods for the analysis of pharmaceuticals included spectrophotometric, titrimetric and polarographic techniques; however, none was specific for benzoyl peroxide.

Typical methods for the analysis of benzoyl peroxide in various matrices are summarized in Table 2.

Table 2. Methods for the analysis of benzoyl peroxide

Sample matrix	Sample preparation	Assay procedure[a]	Limit of detection	Reference
Air	Collect on a cellulose ester membrane filter; desorb with diethyl ether	HPLC/UV	100-150 µg/m^3 (30-litre sample)	Gunderson & Anderson (1980); Purnell et al. (1982)
	For total airborne dust exceeding 5 mg/m^3: collect on glass-fibre filter of a constant weight at 50% relative humidity; weigh; mix filter with a water, potassium iodide and starch solution	Gravimetric detection followed by colorimetry	1-3 µg	National Institute for Occupational Safety and Health (1977)
Dental restorative resins	--	MHPLC and LC-MS	Not given	Miyazaki et al. (1981)
	Not given	Detection by iodometric titration	Not given	Asmussen (1980)
Flour	Repeated diethyl ether extractions to remove interfering impurities	Visible spectrophotometry	Not given	Horwitz (1980)
Cheese	Extract (phosphoric acid, ethanol, diethyl ether and petroleum ether); add cuprous chloride, electrolytic copper and hydrochloric acid; treat with potassium permanganate; transfer to chloroform containing lauric acid	GC/FID	Not given	Karasz et al. (1974)
Blood (human, rabbit and rat)	Shake with water; add acetonitrile and extract; centrifuge and filter supernatant	HPLC/UV	10 ng/20 µl injected	Ehinger & Andermann (1980)

[a]Abbreviations: HPLC/UV, high-performance liquid chromatography/ultraviolet absorption; MHPLC, micro-high-performance liquid chromatography; LC-MS, liquid chromatography-mass spectroscopy; GC/FID, gas chromatography/flame ionization detection

3. Biological Data Relevant to the Evaluation of Carcinogenic Risk to Humans

3.1 Carcinogenicity studies in animals

(a) Oral administration

Mouse: Groups of 25 male and 25 female albino mice [strain and age unspecified] were fed a diet containing different doses of a commercial powder containing 18% benzoyl peroxide [purity unspecified] which were based upon consideration of the use of benzoyl peroxide in bread production (calculated doses of benzoyl peroxide: 0, 28, 280 and 2800 mg/kg of diet), for 80 weeks, at which time the number of survivors was 3, 10, 0 and 2 male mice and 9, 11, 9 and 11 females; all surviving animals were then killed. A few tumours at various sites were observed, but the overall tumour incidence did not differ significantly between treated and control groups (Sharratt et al., 1964). [The Working Group noted the short duration of the experiment and the high mortality.]

Rat: Groups of 25 male and 25 female albino rats [strain and age unspecified] were fed a diet containing different doses of a commercial powder containing 18% benzoyl peroxide [purity unspecified] which were based upon consideration of the use of benzoyl peroxide in bread production (calculated doses of benzoyl peroxide: 0, 28, 280 and 2800 mg/kg of diet) for 120 weeks, at which time they were killed. At 104 weeks, 12, 12, 13 and 11 males and 14, 7, 9 and 11 females survived. A few tumours were observed at various sites, but the overall tumour incidence did not differ significantly between treated and control groups (Sharratt et al., 1964).

(b) *Subcutaneous and/or intramuscular administration*

Mouse: Groups of 25 male and 25 female albino mice [age and strain unspecified] received a single subcutaneous injection of 50 mg of a 20% suspension of benzoyl peroxide [purity unspecified] in starch solution and were observed for 80 weeks, at which time they were killed. An equal number of animals was injected with starch solution only and served as vehicle controls. Nine treated males, seven treated females, no control male and six control females survived. No tumour was found at the site of injection or at any other site in treated or control groups (Sharratt et al., 1964). [The Working Group noted the short duration of the experiment, the high mortality and the low degree of exposure.]

Rat: A group of 25 male and 25 female albino rats [age and strain unspecified] received a single subcutaneous injection of 120 mg of a 20% suspension of benzoyl peroxide [purity unspecified] in starch solution and were observed for 120 weeks, at which time they were killed. An equal number of animals was injected with starch solution only and served as vehicle controls. Ten treated males and 16 control males and nine treated females and 17 control females survived 104 weeks. No tumour was found at the site of injection, and overall tumour incidence did not differ significantly between treated and control groups (Sharratt et al., 1964). [The Working Group noted the low degree of exposure.]

A group of 20 male Charles River CD rats [age unspecified] received subcutaneous injections of 2.9 mg benzoyl peroxide [purity unspecified] in 0.2 ml trioctanoin (tricaprylin) into the right hind leg twice weekly for 12 weeks and were observed for 14 months, at which time all animals were still alive. A further group of 20 male rats received trioctanoin only and served as vehicle controls. No malignant tumour was found at the injection site or in internal organs (Poirier et al., 1967). [The Working Group noted the short duration of the experiment.]

Groups of 20 male and 15 female Bethesda black rats (NIH Black rats) received a subcutaneous implantation of 50 mg benzoyl peroxide [purity unspecified] in a gelatin capsule at the nape of the neck and were observed for 24 months. Mortality was 9/35 [sex unspecified] at week 52 and 22/35 at week 78. None of the rats developed tumours at the site of implantation. Single tumours were found at various other sites, but none of the tumours was considered by the author to be causally related to benzoyl peroxide treatment (Hueper, 1964).

(c) *Skin application*

Mouse: A group of 30 male ICR/Ha mice, eight weeks old, received thrice-weekly applications for life of approximately 100 mg benzoyl peroxide [purity unspecified] dissolved in 5% benzene. Median survival time was 292 days; one mouse developed a skin papilloma. A total of 150 mice (from four different control groups) were treated with benzene alone. Median survival times ranged from 262-412 days; 11 skin tumours, including one carcinoma, were observed (Van Duuren et al., 1963).

Groups of 25 male and 25 female albino mice [age and strain unspecified] received skin applications of one drop (approximately 50 mg) of a 50% suspension of benzoyl peroxide [purity unspecified] in flour paste on the back of the neck on six days per week for 80 weeks. An equal number of mice painted with flour paste only served as controls. No skin tumour was observed, and overall tumour incidence did not differ significantly between treated groups and controls (Sharratt et al., 1964). [The Working Group noted that adequate data on survival were not given.]

A group of 21 female Swiss mice [age unspecified] received twice-weekly skin applications of 0.5% benzoyl peroxide [purity unspecified] in acetone for 80 weeks. No skin tumour was observed. Another group of 20 females received applications of 0.5% benzoyl peroxide in acetone twice weekly for three weeks then, after one week, they were treated with 5% croton oil in mineral oil twice weekly for 67 weeks. No skin tumour was reported (Saffiotti & Shubik, 1963). [The Working Group noted that adequate data on survival were not given.]

Groups of 30 female Sencar mice, aged seven to nine weeks, were used to test the tumour-promoting (A), tumour-initiating (B) and complete carcinogenic (C) activities of benzoyl peroxide [purity unspecified] on the skin. Mice in experiment (A) received a single topical application of 10 nmol 7,12-dimethylbenz[a]anthracene (DMBA) in 0.2 ml acetone, followed one week later by applications of 1, 10, 20 or 40 mg benzoyl peroxide in 0.2 ml acetone twice weekly for 52 weeks. A group receiving a single application of 0.2 ml acetone alone served as controls. The numbers of mice with papillomas at 30 weeks were 1/28 (controls), 9/29 (1 mg), 20/28 (10 mg), 21/27 (20 mg) and 20/24 (40 mg). The numbers of mice with carcinomas at 52 weeks in these groups were 0/28, 1/29, 6/28, 12/27 and 10/24, respectively. Mice in experiment (B) received a single topical application of 0.2 ml acetone alone or 1, 10, 20 or 40 mg benzoyl peroxide in acetone, followed one week later by twice-weekly applications of 2 μg 12-O-tetradecanoylphorbol 13-acetate (TPA) in acetone for 52 weeks. No significant difference was observed in the incidence of papillomas; no carcinoma was observed. Mice in experiment (C) received twice-weekly topical applications of acetone alone or 1, 10, 20 or 40 mg benzoyl peroxide in 0.2 ml acetone for 52 weeks. No significant difference was observed in the incidence of papillomas; no carcinoma was observed. Survival rates in all experiments at 30 weeks ranged from 24-29 mice per group (Slaga et al., 1981). [The Working Group noted the absence of a control group treated with DMBA only, and the short duration of the experiment for complete carcinogenicity.]

Groups of 30-40 female Sencar mice, aged seven to eight weeks, received a single application of 10 nmol DMBA in 0.2 ml acetone and, one week later, were treated twice weekly either with 2 μg TPA in 0.2 ml acetone or 20 mg benzoyl peroxide [purity unspecified] in 0.2 ml acetone for 48 weeks. The incidence of skin papillomas was 100% at 20 weeks in the TPA group and about 80% at 48 weeks in the benzoyl peroxide group. The incidence of squamous-cell carcinomas was 0% in the TPA group and about 50% in the benzoyl peroxide group. Additional groups of 30-40 female C57BL/6 mice, aged seven to eight weeks, received a single skin application of 400 or 800 nmol DMBA or 1000 or 3200 nmol benzo-[a]pyrene (BP) in 0.2 ml acetone, and one week later were treated twice weekly with 4 μg TPA in 0.2 ml acetone or 20 mg benzoyl peroxide in 0.2 ml acetone for 48 weeks. The incidence of squamous-cell carcinomas was about 60% in the DMBA-benzoyl peroxide group, <50% in the DMBA-TPA group, >50% in the BP-benzoyl peroxide group and <50% in the BP-TPA group (Reiners et al., 1984). [Those values were taken from graphs, and actual numbers were not given. The Working Group noted that survival rates were not given, nor were data available on controls treated with DMBA or BP.]

(d) *Other experimental systems*

Mouse: Groups of 25 male and 25 female albino mice [age and strain unspecified] were fed a diet containing 2800 mg/kg benzoyl peroxide [purity unspecified] and received, simultaneously, a subcutaneous injection of 50 mg benzoyl peroxide in 20% starch solution. In addition, the animals were painted on six days per week with one drop (approximately 50 mg) of a 50% suspension of benzoyl peroxide in flour paste. The experiment was terminated after 80 weeks of treatment, when three males and 11 females were still alive. Single tumours occurred at various sites, but the overall tumour incidence did not differ significantly between treated and untreated control groups (Sharratt et al., 1964).

Rat: Groups of 25 male and 25 female albino rats [age and strain unspecified] were fed a diet containing 2800 mg/kg benzoyl peroxide [purity unspecified] and received simultaneously a subcutaneous injection of 120 mg benzoyl peroxide in 20% starch solution. The experiment was terminated after 120 weeks of treatment; 14 males and 10 females were still alive at 104 weeks. Single tumours occurred at various sites, but the overall tumour incidence did not differ significantly between treated and untreated control groups (Sharratt et al., 1964).

3.2 Other relevant biological data

(a) Experimental systems

Toxic effects

The oral LD_{50} of benzoyl peroxide in rats is >950 mg/kg bw (National Institute for Occupational Safety and Health, 1977). The intraperitoneal LD_{50} of benzoyl peroxide in mice was reported to be 17-20 μmol (206-242 mg/kg bw); most of the deaths occurred within hours of the injection (Horgan et al., 1957; Philpot & Roodyn, 1959).

Benzoyl peroxide dust irritates the eyes of albino rabbits if not washed out within five minutes after being placed in the conjunctival sac. Application of a 10% solution of benzoyl peroxide in propylene glycol to the skin of guinea-pigs resulted in slight to moderate erythema (National Institute for Occupational Safety and Health, 1977).

A single application of either 20 or 40 mg benzoyl peroxide to the skin of mice induced marked hyperplasia and a temporary increase in the number of dark basal keratinocytes (Klein-Szanto & Slaga, 1982).

In chronic feeding studies, rats were given a flour-based diet containing a commercial powder (Novadelox), which contains 18% benzoyl peroxide and is used to treat flour (calculated doses of benzoyl peroxide: 28, 280 and 2800 mg/kg of diet). There was a slight depression in growth rate at the high- and medium-dose levels. An increased incidence of testicular atrophy was observed in the high-dose group and in rats receiving diets containing bread prepared from flour that had been treated with the 'usual' and 'ten-times usual' commercially used levels of Novadelox. The authors suggested that destruction of tocopherols in the diet by benzoyl peroxide was responsible for these effects. Similar effects were not observed in mice (Sharratt et al., 1964).

Effects on reproduction and prenatal toxicity

Doses of 0.05, 0.10, 0.21, 0.42, 0.83 and 1.7 μmol benzoyl peroxide (moistened, purum) [purity unspecified] were dissolved in acetone and injected onto the inner shell membrane in the air chamber of three-day-old White Leghorn chicken eggs. There was a dose-related increase in early embryonic deaths at all but the lowest dose level, with an LD_{50} estimated at 0.99 μmol/egg. However, the dose-response curve was flat for the three highest doses, indicating saturation of penetration. Only 1/80 control embryos was malformed, but all doses of benzoyl peroxide increased the malformation rate, although no clear dose-response was evident [perhaps, due to the increasing rate of embryonic death]. The ED_{50} for mortality and malformations was calculated to be 0.27 μmol/egg (Korhonen et al., 1984).

Absorption, distribution, excretion and metabolism

No data were available to the Working Group.

Mutagenicity and other short-term tests (see also 'Appendix: Activity Profiles for Short-term Tests', p. 343)

Benzoyl peroxide was not mutagenic to *Salmonella typhimurium* TA1535, TA1537, TA92, TA94, TA98 or TA100 either in the presence or absence of an exogenous metabolic system (S9 from polychlorinated biphenyl-induced rat liver) (Ishidate *et al.*, 1980).

Benzoyl peroxide did not induce polyploidy or chromosomal aberrations in cultured Chinese hamster lung cells (Ishidate *et al.*, 1980).

No significant increase in dominant lethal mutation rate was observed in ICR/Ha Swiss mice following an intraperitoneal injection of 54 or 62 mg/kg bw benzoyl peroxide (Epstein *et al.*, 1972).

It was reported in an abstract that benzoyl peroxide induced a dose-dependent increase in the incidence of sister chromatid exchanges in Chinese hamster ovary cells in the presence of S9; no such effect was observed in the absence of S9 (Järventaus *et al.*, 1984).

Intercellular communication between Chinese hamster V79 cells, measured by metabolic cooperation between HGPRT$^+$ and HGPRT$^-$ cells (Slaga *et al.*, 1981), and between cultured human epidermal keratinocytes, measured by ^{3}H-uridine transfer (Lawrence *et al.*, 1984), was inhibited by benzoyl peroxide at non-cytotoxic concentrations.

(b) Humans

Toxic effects

Case reports have cited an allergic dermal response to benzoyl peroxide (Eaglstein, 1968; Poole *et al.*, 1970).

No data were available to the Working Group on effects on reproduction and prenatal toxicity, on absorption, distribution, excretion and metabolism, or on mutagenicity and chromosomal effects.

3.3 Case reports and epidemiological studies of carcinogenicity to humans

Sakabe and Fukuda (1977) reported two cases of lung cancer in workers in a small plant in Japan where benzoyl peroxide and benzoyl chloride (see IARC, 1982a) were produced. The factory population from which the two cases were derived varied from 13 individuals in 1952 to 40 in 1963. One of the cases was a 40-year-old male smoker with 17 years of employment in the manufacture of benzoyl peroxide and intermittent exposure to benzoyl chloride over seven years; the other case was a 35-year-old male nonsmoker with a squamous-cell carcinoma, who had had about four years of exposure to the benzoyl peroxide-production process starting about 15 years prior to diagnosis, and had worked for one year in benzoyl chloride production. Both cases would also have been exposed to a number of precursors in the production process, including benzotrichloride (see IARC, 1982c). The authors attributed the two cases to exposure to benzoyl chloride or to benzotrichloride and linked this observation to four cases observed in another plant where benzoyl chloride was manufactured (IARC, 1982a). [The Working Group noted that these cases were also exposed to benzoyl peroxide.]

4. Summary of Data Reported and Evaluation

4.1 Exposure data

Benzoyl peroxide has been produced commercially since 1921. Major sources of exposure are its use in the plastics industry (principally for polyester resin production) and in acne medications. Another potential source is its use in bleaching foods.

4.2 Experimental data

Benzoyl peroxide was tested for carcinogenicity in mice and rats by oral administration in the diet and by subcutaneous administration, and in mice by skin application. In three studies by skin application in mice, benzoyl peroxide was tested for either initiating or promoting activity. All of the studies were inadequate for an evaluation of complete carcinogenicity; two studies indicated that benzoyl peroxide has promoting activity in mouse skin.

The available data are inadequate to evaluate the teratogenic potential of benzoyl peroxide in mammals.

Benzoyl peroxide was not mutagenic to bacteria. It did not induce chromosomal aberrations in mammalian cells *in vitro* and did not induce dominant lethal mutations in mice.

Overall assessment of data from short-term tests: benzoyl peroxide[a]

	Genetic activity			Cell transformation
	DNA damage	Mutation	Chromosomal effects	
Prokaryotes		−		
Fungi/green plants				
Insects				
Mammalian cells (*in vitro*)			−	
Mammals (*in vivo*)			−	
Humans (*in vivo*)				
Degree of evidence in short-term tests for genetic activity: *Inadequate*				Cell transformation: No data

[a]The groups into which the table is divided and the symbol - are defined on pp. 17-18 of the Preamble; the degrees of evidence are defined on p. 18.

4.3 Human data

Among a small factory population, two cases of lung cancer were found in young men who were involved primarily in the production of benzoyl peroxide but were also exposed to benzoyl chloride and other chemicals.

4.4 Evaluation[1]

There is *inadequate evidence* for the carcinogenicity of benzoyl peroxide to experimental animals.

There is *inadequate evidence* for the carcinogenicity of benzoyl peroxide to humans.

No evaluation could be made of the carcinogenicity to humans of benzoyl peroxide.

5. References

American Conference of Governmental Industrial Hygienists (1984) *TLVs Threshold Limit Values for Chemical Substances in Work Environment Adopted by ACGIH for 1984-85*, Cincinnati, OH, p. 10

American Pharmaceutical Association (1982) *Handbook of Nonprescription Drugs*, 7th ed., Washington DC, pp. 552, 554-559, 633, 636

Anon. (1981) Organic peroxides. Lucidol still works at closing the loopholes. *Corpus Market Focus*, 9 February, 5691.1.1-5691.I.2

Asmussen, E. (1980) Quantitative analysis of peroxides in restorative resins. *Acta odontol. scand.*, *38*, 269-272 [*Chem. Abstr.*, *94*, 71612c]

Bloom, T.F. (1975) *Health Hazard Evaluation/Toxicity Determination Report H.H.E. 73-131-196, General Motors Corporation, La Grange, Illinois (PB 249 376)*, Cincinnati, OH, National Institute for Occupational Safety and Health

Burke, D.P. (1979) Catalysts II. Chemicals make it a $1 billion a year market. *Chem. Week*, 4 April, 46-47, 81

Dailey, R. (1978) *Monograph on Benzoyl Peroxide (PB-284 877)*, Prepared for US Food and Drug Administration by Informatics Inc., Springfield, VA, National Technical Information Service, pp. 1, 5-6

Deutsche Forschungsgemeinschaft (1984) *Maximal Work Place Concentrations and Biological Tolerance Values for Compounds in the Work Place* (Ger.), Part XX, Weinheim, Verlag Chemie GmbH, p. 27

Eaglstein, W.H. (1968) Allergic contact dermatitis to benzoyl peroxide. *Arch. Dermatol.*, *97*, 527

Ehinger, C. & Andermann, G. (1980) Simultaneous, rapid, micro-HPLC determination of benzoyl peroxide and benzoic acid in blood. *J. high Resolut. Chromatogr. Chromatogr. Commun.*, *3*, 143-144

---

[1]For definitions of the italicized terms, see the Preamble, pp. 15-16.

Epstein, S.S., Arnold, E., Andrea, J., Bass, W. & Bishop, Y. (1972) Detection of chemical mutagens by the dominant lethal assay in the mouse. *Toxicol. appl. Pharmacol.*, *23*, 288-325

Federation of American Societies for Experimental Biology (1980) *Evaluation of the Health Aspects of Benzoyl Peroxide as a Food Ingredient* (*SCOGS-II-2; PB81-127854*), Springfield, VA, National Technical Information Service, p. 5

Fuller, M. (1982) Analysis of paint-on artificial nails. *J. Soc. cosmet. Chem.*, *33*, 51-74

Gunderson, E.C. & Anderson, C.C. (1980) *Development and Validation of Methods for Sampling and Analysis of Workplace Toxic Substances* (*DHHS (NIOSH) Publ. No. 80-133*), Cincinnati, OH, National Institute for Occupational Safety and Health

Hawley, G.G., ed. (1981) *The Condensed Chemical Dictionary*, 10th ed., New York, Van Nostrand Reinhold Co., p. 120

Horgan, V.J., Philpot, J.St.L., Porter, B.W. & Roodyn, D.B. (1957) Toxicity of autoxidized squalene and linoleic acid, and of simpler peroxides, in relation to toxicity of radiation. *Biochem. J.*, *67*, 551-558

Horwitz, W., ed. (1980) *Official Methods of Analysis of the Association of Official Analytical Chemists*, 13th ed., Washington DC, Association of Official Analytical Chemists, pp. 216-217

Hueper, W.C. (1964) Cancer induction by polyurethan and polysilicone plastics. *J. natl Cancer Inst.*, *33*, 1005-1027

IARC (1979) *IARC Monographs on the Evaluation of the Carcinogenic Risk of Chemicals to Humans, Vol. 19, Some Monomers, Plastics and Synthetic Elastomers, and Acrolein*, Lyon, pp. 245-274

IARC (1982a) *IARC Monographs on the Evaluation of the Carcinogenic Risk of Chemicals to Humans, Vol. 29, Some Industrial Chemicals and Dyestuffs*, Lyon, pp. 83-91

IARC (1982b) *IARC Monographs on the Evaluation of the Carcinogenic Risk of Chemicals to Humans, Vol. 28, The Rubber Industry*, Lyon, p. 316

IARC (1982c) *IARC Monographs on the Evaluation of the Carcinogenic Risk of Chemicals to Humans, Vol. 29, Some Industrial Chemicals and Dyestuffs*, Lyon, pp. 73-82

International Labour Office (1980) *Occupational Exposure Limits for Airborne Toxic Substances*, 2nd (rev.) ed. (*Occupational Safety and Health Series No. 37*), Geneva, pp. 50-51

Ishidate, M., Jr, Sofuni, T. & Yoshikawa, K. (1980) A primary screening for mutagenicity of food additives in Japan (Jpn.). *Hen'lgen to Dokusei* (Mutagens Toxicol.), *3*, 82-80

Järventaus, H., Norppa, H., Linnainmaa, K. & Sorsa, M. (1984) SCE induction in CHO cells by peroxides used in the plastic industry (Abstract II-3C-8). *Mutat. Res.*, *130*, 249

Karasz, A.B., DeCocco, F. & Maxstadt, J.J. (1974) Gas chromatographic measurement of benzoyl peroxide (as benzoic acid) in cheese. *J. Assoc. off. anal. Chem.*, 57, 706-709

Klein-Szanto, A.J.P. & Slaga, T.J. (1982) Effects of peroxides on rodent skin: Epidermal hyperplasia and tumor promotion. *J. invest. Dermatol.*, 79, 30-34

Korhonen, A., Hemminki, K. & Vainio, H. (1984) Embryotoxic effects of eight organic peroxides and hydrogen peroxide on three-day chicken embryos. *Environ. Res.*, 33, 54-61

Lawrence, N.J., Parkinson, E.K. & Emmerson, A. (1984) Benzoyl peroxide interferes with metabolic co-operation between cultured human epidermal keratinocytes. *Carcinogenesis*, 5, 419-421

Luperox GmbH (1979) *Safe Handling of Organic Peroxides (Brochure 30.40/E)*, Günzburg

Mackison, F.W., Stricoff, R.S. & Partridge, L.J., Jr, eds (1981) *NIOSH/OSHA Occupational Health Guidelines for Chemical Hazards: Benzoyl Peroxide (DHHS (NIOSH) Publication No. 81-123)*, Washington DC, US Government Printing Office, pp. 1-4

Mair, R.D. & Hall, R.T. (1971) *Determination of organic peroxides by physical, chemical, and colorimetric methods*. In: Swern, D., ed., *Organic Peroxides*, Vol. II, New York, John Wiley & Sons, pp. 535-635

Markel, H.L., Jr & Jannerfeldt, E. (1981) *Health Hazard Evaluation Report No. HHE-79-156-899, Gulf-Wandes Corporation, Baton Rouge, Louisiana*, Cincinnati, OH, National Institute for Occupational Safety and Health

Markel, H.L., Jr & Wilcox, T. (1981) *Health Hazard Evaluation Report No. HHE-79-104-838, A.O. Smith-Inland, Inc., Little Rock, Arkansas*, Cincinnati, OH, National Institute for Occupational Safety and Health

Miyazaki, K., Horibe, T. & Inoue, H. (1981) Analysis on the resin composition of dental composite materials and bonding agents. *Fukuoka Shika Daigaku Gakki Zasshi*, 8, 277-287 [*Chem. Abstr.*, 98, 132268b]

National Finnish Board of Occupational Safety and Health (1981) *Airborne Contaminants in the Work Places (Safety Bull. No. 3)*, Helsinki, p. 8

National Institute for Occupational Safety and Health (1977) *Criteria for a Recommended Standard...Occupational Exposure to Benzoyl Peroxide (DHEW (NIOSH) Publ. No. 77-166)*, Washington DC, US Government Printing Office

National Institute for Occupational Safety and Health (1980) *Projected Number of Occupational Exposures to Chemical and Physical Hazards*, Cincinnati, OH, p. 13

National Institute for Occupational Safety and Health (1981) *National Occupational Hazard Survey* (microfiche), Cincinnati, OH, pp. 846-847 (generics excluded)

National Institute for Occupational Safety and Health (1983) *NIOSH Criteria Document Review*, Cincinnati, OH, p. 22

National Research Council (1981) *Food Chemicals Codex*, 3rd ed., Washington DC, National Academy Press, pp. 35-36

Paffenbarger, G.C. & Rupp, N.W. (1979) Dental materials. In: Grayson, M., ed., Kirk-Othmer Encyclopedia of Chemical Technology, 3rd ed., Vol. 7, New York, John Wiley & Sons, pp. 466, 504-505, 521

Pennwalt Corporation (1983) Product Bulletin: Diacyl Peroxides, Buffalo, NY, Lucidol Division

Philpot, J.St.L. & Roodyn, D.B. (1959) A comparison between the effects in mice of injected organic peroxides and of whole-body X-irradiation. Int. J. Radiat. Biol., 4, 372-382

Poirier, L.A., Miller, J.A., Miller, E.C. & Sato, K. (1967) N-Benzoyloxy-N-methyl-4-aminoazobenzene: Its carcinogenic activity in the rat and its reactions with proteins and nucleic acids and their constitutents in vitro. Cancer Res., 27, 1600-1613

Poole, R.L., Griffith, J.F. & MacMillan, F.S.K. (1970) Experimental contact sensitization with benzoyl peroxide. Arch. Dermatol., 102, 635-639

Purnell, C.J., Bagon, D.A. & Warwick, C.J. (1982) The determination of organic contaminant concentrations in workplace atmospheres by high-performance liquid chromatography. Pergamon Ser. environ. Sci., 7, 203-219

Reiners, J.J., Jr, Nesnow, S. & Slaga, T.J. (1984) Murine susceptibility to two-stage skin carcinogenesis is influenced by the agent used for promotion. Carcinogenesis, 5, 301-307

Saffiotti, U. & Shubik, P. (1963) Studies on promoting action in skin carcinogenesis. Natl Cancer Inst. Monogr., 10, 489-507

Sakabe, H. & Fukuda, K. (1977) An updating report on cancer among benzoyl chloride manufacturing workers. Ind. Health, 15, 173-174

Sharratt, M., Frazer, A.C. & Forbes, O.C. (1964) Study of the biological effects of benzoyl peroxide. Food Cosmet. Toxicol., 2, 527-538

Sheppard, C.S. & Mageli, O.L. (1982) Peroxides and peroxy compounds, organic. In: Grayson, M., ed., Kirk-Othmer Encyclopedia of Chemical Technology, 3rd ed., Vol. 17, New York, John Wiley & Sons, pp. 63-72, 90

Silbert, L.S. (1971) Physical properties of organic peroxides. In: Swern, D., ed., Organic Peroxides, Vol. II, New York, John Wiley & Sons, pp. 637-639, 678-681, 688-690, 704

Slaga, T.J., Klein-Szanto, A.J.P., Triplett, L.L. & Yotti, L.P. (1981) Skin tumor-promoting activity of benzoyl peroxide, a widely used free radical-generating compound. Science, 213, 1023-1025

US Department of Transportation (1980) Identification numbers, hazardous wastes, hazardous substances, international descriptions, improved descriptions, forbidden materials, and organic peroxides. US Code Fed. Regul., Title 49, Parts 171-174, 176-177; Fed. Regist., 45 (No. 101), pp. 34560, 34596

US Food and Drug Administration (1982a) Benzoyl peroxide; Proposed affirmation of GRAS status. US Code Fed. Regul., Title 21, Part 184; Fed. Regist., 47 (No. 191), pp. 43402-43404

US Food and Drug Administration (1982b) Topical acne drug products for over-the-counter human use; Establishment of a monograph. *US Code Fed. Regul., Title 21*, Part 333; *Fed. Regist., 47* (No. 56), pp. 12430-12433, 12439, 12443-12449, 12474-12477

US International Trade Commission (1976) *Synthetic Organic Chemicals, US Production and Sales, 1974* (*USITC Publication 776*), Washington DC, US Government Printing Office, pp. 197, 205

US International Trade Commission (1977) *Imports of Benzenoid Chemicals and Products, 1975* (*USITC Publication 806*), Washington DC, US Government Printing Office, p. 12

US International Trade Commission (1983a) *Synthetic Organic Chemicals, US Production and Sales, 1982* (*USITC Publication 1422*), Washington DC, US Government Printing Office, pp. 257, 263

US International Trade Commission (1983b) *Imports of Benzenoid Chemicals and Products, 1982* (*USITC Publication 1401*), Washington DC, US Government Printing Office, p. 10

US Pharmacopeial Convention, Inc. (1980) *The United States Pharmacopeia*, 20th rev., Rockville, MD, p. 75

Van Duuren, B.L., Nelson, N., Orris, L., Palmes, E.D. & Schmitt, F.L. (1963) Carcinogenicity of epoxides, lactones, and peroxy compounds. *J. natl Cancer Inst., 31*, 41-55

Verschueren, K. (1977) *Handbook of Environmental Data on Organic Chemicals*, New York, Van Nostrand Reinhold, p. 220

HYDROGEN PEROXIDE

1. Chemical and Physical Data

1.1 Synonyms and trade names

Chem. Abstr. Services Reg. No.: 7722-84-1

Chem. Abstr. Name: Hydrogen peroxide

IUPAC Systematic Name: Hydrogen peroxide

Synonyms: Dihydrogen dioxide; hydrogen dioxide; hydrogen oxide; hydrogen oxide, per-; hydroperoxide; oxydol; peroxide

Trade Names: Albone; Albone DS; Hioxyl; Inhibine; Perhydrol; Perone; Peroxaan; Superoxol; T-Stuff

1.2 Structural and molecular formulae and molecular weight

$$H_2O_2$$

H_2O_2 Mol. wt: 34.0

1.3 Chemical and physical properties of the pure substance

From Kirchner (1981), unless otherwise specified

(a) *Description*: A weakly acidic, clear, colourless liquid

(b) *Boiling-point*: 150.2°C

(c) *Melting-point*: -0.41°C

(d) *Density*: 1.4425 at 25°C

(e) *Spectroscopy data*: A review of spectroscopic data has been published (Silbert, 1971)

(f) *Identity and purity test*: When aqueous solutions are treated with dilute sulphuric acid and diethyl ether, followed by addition of a potassium dichromate solution, a blue colour forms in the aqueous layer and passes into the diethyl ether layer after agitation and standing (National Research Council, 1981).

(g) *Solubility*: Miscible with water; soluble in diethyl ether; insoluble in petroleum ether; decomposed by many organic solvents (Windholz, 1983)

(h) *Viscosity*: 1.245 cP at 20°C

(i) *Stability*: Pure aqueous solutions in clean inert containers are relatively stable (stability increases with increasing concentration and is at a maximum at pH 3.5-4.5), but commercial solutions must be stabilized with additives to prevent possibly violent decomposition due to catalytic impurities.

(j) *Reactivity*: Undergoes a variety of reactions (e.g., molecular additions, substitutions, oxidations and reductions); a strong oxidant; can form free radicals by homolytic cleavage

(k) *Conversion factor*: 1 ppm = 1.39 mg/m^3 at 760 mm Hg and 25°C [calculated by the Working Group]

1.4 Technical products and impurities

Hydrogen peroxide is available only as aqueous solutions containing 3-98% hydrogen peroxide. Grades containing 35, 50, 70, or 90% are the most commonly used for industrial applications and in the laboratory; 3-6% grades are used for cosmetic and medical applications. A variety of stabilizers may be present, depending on the concentration of the solution and its intended use. The quantity of stabilizers required decreases with increasing hydrogen peroxide concentration. Sodium pyrophosphates, sodium stannate, combinations of tin salts and phosphates, and alkali metal silicates are reportedly used as inorganic stabilizers, while certain organic compounds can be used for dilute solutions (Kirchner, 1981). Acetanilide and similar organic compounds have been reported to be used as stabilizers (Windholz, 1983).

In the USA, to meet the requirements of the Food Chemicals Codex, 30-50% aqueous solutions of hydrogen peroxide must pass an identification test and meet the following specifications: purity, as labelled; acidity (as sulphuric acid), 0.03% max; phosphate, 0.005% max; residue on evaporation, 0.006% max; heavy metals (as lead), 0.001% max; tin, 0.001% max; arsenic, 0.0003% max; and iron, 0.00005% max (National Research Council, 1981).

Hydrogen peroxide topical solution USP and hydrogen peroxide concentrate USP must meet certain specifications. Both products must pass an identification test and meet the requirements of tests for acidity, nonvolatile residue and heavy metals, as well as having a content of a suitable preservative or preservatives of 0.05% max. Hydrogen peroxide topical solution USP must contain 2.5-3.5 g hydrogen peroxide per 100 ml, while hydrogen peroxide concentrate USP must contain 29.0-32.0 wt % hydrogen peroxide (US Pharmacopeial Convention, Inc., 1980).

HYDROGEN PEROXIDE

2. Production, Use, Occurrence and Analysis

2.1 Production and use

(a) Production

Hydrogen peroxide was isolated by Thenard in 1818 and has been an item of commerce since the mid-nineteenth century (Kirchner, 1981). Although it has been produced commercially by the following processes, none is used in current commercial practice: (1) reaction of mineral acids with barium peroxide; (2) hydrolysis of organic peracids; (3) autoxidation of various hydrocarbons; (4) reaction of oxygen with secondary alcohols, such as isopropanol; and (5) direct combination of hydrogen and oxygen. Most hydrogen peroxide is made commercially by the hydrogenation of alkyl anthraquinones to the corresponding anthrahydroquinone and oxidation of this to yield hydrogen peroxide and the original alkyl anthraquinone. Much less used is a method in which acid sulphate solutions are subjected to electrolysis to produce persulphates, which are hydrolysed to give hydrogen peroxide and the starting sulphate.

Electrolytic processes for hydrogen peroxide production were introduced in the USA in about 1925, and the first anthraquinone process was used in Germany during the Second World War. All manufacturing facilities built in the USA since 1957 have used the anthraquinone process, and the last US plant using electrolysis closed in 1983. US production has grown steadily from 26 thousand tonnes (100% basis) in 1960 (Kirchner, 1981) to a peak of 105.6 thousand tonnes (including unspecified amounts that were produced but not withdrawn from the system) in 1980; 98.4 thousand tonnes were made in 1982 (US Department of Commerce, 1984) at the seven plants of the five US producing companies. Currently, three US companies produce hydrogen peroxide in four plants.

US imports of hydrogen peroxide (100% active, on the assumption that statistics are for a gross weight of 60% active material) amounted to 2.9 thousand tonnes in 1982 (US Department of Commerce, 1983a), down sharply from the 10.3 thousand tonnes imported in 1979 (US Department of Commerce, 1980a). US exports (100% active, on the assumption that statistics are for a gross weight of 65% active material) amounted to 8.0 thousand tonnes in 1982 (US Department of Commerce, 1983b), up from the 5.4 thousand tonnes exported in 1979 (US Department of Commerce, 1980b).

Hydrogen peroxide is not currently produced in Canada. Two companies have announced plans for plants with a total production capacity of 46 thousand tonnes for 1986.

Hydrogen peroxide is produced by two companies each in Austria, Belgium, France, Italy and Spain and by one company each in Finland, Germany, Greece, the Netherlands, Portugal, Sweden and the UK. Only two of these plants use the electrolysis process; all the others use the anthraquinone process. Total production was about 295 thousand tonnes in 1982; total production capacity in western Europe is estimated to have been about 445 thousand tonnes per year in early 1984.

Hydrogen peroxide has been produced commercially in Japan for over 50 years. Currently, four companies produce it by the anthraquinone process, and total production in 1982 is estimated to have been 85 thousand tonnes, up from 78 thousand tonnes in 1980. Japanese imports were negligible in 1982 and exports totalled about 2.5 thousand tonnes.

Hydrogen peroxide is produced by two companies in the Republic of Korea and one company in Taiwan.

(b) Use

The largest use of hydrogen peroxide is as a chemical intermediate; the next largest are textile bleaching and pulp and paper bleaching. The use pattern for the estimated 94 thousand tonnes of hydrogen peroxide used in the USA in 1982 was as follows: chemical intermediate, 26%; textiles, 18%; pulp and paper, 16%; water treatment, 13%; geothermal energy, 5%; metal cleaning, 3%; mining, 2%; and other uses, 17%.

The largest use as a chemical intermediate in the USA is in the synthesis of inorganic peroxygen compounds, accounting for about 9% of total US usage of hydrogen peroxide. The principal product is sodium perborate, and another is sodium percarbonate; both are used as bleaches. Hydrogen peroxide is also used in large amounts as an intermediate in the synthesis of plasticizers by epoxidizing unsaturated vegetable oils or fatty esters and in the synthesis of organic peroxygen compounds. Important members of the latter group are methyl ethyl ketone peroxide, benzoyl peroxide (see p. 267 of this volume), lauroyl peroxide (see p. 315 of this volume) and various peroxycarbonates. Another group of chemicals made from hydrogen peroxide is the amine oxides (e.g., lauryldimethylamine oxide). Hydrogen peroxide was used in one commercial method for making glycerol, but this process is no longer used.

Hydrogen peroxide is used to bleach fabrics based on natural fibres, and in particular cotton.

In the pulp and paper industry, hydrogen peroxide is principally used to bleach mechanical wood pulps when relatively high quality and brightness are needed. It can also be used in bleaching chemical pulps and in the processing of recycled paper.

Hydrogen peroxide finds use in both industrial and municipal waste-water treatment, principally to eliminate pollutants that are refractory to other water-treatment compounds.

Other uses for hydrogen peroxide in the USA are in the removal of hydrogen sulphide from the steam produced by geothermal power plants, use in various steps during the mining and processing of uranium, pickling of copper and copper alloys, cleaning metals (germanium) and silicon semiconductors used in the electronics industry, and a variety of small-volume applications in photography, cosmetics (e.g., hair bleaches and dyes, mouthwashes, etc.), antiseptics and cleansing agents, food and wine processing, rocket fuels, and treatment of package liners in aseptic packaging.

The International Labour Office (1972) reported that hydrogen peroxide has been used as a foaming agent for foam resins and as a source of oxygen in respiratory protective equipment.

Hydrogen peroxide has also been reported to be a component of non-prescription drugs such as topical anti-infectants, canker-sore treatments, and products used to soften earwax (American Pharmaceutical Association, 1982). A 3% solution of hydrogen peroxide has been used widely as a cleansing and topical antiseptic agent for suppurative wounds and inflammation of the skin and mucous membranes, as well as by the dental profession for irrigation during root-canal therapy and as a mouth rinse for acute necrotizing gingivitis. A 30% solution has been used for bleaching vital and pulpless teeth (US Food and Drug Administration, 1982).

In Canada, about 70-75% of total hydrogen peroxide usage is in pulp and paper bleaching. Textile bleaching and water treatment are also significant markets.

The use pattern for the estimated 290 thousand tonnes of hydrogen peroxide used in western Europe in 1982 was as follows: chemical intermediate (principally for sodium perborate and sodium percarbonate), 58%; pulp and paper, 25%; textiles, 12%; and other uses, 5%.

An estimated 84 thousand tonnes of hydrogen peroxide were used in Japan in 1983, with the following use pattern: pulp and paper, 45%; industrial chemicals, 30%; textiles, 20%; food, 3%; and pharmaceuticals and other uses, 2%.

Occupational exposure to hydrogen peroxide has been limited by regulation or recommended guidelines in at least 11 countries. The standards are listed in Table 1.

Table 1. National occupational exposure limits for hydrogen peroxide[a]

Country	Year	Concentration		Interpretation[b]	Status
		mg/m^3	ppm		
Australia	1978	1.4	1	TWA	Guidelines
Belgium	1978	1.4	1	TWA	Regulation
Bulgaria	1971	1	-	Maximum	Regulation
Finland	1981	1.4	1	TWA	Guideline
		4.2	3	STEL	
Germany, Federal Republic of	1984	1.4	1	TWA	Guideline
Italy	1978	1.4	1	TWA	Guideline
Netherlands	1978	1.4[c]	1	TWA	Guideline
Switzerland	1978	1.4	1	TWA	Regulation
USA[d]					
OSHA	1978	1.4	1	TWA	Regulation
ACGIH	1984-85	-	75	Maximum (30 min)	Guideline
		1.5	1	TWA	
		3	2	STEL	
USSR	1977	2	-	Maximum	Regulation
Yugoslavia	1971	1.4	1	Ceiling	Regulation

[a]From International Labour Office (1980); National Finnish Board of Occupational Safety and Health (1981); American Conference of Governmental Industrial Hygienists (1984); Deutsche Forschungsgemeinschaft (1984)
[b]TWA, time-weighted average; STEL, short-term exposure limit
[c]For 90% hydrogen peroxide
[d]OSHA, Occupational Safety and Health Administration; ACGIH, American Conference of Industrial Hygienists

The US Department of Agriculture (1980) has approved the use of hydrogen peroxide as a bleaching agent for tripe, provided it is removed from the product by rinsing with clear water.

The Bureau of Alcohol, Tobacco and Firearms of the US Department of the Treasury (1982) has authorized the use of hydrogen peroxide in the production and treatment of wine, juice and distilling material. The proposed uses and limitations are summarized in Table 2.

Hydrogen peroxide has been approved for use in a variety of foods and related processing steps by the US Food and Drug Administration (1983a). After a comprehensive safety

Table 2. US Department of the Treasury regulations on use of hydrogen peroxide in wine, juice and distilling material[a]

Product	Use	Reference of limitation
Wine	To facilitate secondary fermentation in the production of sparkling wine	The amount used shall not exceed 3 mg/kg; the finished wine shall contain no residual hydrogen peroxide.
Grape juice	To remove colour from the juice of red and black grapes	The amount used shall not exceed 500 mg/kg; the use of hydrogen peroxide is limited to oxidizing colour pigment in the juice of red and black grapes; wine produced by fermentation of such juice is limited solely to blending with white wines and red wines derived from such juice not subjected to such treatment.
Distilling material	To reduce aldehydes in distilling material	The amount used shall not exceed 200 mg/kg.

[a]From US Department of the Treasury (1982)

review, it has been proposed to affirm that hydrogen peroxide is generally recognized as safe, with specific limitations, as a direct human food ingredient. The proposed uses and limitations are summarized in Table 3.

Table 3. US Food and Drug Administration proposed regulations on direct food use of hydrogen peroxide[a]

Food	Maximum treatment level in food (%)	Functional use	Limitations or restrictions
Milk intended for use during cheese-making, as permitted in the appropriate standards of identity for cheese and related cheese products	0.05	Antimicrobial agent	Residual hydrogen peroxide is removed by addition of catalase.
Whey, during the preparation of modified whey by electrolysis methods	0.04	Antimicrobial agent	Residual hydrogen peroxide is removed by addition of catalase.
Dried eggs, dried egg whites and dried egg yolks	Amount sufficient for the purpose	Oxidizing and reducing agent	Residual hydrogen peroxide is removed by addition of catalase.
Tripe	Amount sufficient for the purpose	Bleaching agent	Residual hydrogen peroxide is removed by a potable water rinse.
Beef feet	Amount sufficient for the purpose (hydrogen peroxide may be in the form of a compound salt, sodium carbonate peroxide)	Bleaching agent	Residual hydrogen peroxide is removed by a potable water rinse.
Herring	Amount sufficient for the purpose	Bleaching agent	No hydrogen peroxide residue is permitted in the final product.
Wine	Amount sufficient for the purpose	Oxidizing and reducing agent	No hydrogen peroxide residue is permitted in the final product.

[a]From US Food and Drug Administration (1983a)

The FAO/WHO Expert Committee on Food Additives (WHO, 1980) concluded that, when no adequate cooling facilities are available, hydrogen peroxide is an acceptable alternative for preserving milk. No acceptable daily intake (ADI) level has been allocated to this compound.

In Japan, hydrogen peroxide has been used for a long time as a food additive for its antiseptic and bleaching properties. Since 1 October 1980, residues of hydrogen peroxide in the final product must not exceed 0.1 mg/kg (Ito et al., 1981a).

An advisory review panel on over-the-counter drugs of the US Food and Drug Administration (1979) concluded that hydrogen peroxide is safe and effective for use in oral wound cleaners but that there was insufficient evidence to decide whether it was safe and effective for use in oral wound-healing agents. Another advisory review panel of the US Food and Drug Administration (1982) concluded that hydrogen peroxide is safe but that there was insufficient evidence to decide whether it was effective for use as an antimicrobial agent in oral health-care drug products for over-the-counter human use. In a tentative final monograph, the US Food and Drug Administration (1983b) proposed a rule requiring that any oral wound-healing products containing hydrogen peroxide not be marketed unless they were the subject of an approved new drug application.

As part of the Hazardous Materials Regulations of the US Department of Transportation (1980), shipments of hydrogen peroxide are subject to a variety of labelling, packaging, quantity and shipping restrictions consistent with its designation as a hazardous material.

2.2 Occurrence

(a) *Natural occurrence*

Gaseous hydrogen peroxide is recognized to be a key component and product of the earth's lower atmospheric photochemical reactions, both in clean and polluted atmospheres. Atmospheric hydrogen peroxide is believed to be generated exclusively by gas-phase photochemical reactions. In the remote troposphere, primary gas-phase photochemical hydrogen peroxide is generated *via* reaction (1),

$$HO_2 + HO_2 \rightarrow H_2O_2 + O_2 \quad (1)$$

while H_2O_2 may be removed by photolysis as in (2),

$$H_2O_2 + h\nu \rightarrow 2\ OH \quad (2)$$

reaction with OH as in (3),

$$H_2O_2 + OH \rightarrow HO_2 + H_2O \quad (3)$$

or by heterogenous loss processes such as rain-out and wash-out (Zika et al., 1982; Das et al., 1983). On the basis of this mechanism, recent photochemical model calculations predict lower and mid-tropospheric hydrogen peroxide levels of the order of 1 ppb by volume (1.4 µg/m^3) (Zika et al., 1982).

In February 1981, the level of hydrogen peroxide in the earth's stratosphere was tentatively estimated to be 1.1 ± 0.5 ppb by volume (1.5 µg/m^3) in a layer between heights of approximately 27 and 35 km from the National Scientific Balloon Facility in Palestine, Texas (Waters et al., 1981).

(b) *Occupational exposure*

On the basis of the 1974 National Occupational Hazard Survey, the National Institute for Occupational Safety and Health (1980, 1981) estimated that 478 000 US workers in 168 industries were exposed to hydrogen peroxide.

(c) Air

Significantly higher hydrogen peroxide concentrations are found in polluted atmospheres, as compared to clean air. They are believed to arise from photochemically-initiated oxidation of reactive hydrocarbons. Hydrogen peroxide concentrations in unpolluted rural air range from 0.3-3 ppb (0.4-4.2 µg/m^3), whereas night-time levels may drop to undetectable limits of 0.01 ppb (Das et al., 1983). However, under severe smog conditions, levels as high as 0.18 ppm (0.25 mg/m^3) have been reported (Bufalini et al., 1972); Kok (1983) reported that atmospheric night-time levels of 2-5 ppb (2.8-7 µg/m^3) did not apparently correlate with smog intensity.

(d) Water

Hydrogen peroxide levels in rain-water have been found to vary according to the level of atmospheric hydrogen peroxide washed out by precipitation (Kok, 1983); and to the formation of hydrogen peroxide cloud-water independently of contamination. Consequent hydrogen peroxide concentrations in rain-water ranged from undetectable to 2.6 mg/l (Zika et al., 1982). Surface-water concentrations have been found to vary between 51-231 mg/l, increasing both with exposure to sunlight and the presence of dissolved organic matter (Draper & Crosby, 1983).

Traces of hydrogen peroxide have been detected in power-plant cooling-water (Zabelin & Karbovnichii, 1983).

(e) Plants

In plant tissues, endogenous hydrogen peroxide has been found at the following levels (mg/kg frozen weight): potato tubers, 7.6; green tomatoes, 3.1; red tomatoes 3.5; and castor beans in water, 4.7 (Warm & Laties, 1982).

(f) Food, beverages and animal feeds

Owing to its use as a food additive for controlling the growth of microorganisms and for bleaching, hydrogen peroxide residues have been detected in foods; at concentrations as low as <0.05 mg/kg in solid foods and 0.02 mg/kg in liquid foods and at least as high as 1.5 mg/kg in solid food (Ito et al., 1981a; Toyoda et al., 1982).

(g) Human tissues and secretions

It is well established that hydrogen peroxide is produced metabolically in intact cells and tissues. It is formed by reduction of oxygen either directly in a two-electron transfer reaction, often catalysed by flavoproteins, or via an initial one-electron step to O_2^- followed by dismutation to hydrogen peroxide (Sies, 1981).

It has been identified in human breath at levels ranging from 1.0 ± 0.5 µg/l to 0.34 ± 0.17 µg/l (Williams et al., 1982) and has also been detected in human serum and liver (Nakane & Kosaka, 1980; Sies, 1981).

(h) Other

Hydrogen peroxide has been found at concentrations of about 10^{-8}M in intact bacterial cells (Micrococcus lysodeikticus) (Chance, 1952).

Trace amounts of hydrogen peroxide occur in polymers or resins, presumably due to its use as a polymerization initiator (Ehrlich & Capone, 1982).

Residues of hydrogen peroxide are present on food packaging surfaces, e.g., polyethylene films used in aseptic packaging systems, as a result of its use as a chemical sterilant (Chin & Cortes, 1983).

2.3 Analysis

Methods for the analysis of hydrogen peroxide in various matrices were reviewed in 1971 (Mair & Hall, 1971).

Typical methods for the analysis of hydrogen peroxide in various matrices are summarized in Table 4. In-situ hydrogen peroxide formation has been noted within sampling impingers (e.g.. Kok et al., 1978).

Table 4. Methods for the analysis of hydrogen peroxide

Sample matrix	Sample preparation	Assay procedure[a]	Limit of detection	Reference
Air	Collect in impingers containing distilled water; react with luminol using alkaline Cu(II) catalyst	Chemiluminescence	0.5 ppb (0.7 ng/l)	Kok et al. (1978); Das et al. (1983)
	Collect in aqueous gas washing traps; react with scopoletin and horseradish peroxidase	Fluorescence decay (365 nm and 490 nm)	At least 70 ng/l in trap water	Zika & Saltzman (1982)
Cooling water	Mix with phenolphthalein leuco base	Photometry	Not given	Zabelin & Karbovnichii (1983)
Water	Add leuco crystal violet, horseradish peroxidase and acetate buffer	Spectrophotometry (596 nm)	50 µg/l	Draper & Crosby (1983)
	Boil, cool, filter and irradiate	Thin-layer chromatography using peroxidase-catalysed leuco crystal violet spray	50 µg/l	Draper & Crosby (1983)
Rain and cloud water	React with alkaline luminol	Chemiluminescence	1 µg/l	Kok et al. (1978)
Cleaning antiseptic solutions	Place in closed gas collection system; add potassium permanganate and sulphuric acid	Determination of volume of oxygen released	Not given	Worley (1983)
Treated packaging films, e.g., polyethylene	Fill package with distilled water	Potentiometric titration	Upper limit, 0.20 mg/kg	Chin & Cortes (1983)
	Add distilled water, purified leuco crystal violet, horseradish peroxidase and acetate buffer	Spectrophotometry (596 nm)	0.02 mg/kg	Chin & Cortes (1983)
Pickling baths for copper and copper alloys	Dilute in a dispersion coil	Amperometric detection	3 mg for 5-µl samples; 30 µg/l for 50-µl samples	Lundbaeck (1983)
Industrial solution from manufacture of thiourea dioxide	React with titanium(4+) in dilute sulphuric acid	Spectrophotometry (410 nm)	Not given	Yakovleva et al. (1983)
Polymer solutions	Extract (aqueous chelating agent); add luminol	Chemiluminescence	Not given	Ehrlich & Capone (1982)

Sample matrix	Sample preparation	Assay procedure[a]	Limit of detection	Reference
Various Japanese foods	Homogenize; extract (potassium bromate-phosphate buffer); filter; purge with nitrogen; add catalase	Oxygen detection by oxygen electrode	0.1 mg/kg for solid food; 0.01 mg/kg for liquid food	Toyoda et al. (1982); Kawamura et al. (1983)
	Homogenize with aqueous acetic acid; centrifuge; filter; treat with acidic sulphate and ammonium thiocyanate	Spectrophotometry (480 nm)	0.05 mg/kg	Asai et al. (1982)
	Homogenize, extract (aqueous methanol); centrifuge; mix supernatant with phosphate buffer, hydrogen peroxide peroxidase and sesamol dimer; extract (chloroform)	Spectrophotometry (550 nm)	0.5 mg/kg	Kikugawa et al. (1982)
	Add potassium bromate; extract (methanol); add phosphate buffer and zinc sulphate; react with phenol, 4-aminoantipyrine and peroxidase; purify with Florisil and concentrate	Colorimetry (505 nm)	0.05 mg/kg	Ito et al. (1981a)
Cooked noodles	Homogenize in water; centrifuge; mix supernatant with sodium phosphate buffer containing 2,2'-azinobis(3-ethylbenzothiazoline-6-sulphonic acid) ammonium salt and peroxidase	Colorimetry (420, 660 or 740 nm)	At least 1 mg/kg	Hasegawa & Sugawara (1981)
	Extract (methanol); add catalase; add 3-methyl-2-benzothiazolone hydrazone and ferric chloride	Photometry	Not given	Nanase et al. (1981)
Milk	Dialyse; react with gum guaiac or guaiacol in the presence of peroxidase	Spectrophotometry	Not given	Iwaida et al. (1981)
Plant tissues	Transfer frozen sliced tissue into 5% thyrocalcitonin; homogenize; centrifuge; pass over anion exchange resin; add to ammoniacal luminol; add potassium ferricyanide	Chemiluminescence	At least 1 ng (corresponding to 0.1-1 g fresh material)	Warm & Laties (1982)
Intact animal cells and tissue	Mix tissue homogenates or subcellular fractions with catalase	Spectrophotometry (640-660 nm)	Not given	Sies (1981)
Blood serum	Add sodium azide, ascorbate oxidase, catalase and 1,4-piperazinediethanesulphonic acid or phosphate buffer	Hydrogen peroxide-selective electrode with an oxidase meter	Not given	Nakane & Kosaka (1980)

3. Biological Data Relevant to the Evaluation of Carcinogenic Risk to Humans

3.1 Carcinogenicity studies in animals[1]

(a) *Oral administration*

Mouse: Groups of 98, 101 and 99 inbred C57BL/6J mice of both sexes, eight weeks of age, were given 0, 0.1 and 0.4% hydrogen peroxide (a solution of 30% for food-additive use)

---

[1]The Working Group was aware of a recently-completed study in rats administered hydrogen peroxide in drinking-water (IARC, 1984).

in distilled water, as drinking-water, for 100 weeks. One adenoma of the duodenum was noted in controls; six adenomas and one carcinoma of the duodenum were seen in mice receiving 0.1%, and two adenomas and five carcinomas of the duodenum occurred in mice given 0.4% hydrogen peroxide ($p < 0.05$ compared with controls) (Ito et al., 1981b). [The Working Group noted that data on survival were not given.]

A group of 138 male and female C57BL/6N mice were treated with 0.4% hydrogen peroxide (food grade, see above) in the drinking-water. Groups of 5-17 mice were killed sequentially at 30-day intervals up to 210 days and then every 60, 70 or 90 days up to 630 days; 29 mice were killed on day 700, when the experiment was terminated. Gastric erosions and duodenal 'plaques' (round, flat, avillous areas) appeared at the first kill (30 days) and were present consistently at each subsequent kill. 'Nodules' (hyperplastic lesions, adenomas and carcinomas) were found in both duodenum and stomach from 90 days until the end of the experiment, but not on days 210 and 360 in the stomach. The lesions did not appear to increase in frequency, but atypical hyperplasia appeared late in the experiment and 5% of the animals developed duodenal adenocarcinoma. No such lesion was observed in controls. The reversibility of the lesions was investigated in groups of mice treated with 0.4% hydrogen peroxide for 120, 140, 150 or 180 days after a treatment-free period of 10-30 days. The stomach lesions regressed completely, irrespective of length of treatment, but some of the duodenal lesions persisted. Groups of 22 DBA/2N, 39 BALB/cAnN and 34 C57BL/6N mice of both sexes were given 0.4% hydrogen peroxide (food grade) in distilled water as drinking-water. The mice were examined sequentially from 90 to 210 days of treatment for strain differences in the development of gastric and duodenal 'nodules' (hyperplastic lesions, adenomas and carcinomas). The incidences of gastric nodules were 2/22, 1/39 and 12/34 and those of duodenal nodules were 14/22, 7/39 and 22/34 in DBA, BALB/c and C57BL mice, respectively. The duodenal nodules appeared at 90 days in all three strains (Ito et al., 1982).

Groups of 18-24 female C3H/HeN, B6C3F1, C57BL/6N and C3H/C$_s^b$ mice with different levels of catalase activities in the duodenal mucosa (5.3, 1.7, 0.7 and 0.4 $\times 10^{-4}$ k/mg protein, respectively) were given 0.4% hydrogen peroxide (food grade) in distilled water as drinking-water for six or seven months. The incidences of duodenal 'nodules' (hyperplastic lesions, adenomas and carcinomas) were 2/18, 7/22, 21/21 and 22/24, respectively (Ito et al., 1984). [The Working Group noted that the pathology of the tumours was not well documented.]

(b) Subcutaneous and/or intramuscular administration

Mouse: Hydrogen peroxide (0.1 ml of a 0.5% solution in saline) was tested by subcutaneous injection in mice [strain and age, number of animals and sex unspecified] either alone or as a mixture with diatomaceous earth, hydroxylamine hydrochloride (a catalase inhibitor) or ferrous sulphate. No tumour was found in the few animals reported to have survived longer than 200 days (Nakahara & Fukuoka, 1959). [The Working Group noted that the study was inadequate for evaluating the carcinogenicity of hydrogen peroxide.]

A total of 649 male and female newborn mice of strains AB and C57BlxA/JAX received subcutaneous injections of 0.1 ml of 0.6% hydrogen peroxide [purity unspecified] diluted in Ringer's solution; 303 animals received one injection and 346 animals were injected three times. Of those that received the single injection, 12 males and 18 females survived longer than six months; of these, six developed tumours at various sites. Of those that received the repeated injections, 21 males and 21 females survived longer than six months; of these, 14 developed tumours at various sites (Schmidt, 1964). [The Working Group could not evaluate the results of this study because of the absence of a proper control group.]

Groups of female ddN mice, 30 days of age, received subcutaneous injections of hydrogen peroxide [purity and stability unspecified] in a study of its effect on benzo[a]pyrene (BP) carcinogenesis. (1) Groups of 30 mice received subcutaneous injections of 0.1 ml tricaprylin followed immediately by 0.05 ml of 0.6% hydrogen peroxide; 0.3 mg BP in 0.1 ml tricaprylin; or 0.3 mg BP in 0.1 ml tricaprylin followed immediately by 0.05 ml of 0.6% hydrogen peroxide. After 437 days, the incidences of fibrosarcomas at the site of injection were 0/30, 19/30 and 22/30, respectively. All mice were alive at the appearance of the first tumour. (2) Groups of 30 mice received subcutaneous injections of 0.3 mg BP in 0.1 ml tricaprylin followed immediately by 0.05 ml distilled water; 0.3 mg BP in 0.1 ml tricaprylin followed immediately by 0.05 ml of 0.6% hydrogen peroxide; or 0.3 mg BP in 0.1 ml tricaprylin followed immediately by 0.05 ml Fenton's reagent (0.6% hydrogen peroxide + 0.1 mg ferrous chloride). At 480 days, when the study was terminated, the incidences of fibrosarcomas at the injection site were 19/29, 18/29 and 20/29, respectively, the denominator being the number of mice alive at the appearance of the first tumour. (3) Groups of 31-32 mice received a subcutaneous injection every two days, for 12 days, of 0.3 mg BP in 0.1 ml tricaprylin followed immediately by 0.05 ml of 0.6% hydrogen peroxide; 0.2 mg BP in 0.1 ml tricaprylin followed immediately by 0.05 ml distilled water; 0.2 mg BP in 0.1 ml tricaprylin followed immediately by 0.05 ml of 0.6% hydrogen peroxide. At 480 days, the incidences of fibrosarcomas at the injection site were 17/31, 21/26 and 13/27, respectively. These results indicate that single or repeated subcutaneous injections of hydrogen peroxide at the site of a subcutaneous injection of BP did not alter the number or time of onset of the induced fibrosarcomas (Nagata et al., 1973).

(c) *Skin application*

Mouse: In a two-stage mouse-skin assay, a group of 30 female ICR Swiss mice, eight weeks of age, received a single application of 125 μg 7,12-dimethylbenz[a]anthracene (DMBA) in 0.25 ml acetone. Three weeks later, the mice received applications of 0.2 ml 3% hydrogen peroxide [purity and stability unspecified] in water five times weekly for 56 weeks, at which time the experiment was terminated. No skin tumour was found (Bock et al., 1975). [The Working Group noted the short duration of the experiment and that survival data were not given.]

Groups of 60 female Sencar mice, aged seven to nine weeks, were used to test the tumour-promoting (A), tumour-initiating (B) and complete carcinogenic (C) activities of hydrogen peroxide [purity unspecified] on the skin. Mice in experiment (A) received a single topical application of 10 nmol DMBA in 0.2 ml acetone, followed one week later by applications of a 30% solution of hydrogen peroxide diluted 1:1 (once and twice weekly), 1:2 or 1:5 in 0.2 ml acetone twice weekly for 25 weeks. Controls received acetone alone. The proportions of mice with papillomas at 25 weeks were 0/60 (controls), 3/58, 5/59, 6/59 and 6/60, respectively. Mice in experiment (B) received a single topical application of hydrogen peroxide diluted 1:1 in 0.2 ml acetone, or acetone alone (controls), followed one week later by twice-weekly applications of 2 μg 12-*O*-tetradecanoylphorbol 13-acetate (TPA) in acetone for 25 weeks. Papillomas were found after 25 weeks in 3/56 and 6/58 control and hydrogen peroxide-treated animals, respectively. Mice in experiment (C) received twice-weekly topical applications of hydrogen peroxide diluted 1:1 in 0.2 ml acetone for 25 weeks; 3/57 had papillomas at that time. No squamous-cell carcinoma was found when these animals were observed up to 50 weeks (Klein-Szanto & Slaga, 1982). [The Working Group noted the absence of a DMBA-treated control group for the promotion experiment and the short duration of the experiment for complete carcinogenicity.]

3.2 Other relevant biological data

(a) Experimental systems

Toxic effects

The intravenous LD_{50} of hydrogen peroxide in rats was reported to be 21 mg/kg bw (Spector, 1956). In rabbits, an inverse relationship between the intravenous LD_{50} (3-19 mg/kg bw) and the administered concentration of hydrogen peroxide (3.6-90%) was observed: intravascular oxygen bubbles appear at the site of injection and mechanically hinder further access of hydrogen peroxide to the general circulation; the more concentrated the solution, the more marked are these effects and the less the quantity of hydrogen peroxide that actually reaches the systemic circulation. The LD_{50} for percutaneous application was much higher, and there were marked species and strain differences in susceptibility (rabbit, 630 mg/kg bw; rat, 700->7500 mg/kg bw) (US Food and Drug Administration, 1983a).

In rabbits and cats that died after intravenous administration of hydrogen peroxide, the lungs were found to be pale and emphysematous, with considerable amounts of gas in the great veins and in the right side of the heart (Lorincz et al., 1948). After intraperitoneal injection of 0.5 ml of 5% hydrogen peroxide into adult mice, a radiation-like effect was observed; pyknotic nuclei were induced in the intestine and thymus within two hours and persisted for up to 24 hours (Dustin & Gompel, 1949).

Rats receiving 2.5% hydrogen peroxide in their drinking-water died within 43 days (US Food and Drug Administration, 1983a). Dose-related growth retardation, induction of dental caries and pathological changes in the periodontium were observed in young male rats receiving 1.5% hydrogen peroxide as their drinking-fluid for eight weeks (Shapiro et al., 1960). Aoki and Tani (1972) reported that treatment of 16 mice for 35 weeks with 0.15% hydrogen peroxide resulted in hydropic degeneration of hepatic and renal tubular epithelial tissues, necrosis, inflammation, irregularities of tissue structure of the stomach wall and hypertrophy of the lymphatic tissue of the small intestinal wall; concentrations in excess of 1% resulted in pronounced loss of body weight and death within two weeks. In a sequential study of male and female mice treated with 0.4% hydrogen peroxide in the drinking-water, gastric erosion appeared at the time the first animals were killed (30 days) and was present consistently throughout the study period (108 weeks) (Ito et al., 1982).

The characteristic whitening of the skin after topical application of hydrogen peroxide is believed to be the result of ischaemia of the skin produced by oxygen bubbles acting microembolically in the capillaries (Ludewig, 1964a). A single topical application of 15 or 30% hydrogen peroxide to the dorsal skin of mice produced extensive epidermolysis, inflammation and vascular injury, similar to that produced by tumour promoters, followed by quick regeneration and epidermal hyperplasia, with a temporary increase in the number of dark basal keratinocytes. Extensive endothelial damage to the dermal blood vessels also occurred (Klein-Szanto & Slaga, 1982).

Concentrations of 0.25-2 mM [8.5-68 μg/ml] hydrogen peroxide in primary cultures of rat hepatocytes killed 30-90%, respectively, of the cells within three hours. Morphologically, the cells exhibited cytoplasmic vacuolization and prominant surface blebs. Lipid peroxidation preceded cell death. Addition of antioxidants prevented both lipid peroxidation and cell death (Rubin & Farber, 1984).

Endothelial cells, fibroblasts and several tumour-cell lines were up to 10 times more sensitive to enzymatically-generated hydrogen peroxide than rat hepatocytes (Sacks et al., 1978; Nathan et al., 1980; Simon et al., 1981; Weiss et al., 1981; Suttorp & Simon, 1982).

Phenanthroline, a strong iron chelator, prevented the death of mouse fibroblast 3T3 cells and the formation of DNA single-strand breaks produced by hydrogen peroxide. It was concluded that hydroxyl radicals formed when hydrogen peroxide reacts with chromatin-bound Fe^{+2} were responsible for the breaks that resulted in cell death (Mello Filho et al., 1984). Hydroxy radicals are considered to be the species primarily responsible for radiation effects on DNA (Sasaki & Matsubara, 1977). DNA-strand breaks have also been observed in intact leucocytes following activation of endogenous hydrogen peroxide formation by phorbol myristyl acetate (Birnboim, 1982).

Hydrogen peroxide induces liver microsomal lipid peroxidation in the presence of haematin (Ursini et al., 1981) but not of iron ions (Morehouse et al., 1983). Microsomal lipid peroxidation results in the formation of high-molecular-weight protein (Koster et al., 1982), the destruction of cytochrome-P450 (Levin et al., 1973) and inactivation of glucose-6-phosphatase (Wills, 1971). The induction of liver peroxisomes in rats in vivo with peroxisome proliferators resulted in increased steady-state levels of hydrogen peroxide production and lipofuscin, indicating an increase in lipid peroxidation (Reddy & Lalwai, 1983).

Hydrogen peroxide formed intracellularly- or added exogenously-caused methaemoglobinaemia and haemolysis when incubated with erythrocytes (Cohen & Hochstein, 1964). Incubation of 10 mM [34 µg/ml] hydrogen peroxide with erythrocytes resulted in the release of volatile alkanes believed to be formed as a result of lipid peroxidation (Clemens et al., 1983). Cu^{+2} readily catalysed the peroxidation of rat-erythrocyte membrane lipid by hydrogen peroxide (Chan et al., 1982).

Effects on reproduction and prenatal toxicity

Doses of 1.4, 2.8, 5.5 and 11 µmol/egg hydrogen peroxide (purity, 30%) dissolved in water were injected into the airspace of groups of 20-30 White Leghorn chicken eggs on day 3 of incubation. Embryos were examined on day 14. The incidence of embryonic deaths and malformations was dose-related and detected at doses of 2.8 µmol/egg and above; the combined ED_{50} was 2.7 µmol/egg (Korhonen et al., 1984).

Female rats that received 0.45% hydrogen peroxide as the sole drinking-fluid for five weeks produced normal litters when mated with untreated males (Hankin, 1958). Furthermore, 1% hydrogen peroxide given as the sole drinking-fluid to three-month-old male mice for 7-28 days did not cause infertility (Wales et al., 1959).

Absorption, distribution, excretion and metabolism

Meaningful information on the rate of absorption, distribution and excretion of hydrogen peroxide is difficult to obtain. Firstly, hydrogen peroxide is decomposed in the bowel before absorption; the large volumes of oxygen released as a result of catalase action (ten times the volume of solution) can cause rupture of the colon, proctitis and ulcerative colitis. Furthermore at these concentrations of hydrogen peroxide (3%), oesophagitis and gastritis occur as a result of ischaemia produced by oxygen bubbles acting as microemboli in the tissues and capillaries, as well as by its inflammatory irritating effect (Sheehan & Brynjolfsson, 1960). Secondly, hydrogen peroxide is a normal product of aerobic metabolism and may result from a number of oxidase-catalysed reactions (e.g., D-amino acid oxidase, urate oxidase, glycolate oxidase, fatty acyl-CoA dehydrogenase), or by the breakdown of superoxide by superoxide dismutase (Fridovich, 1978; Reddy & Lalwai, 1983).

Intracellular hydrogen peroxide levels are markedly increased in liver homogenates obtained from rats after in-vivo administration of peroxisome proliferators, e.g., certain hypolipidaemic drugs (Reddy & Lalwai, 1983). Peroxisomal-catalysed β-oxidation of fatty acids was also markedly increased. The authors suggested that hepatocellular genetic damage could occur as a result of hydrogen peroxide released from peroxisomes during fatty acid oxidation.

Studies conducted by Ludewig (1964a) show that hydrogen peroxide can penetrate the epidermis and mucous membranes and decompose in the underlying tissues, producing emphysemae and emboli. The presence of oxygen bubbles in the tongue and jugular veins following sublingual application of 3-30% hydrogen peroxide to dogs, cats and rabbits suggests that significant amounts of hydrogen peroxide were absorbed (Ludewig, 1959). Within one hour, 33% of the ^{18}O of a 19% solution of $H_2^{18}O_2$ was present in the expired air following sublingual application to cats (Ludewig, 1964b). Perfusion of the large intestine of dogs with dilute hydrogen peroxide has been shown to elevate the oxygen content of the blood, indicating that the compound was absorbed by the intestine (Urschel, 1967).

Hydrogen peroxide is formed intracellularly by mitochondria, endoplasmic reticulum, peroxisomes and soluble enzymes as a by-product during metabolism (Chance et al., 1979) (see Fig. 1).

Data on hydrogen peroxide formation in various subcellular fractions vary, reflecting different assay conditions (Chance et al., 1979). Boveris et al. (1972) studied intracellular pro-

Fig. 1. General scheme of roles of catalase, glutathione peroxidase and superoxide dismutase in different subcellular locations[a]

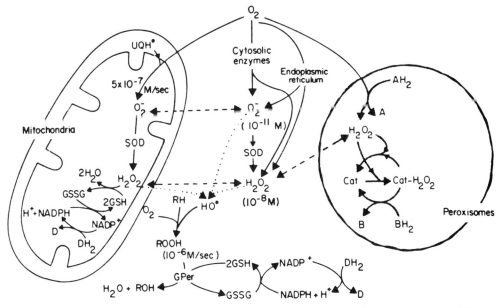

[a]From Chance et al. (1979). Concentrations and formation rates of oxygen metabolites are estimated. Abbreviations: UQH·, ubiquinone radical; GSSG, oxidized glutathione; GSH, reduced glutathione; DH_2 and D, a nonspecified NADP-reducing system; SOD, superoxide dismutase; NADPH and NADP, nicotinamide adenine dinucleotide phosphate; O_2^-, superoxide anion; HO·, hydroxyl radical; ROOH, an alkyl hydroperoxide; GPer, glutathione peroxidase; Cat, catalase; B and BH_2, hydrogen donors of a specificity appropriate to catalase, such as ethanol.

duction of hydrogen peroxide in various subcellular fractions under optimized physiological conditions; by summing the levels of hydrogen peroxide production in each fraction, they estimated a total production of 88 nmol/min per g wet-weight of rat liver. In rat livers perfused in vitro, highly variable results have been obtained with regard to hydrogen peroxide production, depending on the substrate used (Chance et al., 1979). During decanoate oxidation, 80 nmol/min per g wet-weight of perfused rat liver were formed (Sies, 1981). When uric acid was used as the substrate and rat livers were perfused in situ, somewhat higher rates of hydrogen peroxide production were measured (100-300 nmol/min per g wet-weight) (Chance et al., 1979).

Hydrogen peroxide levels are particularly high in rat kidney (25-50 µg/17 mg dry-weight: 1.5-3 µg/g) and may reflect the high peroxisomal content (Rondoni & Cudkowicz, 1953). Polymorphonuclear leucocytes are particularly effective at forming hydrogen peroxide in response to phagocytosable particles or soluble agonists and is associated with their bactericidal function. Rates of hydrogen peroxide formation of about 20 nmol/min/10^7 cells have been found with guinea-pig leucocytes during phagocytosis (Dri et al., 1979).

In studies on the metabolism of hydrogen peroxide by erythrocytes, glutathione peroxidase was found to be responsible for decomposition of low concentrations of hydrogen peroxide, whereas catalase was the primary catalyst for decomposition of higher concentrations (Cohen & Hochstein, 1963; Nicholls, 1972). With hepatocytes, glutathione peroxidase was responsible for the decomposition of cytosolic hydrogen peroxide formed during drug metabolism in the endoplasmic reticulum, whereas catalase in the peroxisomes was responsible for the decomposition of peroxisomal hydrogen peroxide formed on incubation of hepatocytes with 10 mM [34 µg/ml] glycolate (Jones et al., 1981).

The hexose-monophosphate shunt of erythrocytes participates in the removal of hydrogen peroxide as a result of the oxidation of intracellular NADPH and glutathione (GSH) (Sullivan & Stern, 1980). Incubation of isolated hepatocytes with 0.5 nmol hydrogen peroxide added per ml per min resulted in the oxidation of cellular GSH and the release of GSSG (oxidized glutathione) into the medium, but had only a minor effect on the cellular NADPH:-NADP$^+$ ratio (Eklöw et al., 1981).

In-vitro studies show that the interactions of haem proteins with hydrogen peroxide result in protein cross-linking (Rice et al., 1983) and modification of the haem-protein bonds and of certain amino acids, depending on the pH of the hydrolysate (O'Brien, 1966).

Base destruction, single-strand breakage and cross-linking have been detected in isolated DNA upon interaction with hydrogen peroxide in the presence of Cu^{+2} and Fe^{+2} (Massie et al., 1972). Pyrimidine deoxyribonucleotides were more susceptible to oxidation than purine deoxyribonucleotides (Melzer & Tomlinson, 1966). Hydrogen peroxide also liberates all bases from DNA as a result of oxidation of the C-1 carbon of deoxyribose (Uchida et al., 1965; Rhaese & Freese, 1968). At low concentrations of hydrogen peroxide (0.3 mM [10 µg/ml]), thymine bases of DNA are converted to 5,6-glycols, with relatively few accompanying strand breaks or apurinic/apyrimidinic bases (Demple & Linn, 1982).

Mutagenicity and other short-term tests (see also 'Appendix: Activity Profiles for Short-term Tests', p. 344)

Prophages are induced by treatment of lysogenized bacteria with hydrogen peroxide (Northrop, 1958). In *Escherichia coli*, hydrogen peroxide induced single-strand breaks in DNA (Ananthaswamy & Eisenstark, 1977) and was positive in DNA-repair assays (Rosenkranz, 1973; Ananthaswamy & Eisenstark, 1977; Hartman & Eisenstark, 1978).

Hydrogen peroxide was mutagenic to *Salmonella typhimurium* TA92 (Ames *et al.*, 1981) and TA102 (Levin *et al.*, 1982) and was positive in a forward-mutation test in *S. typhimurium* SV50 (Xu *et al.*, 1984). Both positive (Norkus *et al.*, 1983) and negative results have been reported in strain TA100 and negative results in strain TA98 (Stich *et al.*, 1978; Yamaguchi & Yamashita, 1980). Hydrogen peroxide was mutagenic to *Micrococcus aureus* (Clark, 1953), *Haemophilus influenzae* (Kimball & Hirsch, 1975), *Bacillus subtilis* (Sacks & MacGregor, 1982) and *E. coli* (Demerec *et al.*, 1951).

Hydrogen peroxide was mutagenic to *Saccharomyces cerevisiae* (Thacker, 1976; Thacker & Parker, 1976), *Neurospora crassa* (Dickey *et al.*, 1949; Jensen *et al.*, 1951) and *Aspergillus chevalieri* (Nanda *et al.*, 1975) but not to *Streptomyces griseoflavus* (Mashima & Ikeda, 1958).

Hydrogen peroxide did not induce sex-linked recessive lethal mutations in *Drosophila melanogaster* (DiPaolo, 1952).

Hydrogen peroxide induced single-strand breaks in DNA of V79-4 Chinese hamster cells (Bradley *et al.*, 1979) and human cells (Hoffmann & Meneghini, 1979; Taylor *et al.*, 1979; Wang *et al.*, 1980). No DNA-DNA or DNA-protein cross-link was found after treatment of V79-4 Chinese hamster cells with hydrogen peroxide (Bradley *et al.*, 1979).

Unscheduled DNA synthesis was induced by treating cultured human fibroblasts with hydrogen peroxide (Stich *et al.*, 1978; Coppinger *et al.*, 1983).

Hydrogen peroxide did not induce resistance to 6-thioguanine (Bradley *et al.*, 1979; Bradley & Erickson, 1981), 8-azaguanine or ouabain in V79 Chinese hamster cells (Tsuda, 1981).

Hydrogen peroxide induced sister chromatid exchanges in Chinese hamster ovary cells (MacRae & Stich, 1979; Wilmer & Natarajan, 1981), V79 Chinese hamster cells (Bradley *et al.*, 1979; Speit *et al.*, 1982) and cultured human cells (Estervig & Wang, 1979 [abstract, details not given]). Chromosomal aberrations were induced by exposure to hydrogen peroxide of Chinese hamster ovary cells (Stich *et al.*, 1978; Tsuda, 1981; Wilmer & Natarajan, 1981; Hanham *et al.*, 1983), Chinese hamster DON-6 cells (Sasaki *et al.*, 1980), primary cultures of Syrian hamster embryo cells and newborn Balb/c mouse back-skin cells (Tsuda, 1981), and human fibroblasts (Parshad *et al.*, 1980).

Exposure of mice carrying ascite-tumour cells to hydrogen peroxide *in vivo* resulted in an increased number of chromosomal aberrations in the tumour cells (Schöneich, 1967; Schöneich *et al.*, 1970). Hydrogen peroxide did not induce chromosomal aberrations in bone-marrow cells of rats treated *in vivo* (Kawachi *et al.*, 1980). [No detail was given.]

(b) Humans

Toxic effects

A characteristic whitening of the skin occurs after topical application of hydrogen peroxide (1-30%), which is believed to be the result of avascularity of the skin produced by oxygen bubbles acting microembolically in the capillaries (Hauschild *et al.*, 1958).

In five persons who accidentally drank about 50 ml of 33% hydrogen peroxide solution, stomach and chest pain, retention of breath, foaming at the mouth and loss of consciousness ensued. Later, motor and sensory disorders, fever, microhaemorrhages and moderate leucocytosis were noted. One patient developed pneumonia. All recovered completely within two to three weeks (Budagovskiya et al., 1971).

Cases of rupture of the colon, inflammation of the anus or rectum and ulcerative colitis have been reported following hydrogen peroxide enemas (Pumphrey, 1951; Ludington et al., 1958; Schechan & Brynjolfsson, 1960).

Human erythrocytes exhibit increased osmotic fragility when incubated with hydrogen peroxide (Cohen & Hochstein, 1963). Erythrocytes from vitamin E-deficient individuals (Younkin et al., 1971) or glucose-6-phosphate dehydrogenase-deficient individuals (Cohen & Hochstein, 1964) are more susceptible to haemolysis by hydrogen peroxide than those of normal individuals; the mechanism was associated with membrane lipid peroxidation. Vitamin E deficiency has been reported in patients not only with general malnutrition, but also with steatorrhea, acanthocytosis, sickle-cell anaemia and in premature infants; lipid peroxidation of their erythrocytes readily occurs on incubation with hydrogen peroxide (Chiu et al., 1982). Hereditary acatalasia is a rare enzyme defect characterized by catalase deficiency; erythrocytes from individuals with this anomaly are also particularly susceptible to the above peroxidative changes with hydrogen peroxide (Jacob et al., 1965). Increased sensitivity to hydrogen peroxide has also been found in erythrocytes isolated from persons deficient in enzymes of glutathione metabolism, persons with nocturnal haemoglobinuria, erythropoietic protoporphyria or thalassaemia syndromes (Chiu et al., 1982).

Incubation of erythrocytes *in vitro* with oxidative drugs, such as primaquine, quinine, phenylhydrazine and dapsone, resulted in the intracellular formation of hydrogen peroxide through interactions of the drug with haemoglobin. This led to the formation of methaemoglobin and to haemoglobin denaturation, followed by precipitation inside the red cells as Heinz bodies (Cohen & Hochstein, 1964; Chiu et al., 1982).

Leucocytes from children and young adults with the genetic disorder, chronic granulomatous disease, fail to generate adequate amounts of oxygen and hydrogen peroxide that are required for the effective destruction of ingested microbes, resulting in recurrent infections (Baehner et al., 1982).

Cultured human $D98/AH_2$ cells lose their viability in the presence of 1 µg/ml hydrogen peroxide (Wang & Nixon, 1978; Wang et al., 1980).

Effects on reproduction and prenatal toxicity

No data were available to the Working Group.

Absorption, distribution, excretion and metabolism

Glutathione peroxidase, responsible for decomposing hydrogen peroxide, is present in normal human tissues (Flohé, 1982). Hydrogen peroxide has been detected in serum and in intact liver (Nakane & Kosaka, 1980; Sies, 1981).

Mutagenicity and chromosomal effects

No data were available to the Working Group.

3.3 Case reports and epidemiological studies of carcinogenicity to humans

No data were available to the Working Group.

4. Summary of Data Reported and Evaluation

4.1 Exposure data

Hydrogen peroxide has been produced commercially since 1925 and is employed principally as a chemical intermediate and for textile, paper and pulp bleaching. Large quantities are also used in industrial and municipal waste treatment and as a component of non-prescription drugs and topical anti-infectants. It has been found in many samples of air, water and food. It is also formed endogenously in intact human cells and tissues.

4.2 Experimental data

Hydrogen peroxide has been tested for carcinogenicity only in mice, by oral administration in drinking-water, by skin application and by subcutaneous administration. Adenomas and carcinomas of the duodenum were reported following its oral administration. The other studies were inadequate for an evaluation of carcinogenicity. One study by skin application indicated that hydrogen peroxide has no promoting activity.

The available data are inadequate to evaluate the teratogenic potential of hydrogen peroxide in mammals.

Hydrogen peroxide induced DNA damage in bacteria and was mutagenic to bacteria and fungi. It was not mutagenic to insects or to mammalian cells *in vitro*. Hydrogen peroxide induced DNA damage, sister chromatid exchanges and chromosomal aberrations in mammalian cells *in vitro*.

Overall assessment of data from short-term tests: hydrogen peroxide[a]

	Genetic activity			Cell transformation
	DNA damage	Mutation	Chromosomal effects	
Prokaryotes	+	+		
Fungi/green plants		+		
Insects		−		
Mammalian cells (*in vitro*)	+	−	+	
Mammals (*in vivo*)				
Humans (*in vivo*)				
Degree of evidence in short-term tests for genetic activity: *Sufficient*				Cell transformation: No data

[a]The groups into which the table is divided and the symbols + and − are defined on pp 17-18 of the Preamble, the degrees of evidence are defined on p. 18.

4.3 Human data

No case report or epidemiological study of the carcinogenicity of hydrogen peroxide to humans was available to the Working Group.

4.4 Evaluation[1]

There is *limited evidence* of the carcinogenicity of hydrogen peroxide to experimental animals.

In the absence of epidemiological data, no evaluation could be made of the carcinogenicity of hydrogen peroxide to humans.

5. References

American Conference of Governmental Industrial Hygienists (1984) *TLVs Threshold Limit Values for Chemical Substances in the Work Environment Adopted by ACGIH for 1984-85*, Cincinnati, OH, p. 21

American Pharmaceutical Association (1982) *Handbook of Nonprescription Drugs*, Washington DC, pp. 412, 463, 538

Ames, B.N., Hollstein, M.C. & Cathcart, R. (1981) Lipid peroxidation and oxidative damage to DNA. In: Yagi, K., ed., *Lipid Peroxide in Biology and Medicine*, New York, Academic Press, pp. 339-351

Ananthaswamy, H.N. & Eisenstark, A. (1977) Repair of hydrogen peroxide induced single-strand breaks in *Escherichia coli* deoxyribonucleic acid. *J. Bacteriol.*, 130, 187-191

Aoki, M. & Tani, Y. (1972) Growth and histopathologic changes in mice fed with hydrogen peroxide solution instead of water (Jpn.). *Igaku to Seibutsugak* (Med. Biol), 84, 159-162

Asai, Y., Kuwahira, H., Shimoda, K. & Sato, K. (1982) Rapid microanalysis of residual hydrogen peroxide in foods by use of ammonium thiocyanate (Jpn.). *Shokuhin Eiseigaku Zasshi*, 23, 438-443 [*Chem. Abstr.*, 98, 105821v]

Baehner, R.L., Boxer, L.A. & Ingraham, L.M. (1982) *Reduced oxygen by-products and white blood cells*. In: Pryor, W., ed., *Free Radicals in Biology*, Vol. 5, New York, Academic Press, pp. 91-113

Birnboim, H.C. (1982) DNA strand breakage in human leukocytes exposed to a tumor promoter, phorbol myristate acetate. *Science*, 215, 1247-1249

Bock, F.G., Myers, H.K. & Fox, H.W. (1975) Cocarcinogenic activity of peroxy compounds. *J. natl Cancer Inst.*, 55, 1359-1361

---

[1]For definitions of the italicized terms, see the Preamble, pp. 15-16.

Boveris, A., Oshino, N. & Chance, B. (1972) The cellular production of hydrogen peroxide. *Biochem. J.*, *128*, 617-630

Bradley, M.O. & Erickson, L.C. (1981) Comparison of the effects of hydrogen peroxide and X-ray irradiation on toxicity, mutation, and DNA damage/repair in mammalian cells (V-79). *Biochim. biophys. Acta*, *654*, 135-141

Bradley, M.O., Hsu, I.C. & Harris, C.C. (1979) Relationships between sister chromatid exchange and mutagenicity, toxicity, and DNA damage. *Nature*, *282*, 318-320

Budagovskiya, M.T., Vadachkoriya, M.K. & Desyatov, A.I. (1971) Poisoning with hydrogen peroxide (Russ.). *Otravl. peridrolem. Voen. Med. Zh.*, *9*, 79-81

Bufalini, J.J., Gay, B.W. & Brubaker, K.L. (1972) Hydrogen peroxide formation from formaldehyde photooxidation and its presence in urban atmospheres. *Environ. Sci. Technol.*, *6*, 816-821

Chan, P.C., Peller, O.G. & Kesner, L. (1982) Copper (II)-catalysed lipid peroxidation in liposomes and erythrocyte membranes. *Lipids*, *17*, 331-337

Chance, B. (1952) The state of catalase in the respiring bacterial cell. *Science*, *116*, 202-203

Chance, B., Sies, H. & Boveris, A. (1979) Hydroperoxide metabolism in mammalian organs. *Physiol. Rev.*, *59*, 527-605

Chin, H.B. & Cortes, A. (1983) Comparison between potentiometric titration and enzyme-catalyzed determination of hydrogen peroxide. *J. Assoc. off. anal. Chem.*, *66*, 199-202

Chiu, D., Lubin, B. & Shohet, S.B. (1982) *Peroxidative reactions in red cell biology*. In: Pryor, W., ed., *Free Radicals in Biology*, Vol. 5, New York, Academic Press, pp. 115-160

Clark, J.B. (1953) The mutagenic action of various chemicals on *Micrococcus aureus*. *Proc. Oklahoma Acad. Sci.*, *34*, 114-118

Clemens, M.R., Einsele, H., Frank, H., Remmer, H. & Waller, H.D. (1983) Volatile hydrocarbons from hydrogen-peroxide induced lipid peroxidation of erythrocytes and their cell components. *Biochem. Pharmacol.*, *32*, 3877-3878

Cohen, G. & Hochstein, P. (1963) Glutathione peroxidase: The primary agent for the elimination of hydrogen peroxide in erythrocytes. *Biochemistry*, *2*, 1420-1428

Cohen, G. & Hochstein, P. (1964) Generation of hydrogen peroxide in erythrocytes by hemolytic agents. *Biochemistry*, *3*, 895-900

Coppinger, W.J., Wong, T.K. & Thompson, E.D. (1983) Unscheduled DNA synthesis and DNA repair studies of peroxyacetic and monoperoxydecanoic acids. *Environ. Mutagenesis*, *5*, 177-192

Das, T.N., Moorthy, P.N. & Rao, K.N. (1983) Chemiluminescent measurement of atmospheric hydrogen peroxide in the Bombay area. *Atmos. Environ.*, *17*, 79-82

Demerec, M., Bertani, G. & Flint, J. (1951) A survey of chemicals for mutagenic action on *E. coli. Am. Nat.*, *85*, 119-136

Demple, B. & Linn, S. (1982) 5,6-Saturated thymine lesions in DNA: Production by ultraviolet light or hydrogen peroxide. *Nucleic Acids Res.*, *10*, 3781-3789

Deutsche Forschungsgemeinschaft (1984) *Maximal Work Place Concentrations and Biological Tolerance Values for Compounds in the Work Place* (Ger.)., Part XX, Weinheim, Verlag Chemie GmbH, p. 55

Dickey, F.H., Cleland, G.H. & Lotz, C. (1949) The role of organic peroxides in the induction of mutations. *Proc. natl Acad. Sci. USA*, *35*, 581-586

DiPaolo, J.A. (1952) Studies on chemical mutagenesis utilizing nucleic acid components, urethane, and hydrogen peroxide. *Am. Nat.*, *86*, 49-56

Draper, W.M. & Crosby, D.G. (1983) The photochemical generation of hydrogen peroxide in natural waters. *Arch. environ. Contam. Toxicol.*, *12*, 121-126

Dri, P., Bellavite, P., Berton, G. & Rossi, F. (1979) Interrelationship between oxygen consumption, superoxide anion and hydrogen peroxide formation in phagocytosing guinea pig polymorphonuclear leukocytes. *Molec. cell. Biochem.*, *23*, 109-122

Dustin, P. & Gompel, C. (1949) Effect of hydrogen peroxide on intestinal mitoses in the mouse (Fr.). *C.R. Soc. belge Biol.*, *143*, 874-875

Ehrlich, S.H. & Capone, S.M. (1982) Method for determining hydrogen peroxide concentration in polymers or resins. *Res. Discl.*, *215*, 64 [*Chem. Abstr.*, *97*, 163763t]

Eklöw, L., Thor, H. & Orrenius, S. (1981) Formation and efflux of glutathione disulfide studied in isolated rat hepatocytes. *FEBS Lett.*, *127*, 125-128

Estervig, N. & Wang, R.J. (1979) Chromosomal damage induced by fluorescent light photoproducts (Abstract No. CH744). *J. Cell Biol.*, *83*, 158a

Flohé, L. (1982) *Glutathione peroxidase brought into focus.* In: Pryor, W., ed., *Free Radicals in Biology*, Vol. 5, New York, Academic Press, pp. 223-254

Fridovich, I. (1978) The biology of oxygen radicals. *Science*, *201*, 875-880

Hanham, A.F., Dunn, B.P. & Stich, H.F. (1983) Clastogenic activity of caffeic acid and its relationship to hydrogen peroxide generated during autooxidation. *Mutat. Res.*, *116*, 333-339

Hankin, L. (1958) Hydrogen peroxide ingestion and the growth of rats. *Nature*, *182*, 1453

Hartman, P.S. & Eisenstark, A. (1978) Synergistic killing of *Escherichia coli* by near-UV radiation and hydrogen peroxide: Distinction between *recA*-repairable and *recA*-nonrepairable damage. *J. Bacteriol.*, *133*, 769-774

Hasegawa, N. & Sugawara, E. (1981) Determination of hydrogen peroxide using 2,2'-azinobis(3-ethylbenzthiazoline-6-sulfonic acid) ammonium salt (ABTS)-peroxidase system (Jpn.). *Kankyo Igaku Kenkyusho Nenpo (Nagoya Daigaku)*, *32*, 182-186 [*Chem. Abstr.*, *96*, 4997k]

Hauschild, F., Ludewig, R. & Mühlberg, H. (1958) Corrosive action of hydrogen peroxide (Ger.). *Naunyn-Schmiedeberg's Arch. exp. Pathol. Pharmakol., 235*, 51-62

Hoffmann, M.E. & Meneghini, R. (1979) Action of hydrogen peroxide on human fibroblast in culture. *Photochem. Photobiol., 30*, 151-155

IARC (1984) *Information Bulletin on the Survey of Chemicals Being Tested for Carcinogenicity*, No. 11, Lyon, p. 86

International Labour Office (1972) *Encyclopedia of Occupational Health and Safety*, Vol. 1 (A-K), New York, McGraw-Hill, pp. 696, 697

International Labour Office (1980) *Occupational Exposure Limits for Airborne Toxic Substances*, 2nd (rev.) ed. (*Occupational Safety and Health Series No. 37*), Geneva, pp. 128-129

Ito, Y., Tonogai, Y., Suzuki, H., Ogawa, S., Yokoyama, T., Hashizume, T., Santo, H., Tanaka, K.I., Nishigaki, K. & Iwaida, M. (1981a) Improved 4-aminoantipyrine colorimetry for detection of residual hydrogen peroxide in noodles, fish paste, dried fish, and herring roe. *J. Assoc. off. anal. Chem., 64*, 1448-1452

Ito, A., Watanabe, H., Naito, M. & Naito, Y. (1981b) Induction of duodenal tumors in mice by oral administration of hydrogen peroxide. *Gann, 72*, 174-175

Ito, A., Naito, M., Naito, Y. & Watanabe, H. (1982) Induction and characterization of gastro-duodenal lesions in mice given continuous oral administration of hydrogen peroxide. *Gann, 73*, 315-322

Ito, A., Watanabe, H., Naito, M., Naito, Y. & Kawashima, K. (1984) Correlation between induction of duodenal tumor by hydrogen peroxide and catalase activity in mice. *Gann, 75*, 17-21

Iwaida, M., Ito, Y., Toyoda, M. & Masaki, A. (1981) Rapid estimation of hydrogen peroxide in milk by the use of guaiac test solution (Jpn.). *Shokuhin Eiseigaku Zasshi, 22*, 432-435 [*Chem. Abstr., 96*, 33460k]

Jacob, H.S., Ingbar, S.H. & Jandl, J.H. (1965) Oxidative hemolysis and erythrocyte metabolism in hereditary acatalasia. *J. clin. Invest., 44*, 1187-1199

Jensen, K.A., Kirk, I., Kølmark, G. & Westergaard, M. (1951) *Chemically induced mutations in Neurospora*. In: *Cold Spring Harbor Symposia on Quantitative Biology*, Vol. 16, *Genes and Mutations*, Cold Spring Harbor, NY, The Biological Laboratory, pp. 245-261

Jones, D.P., Eklöw, L., Thor, H. & Orrenius, S. (1981) Metabolism of hydrogen peroxide in isolated hepatocytes: Relative contribution of catalase and glutathione peroxidase in decomposition of endogenously generated H_2O_2. *Arch. Biochem. Biophys., 210*, 505-516

Kawachi, T., Yahagi, T., Kada, T., Tazima, Y., Ishidate, M., Sasaki, M. & Sugiyama, T. (1980) *Cooperative programme on short-term assays for carcinogenicity in Japan*. In: Montesano, R., Bartsch, H. & Tomatis, L., eds, *Molecular and Cellular Aspects of Carcinogen Screening Tests* (*IARC Scientific Publications No. 27*), Lyon, International Agency for Research on Cancer, pp. 323-330

Kawamura, N., Saito, I., Ohshima, H. & Uno, K. (1983) Determination of residual hydrogen peroxide in boiled and semidried white bait (*shirasuboshi*)by an oxygen electrode method (Jpn.). *Aichi-ken Eisei Kenkyusho Ho, 33,* 17-22 [*Chem. Abstr., 98,* 177582r]

Kikugawa, K., Ohhashi, Y. & Kurechi, T. (1982) Determination of hydrogen peroxide by use of sesamol dimer (Jpn.). *Shokuhin Eiseigaku Zasshi, 23,* 462-467 [*Chem. Abstr., 98,* 124305w]

Kimball, R.F. & Hirsch, B.F. (1975) Tests for the mutagenic action of a number of chemicals on *Haemophilus influenzae* with special emphasis on hydrazine. *Mutat. Res., 30,* 9-20

Kirchner, J.R. (1981) *Hydrogen peroxide.* In: Grayson, M., ed., *Kirk-Othmer Encyclopedia of Chemical Technology,* 3rd ed., Vol. 13, New York, John Wiley & Sons, pp. 12-38

Klein-Szanto, A.J.P. & Slaga, T.J. (1982) Effects of peroxides on rodent skin: Epidermal hyperplasia and tumor promotion. *J. invest. Dermatol., 79,* 30-34

Kok, G.L. (1983) *Measurements of Formaldehyde and Hydrogen Peroxide in the California South Coast Air Basin (EPA-600/3-83-030; PB83-196725),* Research Triangle Park, NC, Environmental Protection Agency, pp. 1-125

Kok, G.L., Holler, T.P., Lopez, M.B., Nachtrieb, H.A. & Yuan, M. (1978) Chemiluminescent method for determination of hydrogen peroxide in the ambient atmosphere. *Environ. Sci. Technol., 12,* 1072-1076

Korhonen, A., Hemminki, K. & Vainio, H. (1984) Embryotoxic effects of eight organic peroxides and hydrogen peroxide on three-day chicken embryos. *Environ. Res., 33,* 54-61

Koster, J.F., Slee, R.G. & Van Berkel, T.J.C. (1982) On the lipid peroxidation of rat liver hepatocytes, the formation of fluorescent chromolipids and high molecular weight protein. *Biochim. biophys. Acta, 710,* 230-235

Levin, D.E., Hollstein, M., Christman, M.F., Schwiers, E.A. & Ames, B.N. (1982) A new *Salmonella* tester strain (TA102) with A.T base pairs at the site of mutation detects oxidative mutagens. *Proc. natl Acad. Sci. USA, 79,* 7445-7449

Levin, W., Lu, A.Y.H., Jacobson, M. & Kuntzman, R. (1973) Lipid peroxidation and the degradation of cytochrome P-450 heme. *Arch. Biochem. Biophys., 158,* 842-852

Lorincz, A.L., Jacoby, J.J. & Livingstone, H.M. (1948) Studies on the parenteral administration of hydrogen peroxide. *Anesthesiology, 9,* 162-174

Ludewig, R. (1959) Intraoral application of hydrogen peroxide (Ger.). *Z. ges. exp. Med., 131,* 452-465

Ludewig, R. (1964a) Distribution of skin hydroperoxidase and transepidermal penetration of hydrogen peroxide following its epicutaneous application (Ger.). *Acta histochem., 19,* 303-315

Ludewig, R. (1964b) The detection of ^{18}O in the expired air and in the blood during sublingual application of hydrogen peroxide (Ger.). *Abhandl. Dtsch. Akad. Wiss., Berlin, Kl. Chem., Geol. Biol., 7,* 549-552

Ludington, L.G., Hartman, S.W., Keplinger, J.E. & Williams, F.S. (1958) Incomplete rupture of the colon following hydrogen peroxide enema. *Arch. Surg.*, *76*, 658-660

Lundbaeck, H. (1983) Amperometric determination of hydrogen peroxide in pickling baths for copper and copper alloys by flow injection analysis. *Anal. chim. Acta*, *145*, 189-196 [*Chem. Abstr.*, *98*, 118751x]

MacRae, W.D. & Stich, H.F. (1979) Induction of sister-chromatid exchanges in Chinese hamster ovary cells by thiol and hydrazine compounds. *Mutat. Res.*, *68*, 351-365

Mair, R.D. & Hall, R.T. (1971) *Determination of organic peroxides by physical, chemical, and colorimetric methods*. In: Swern, D., ed., *Organic Peroxides*, Vol. II, New York, John Wiley & Sons, pp. 535-635

Mashima, S. & Ikeda, Y. (1958) Selection of mutagenic agents by the *Streptomyces* reverse mutation test. *Appl. Microbiol.*, *6*, 45-49

Massie, H.R., Samis, H.V. & Baird, M.B. (1972) The kinetics of degradation of DNA and RNA by H_2O_2. *Biochim. biophys. Acta*, *272*, 539-548

Melzer, M.S. & Tomlinson, R.V. (1966) Antioxidative effects of purine bases on hydrogen peroxide oxidation of pyrimidine bases. *Arch. Biochem. Biophys.*, *115*, 226-229

Morehouse, L.A., Tien, M., Bucher, J.R. & Aust, S.D. (1983) Effect of hydrogen peroxide on the initiation of microsomal lipid peroxidation. *Biochem. Pharmacol.*, *32*, 123-127

Nagata, C., Tagashira, T., Kodama, M., Ioki, Y. & Oboshi, S. (1973) Effect of hydrogen peroxide, Fenton's reagent, and iron ions on the carcinogenicity of 3,4-benzopyrene. *Gann*, *64*, 277-285

Nakahara, W. & Fukuoka, F. (1959) On the mechanism of radiation carcinogenesis. *Gann*, *50*, 17-21

Nakano, K. & Kooaka, A. (1900) Applications of a hydrogen peroxide electrode in clinical chemistry (Jpn.). *Rinsho Kagaku Shimpojumu*, *20*, 85-89 [*Chem. Abstr.*, *96*, 31123d]

Nanase, Y., Takata, K. & Yamada, T. (1981) Hydrogen peroxide in Japanese noodles (Jpn.). *Kenkyu Hokoku-Hiroshima-ken Eisei Kenkyusho*, *28*, 31-34 [*Chem. Abstr.*, *96*, 160956b]

Nanda, G., Nandi, P. & Mishra, A.K. (1975) Studies on induced reversions at the arginine locus of *Aspergillus chevalieri* (Mangin). *Zbl. Bakt. Abt. II*, *130*, 105-108

Nathan, C.F., Arrick, B.A., Murray, H.W., DeSantis, N.M. & Cohn, Z.A. (1980) Tumor cell anti-oxidant defenses. Inhibition of the glutathione redox cycle enhances macrophage-mediated cytolysis. *J. exp. Med.*, *153*, 766-782

National Finnish Board of Occupational Safety and Health (1981) *Airborne Contaminants in the Workplaces (Safety Bulletin No. 3)*, Helsinki, p. 27

National Institute for Occupational Safety and Health (1980) *Projected Number of Occupational Exposures to Chemical and Physical Hazards*, Cincinnati, OH, p. 73

National Institute for Occupational Safety and Health (1981) *National Occupational Hazard Survey Data* (microfiche), Cincinnati, OH, pp. 4769-4772 (generics excluded)

National Research Council (1981) *Food Chemicals Codex*, 3rd ed., Washington DC, National Academy Press, pp. 146-147

Nicholls, P. (1972) Contributions of catalase and glutathione peroxidase to red cell peroxide removal. *Biochim. biophys. Acta*, *279*, 306-309

Norkus, E.P., Kuenzig, W. & Conney, A.H. (1983) Studies on the mutagenic activity of ascorbic acid in vitro and in vivo. *Mutat. Res.*, *117*, 183-191

Northrop, J.H. (1958) Studies of the origin of bacterial viruses. I. The incidence of phage-producing cells in various *B. megatherium* cultures. II. The effect of ultraviolet light on the incidence of phage-producing cells and of terramycin-resistant cells. III. The effect of hydrogen peroxide on the incidence of phage-producing cells and of terramycin-resistant cells. IV. Calculation of the incidence of phage-producing cells. *J. gen. Physiol.*, *42*, 109-136

O'Brien, P.J. (1966) The effects of hydrogen peroxide or lipid peroxide on cytochrome *c* (Abstract). *Biochem. J.*, *101*, 12P-13P

O'Brien, P.J. (1984) *Multiple mechanisms of metabolic activation of aromatic amine carcinogens*. In: Pryor, W., ed., *Free Radicals in Biology*, Vol. IV, New York, Academic Press, pp. 289-322

Parshad, R., Taylor, W.G., Sanford, K.K., Camalier, R.F., Gantt, R. & Tarone, R.E. (1980) Fluorescent light-induced chromosome damage in human IMR-90 fibroblasts. Role of hydrogen peroxide and related free radicals. *Mutat. Res.*, *73*, 115-124

Pumphrey, R.E. (1951) Hydrogen peroxide proctitis. *Am. J. Surg.*, *81*, 60-62

Reddy, J.K. & Lalwai, N.D. (1983) Carcinogenesis by hepatic peroxisome proliferators: Evaluation of the risk of hypolipidemic drugs and industrial plasticizers to humans. *C.R.C. crit. Rev. Toxicol.*, *12*, 1-58

Rhaese, H.-J. & Freese, E. (1968) Chemical analysis of DNA alterations. I. Base liberation and backbone breakage of DNA and oligodeoxyadenylic acid induced by hydrogen peroxide and hydroxylamine. *Biochim. biophys. Acta*, *155*, 476-490

Rice, R.H., Lee, Y.M. & Brown, W.D. (1983) Interactions of heme proteins with hydrogen peroxide: Protein crosslinking and covalent binding of benzo[a]pyrene and 17β-estradiol. *Arch. Biochem. Biophys.*, *221*, 417-427

Rondoni, P. & Cudkowicz, G. (1953) Hydrogen peroxide in tumours. Its possible significance in carcinogenesis. *Experientia*, *9*, 348-349

Rosenkranz, H. (1973) Sodium hypochlorite and sodium perborate: Preferential inhibitors of DNA polynuclease-deficient bacteria. *Mutat. Res.*, *21*, 171-174

Rubin, R. & Farber, J.L. (1984) Mechanisms of the killing of cultured hepatocytes by hydrogen peroxide. *Arch. Biochem. Biophys.*, *228*, 450-459

Sacks, L.E. & MacGregor, J.T. (1982) The *B. subtilis* multigene sporulation test for mutagens: Detection of mutagens inactive in the Salmonella *his* reversion test. *Mutat. Res.*, 95, 191-202

Sacks, T., Moldow, C.F., Craddock, P.R., Bowers, T.K. & Jacob, H.S. (1978) Oxygen radicals mediate endothelial cell damage by complement-stimulated granulocytes. An in vitro model of immune vascular damage. *J. clin. Invest.*, 61, 1161-1167

Sasaki, M.S. & Matsubara, S. (1977) Free radical scavenging in protection of human lymphocytes against chromosome aberration formation by gamma-ray irradiation. *Int. J. Radiat. Biol.*, 32, 439-445

Sasaki, M., Sugimura, K., Yoshida, M.A. & Abe, S. (1980) Cytogenetic effects of 60 chemicals on cultured human and Chinese hamster cells. *Kromosomo*, 20, 574-584

Schmidt, F. (1964) Experiments on the carcinogenic effect of hydrogen peroxide and on the mechanisms of radiation carcinogenesis (Ger.). *Acta biol. med. ger.*, 13, 74-85

Schöneich, J. (1967) The induction of chromosomal aberrations by hydrogen peroxide in strains of ascites tumors in mice. *Mutat. Res.*, 4, 384-388

Schöneich, J., Michaelis, A. & Rieger, R. (1970) Caffeine and the chemical induction of chromatid aberrations in *Vicia faba* and ascite tumours of the mouse (Ger.). *Biol. Zbl.*, 88, 49-63

Shapiro, M., Brat, V. & Ershoff, B.H. (1960) Induction of dental caries and pathological changes in periodontium of rat with hydrogen peroxide and other oxidising agents. *J. dent. Res.*, 39, 332-343

Sheehan, J.F. & Brynjolfsson, G. (1960) Ulcerative colitis following hydrogen peroxide enema: Case report and experimental production with transient emphysema of colonic wall and gas embolism. *Lab. Invest.*, 9, 150-168

Sies, H. (1981) *Measurement of hydrogen peroxide formation* in situ. In: Jakoby, W.B., ed., *Methods in Enzymology*, Vol. 77, *Detoxification and Drug Metabolism: Conjugation and Related Systems*, New York, Academic Press, pp. 15-20

Silbert, L.S. (1971) *Physical properties of organic peroxides*. In: Swern, D., ed., *Organic Peroxides*, Vol. II, New York, John Wiley & Sons, pp. 637-639, 678, 684-686

Simon, R.H., Scoggins, C.H. & Patterson, D. (1981) Hydrogen peroxide causes the fatal injury to human fibroblasts exposed to oxygen radicals. *J. biol. Chem.*, 256, 7181-7186

Spector, W.S. (1956) *Acute toxicities of solids, liquids and gases to laboratory animals*. In: Spector, W.S., ed., *Handbook of Toxicology*, Vol. 1, Philadelphia, PA, W.B. Saunders, p. 160

Speit, G., Vogel, W. & Wolf, M. (1982) Characterization of sister chromatid exchange induction of hydrogen peroxide. *Environ. Mutagenesis*, 4, 135-142

Stich, H.F., Wei, L. & Lam, P. (1978) The need for a mammalian test system for mutagens: Action of some reducing agents. *Cancer Lett.*, 5, 199-204

Sullivan, S.G. & Stern, A. (1980) Interdependence of hemoglobin, catalase and the hexose monophosphate shunt in red blood cells exposed to oxidative agents. *Biochem. Pharmacol.*, *29*, 2351-2359

Suttorp, N. & Simon, L.M. (1982) Lung cell oxidant injury. Enhancement of polymorphonuclear leukocyte-mediated cytotoxicity in lung cells exposed to sustained in vitro hyperoxia. *J. clin. Invest.*, *70*, 342-350

Taylor, W.G., Camalier, R.F., Baeck, A.E. & Gantt, R. (1979) Type-specific cell killing and DNA strand breaks after exposure to visible light or hydrogen peroxide (Abstract No. CU463). *J. Cell Biol.*, *83*, 111a

Thacker, J. (1976) Radiomimetic effects of hydrogen peroxide in the inactivation and mutation of yeast. *Radiat. Res.*, *68*, 371-380

Thacker, J. & Parker, W.F. (1976) The induction of mutation in yeast by hydrogen peroxide. *Mutat. Res.*, *38*, 43-52

Toyoda, M., Ito, Y., Iwaida, M. & Fujii, M. (1982) Rapid procedure for the determination of minute quantities of residual hydrogen peroxide in food by using a sensitive oxygen electrode. *J. agric. Food Chem.*, *30*, 346-349

Tsuda, H. (1981) Chromosomal aberrations induced by hydrogen peroxide in cultured mammalian cells. *Jpn. J. Genet.*, *56*, 1-8

Uchida, Y., Shigematu, H. & Yamafuji, K. (1965) The mode of action of hydrogen peroxide on deoxyribonucleic acid. *Enzymologia*, *29*, 369-376

Urschel, H.C., Jr (1967) Cardiovascular effects of hydrogen peroxide: Current status. *Dis. Chest*, *51*, 180-192

Ursini, F., Maiorino, M., Ferri, L., Valente, M. & Gregolin, C. (1981) Hydrogen peroxide and hematin in microsomal lipid peroxidation. *J. inorg. Biochem.*, *15*, 163-169

US Department of Agriculture (1980) Animals and animal products. *US Code Fed. Regul., Title 9*, Part 318.7

US Department of Commerce (1980a) *US Imports for Consumption and General Imports (FT246/Annual 1979)*, Bureau of the Census, Washington DC, US Government Printing Office, p. 1-247

US Department of Commerce (1980b) *US Exports, Schedule E Commodity by Country (FT410/December 1979)*, Bureau of the Census, Washington DC, US Government Printing Office, p. 2-93

US Department of Commerce (1983a) *US Imports for Consumption and General Imports (FT246/Annual 1982)*, Bureau of the Census, Washington DC, US Government Printing Office, p. 1-284

US Department of Commerce (1983b) *US Exports, Schedule E Commodity by Country (FT410/December 1982)*, Bureau of the Census, Washington DC, US Government Printing Office, p. 2-99

US Department of Commerce (1984) *Current Industrial Reports, Inorganic Chemicals, 1982 (Series MA28A(82)-1)*, Washington DC, Bureau of the Census, pp. 10, 12

US Department of Transportation (1980) Identification numbers, hazardous wastes, hazardous substances, international descriptions, improved descriptions, forbidden materials, and organic peroxides. *US Code Fed. Regul., Title 49*, Parts 171-174, 176-177; *Fed. Regist., 45* (No. 101), pp. 34560, 34620

US Department of the Treasury (1982) Materials and processes for the production and treatment of wine, juice, and distilling material. *US Code Fed. Regul., Title 27*, Part 240; *Fed. Regist., 47* (No. 118), pp. 26399-26405

US Food and Drug Administration (1979) Oral mucosal injury drug products for over-the-counter human use, establishment of a monograph; Proposed rulemaking. *US Code Fed. Regul., Title 21*, Part 353; *Fed. Regist., 44* (No. 214), pp. 63270-63273, 63280-63281, 63289-63290

US Food and Drug Administration (1982) Oral health care drug products for over-the-counter human use; Establishment of a monograph. *US Code Fed. Regul., Title 21*, Part 356; *Fed. Regist., 47* (No. 101), pp. 22760-22763, 22858, 22875-22877

US Food and Drug Administration (1983a) Hydrogen peroxide; Proposed affirmation of GRAS status as a direct human food ingredient with specific limitations. *US Code Fed. Regul., Title 21*, Parts 182, 184; *Fed. Regist., 48* (No. 223), pp. 52323-52333

US Food and Drug Administration (1983b) Oral mucosal injury drug products for over-the-counter human use; Tentative final monograph. *US Code Fed. Regul., Title 21*, Part 353; *Fed. Regist., 48* (No. 144), pp. 33984-33994

US Pharmacopeial Convention, Inc. (1980) *The United States Pharmacopeia*, 20th rev., Rockville, MD, p. 381

Wales, R.G., White, I.G. & Lamond, D.R. (1959) The spermicidal activity of hydrogen peroxide in vitro and in vivo. *J. Endocrinol., 18*, 236-244

Wang, R.J. & Nixon, B.T. (1978) Identification of hydrogen peroxide as a photoproduct toxic to human cells in tissue-culture medium irradiated with 'daylight' fluorescent light. *In Vitro, 14*, 715-722

Wang, R.J., Ananthaswamy, H.N., Nixon, B.T., Hartman, P.S. & Eisenstark, A. (1980) Induction of single-strand DNA breaks in human cells by H_2O_2 formed in near-UV (black light)-irradiated medium. *Radiat. Res., 82*, 269-276

Warm, E. & Laties, G.G. (1982) Quantification of hydrogen peroxide in plant extracts by the chemiluminescence reaction with luminol. *Phytochemistry, 21*, 827-831

Waters, J.W., Hardy, J.C., Jarnot, R.F. & Pickett, H.M. (1981) Chlorine monoxide radical, ozone, and hydrogen peroxide: Stratospheric measurements by microwave limb sounding. *Science, 214*, 61-64

Weiss, S.J., Young, J., LoBuglio, A.F. & Slivka, A. (1981) Role of hydrogen peroxide in neutrophil-mediated destruction of cultured endothelial cells. *J. clin. Invest., 68*, 714-721

WHO (1980) Evaluation of certain food additives. Twenty-fourth Report of the Joint FAO/WHO Expert Committee on Food Additives. *WHO tech. Rep. Ser. No. 653*, Geneva

Williams, M.D., Leigh, J.S., Jr & Chance, B. (1982) Hydrogen peroxide in human breath and its probable role in spontaneous breath luminescence. *Ann. N.Y. Acad. Sci., 386*, 478-483

Wills, E.D. (1971) Effects of lipid peroxidation on membrane-bound enzymes of the endoplasmic reticulum. *Biochem. J., 123*, 983-991

Wilmer, J.W.G.M. & Natarajan, A.T. (1981) Induction of sister-chromatid exchanges and chromosome aberrations by γ-irradiated nucleic acid constituents in CHO cells. *Mutat. Res., 88*, 99-107

Windholz, M., ed. (1983) *The Merck Index*, 10th ed., Rahway, NJ, Merck & Co., p. 697

Worley, J.D. (1983) Hydrogen peroxide in cleansing antiseptics. *J. chem. Educ., 60*, 678

Xu, J., Whong, W.-Z. & Ong, T.-M. (1984) Validation of the *Salmonella* (SV50)/arabinose-resistant forward mutation assay system with 26 compounds. *Mutat. Res., 130*, 79-86

Yakovleva, E.N., Nikonova, N.P. & Kreingol'd, S.U. (1983) Method of analyzing industrial solutions in the production of thiourea dioxide (Russ.). *Zavod. Lab., 49*, 24-25 [*Chem. Abstr., 98*, 172272e]

Yamaguchi, T. & Yamashita, Y. (1980) Mutagenicity of hydroperoxides of fatty acids and some hydrocarbons. *Agric. biol. Chem., 44*, 1675-1678

Younkin, S., Oski, F.A. & Barness, L.A. (1971) Mechanism of the hydrogen peroxide hemolysis test and its reversal with phenols. *Am. J. clin. Nutr., 24*, 7-13

Zabelin, A.I. & Karbovnichii, P.N. (1983) Optimum conditions for the determination of trace amounts of hydrogen peroxide with a phenolphthalein leuco base (Russ.). *Energetik, 6*, 26-27 [*Chem. Abstr., 99*, 76526r]

Zika, R.G. & Saltzman, E.S. (1982) Interaction of ozone and hydrogen peroxide in water: Implications for analysis of H_2O_2 in air. *Geophys. Res. Lett., 9*, 231-234

Zika, R., Saltzman, E., Chameides, W.L. & Davis, D.D. (1982) H_2O_2 levels in rainwater collected in south Florida and the Bahama Islands. *J. geophys. Res., 87*, 5015-5017

LAUROYL PEROXIDE

1. Chemical and Physical Data

1.1 Synonyms and trade names

Chem. Abstr. Services Reg. No.: 105-74-8

Chem. Abstr. Name: Peroxide, bis(1-oxododecyl)

IUPAC Systematic Name: Lauroyl peroxide

Synonyms: Dilauroyl peroxide; dodecanoyl peroxide

Trade Names: Alperox C, C/S,F and S-35; DYP-97F; Laurox Q and W40; Laurydol; LYP 97 and 97F

1.2 Structural and molecular formulae and molecular weight

$$CH_3-(CH_2)_{10}-\overset{O}{\underset{\|}{C}}-O-O-\overset{O}{\underset{\|}{C}}-(CH_2)_{10}-CH_3$$

$C_{24}H_{46}O_4$ Mol. wt: 398.6

1.3 Chemical and physical properties of the pure substance

From Hawley (1981), unless otherwise specified

(a) *Description:* White, coarse powder

(b) *Melting-point:* 54.7-55°C (Sheppard & Mageli, 1982); 49°C

(c) *Solubility:* Insoluble in water; slightly soluble in alcohols; soluble in oils and most organic solvents

(d) *Spectroscopy data:* A review on spectroscopy data has been published (Silbert, 1971)

(e) *Stability:* Relatively stable compared to other commercially used peroxides: decomposition half-life is about 1 h at 80°C (Mair & Hall, 1971); inflammable and explosive

(f) *Reactivity*: Strong oxidizing agent

(g) *Conversion factor*: 1 ppm = 16.3 mg/m^3 at 760 mm Hg and 25°C [calculated by the Working Group]

1.4 Technical products and impurities

Lauroyl peroxide is available in the USA and in western Europe as a flaked product containing 98% min lauroyl peroxide (Luperox GmbH, 1979; Pennwalt Corporation, 1983); it is also available in Europe as a 35% active suspension in water (Luperox GmbH, 1979).

2. Production, Use, Occurrence and Analysis

2.1 Production and use

(a) *Production*

Lauroyl peroxide can be prepared by the reaction of lauroyl chloride with sodium hydroxide and hydrogen peroxide (i.e., sodium peroxide) (Sheppard & Mageli, 1982), and this is the method used for its commercial production.

Commercial production of lauroyl peroxide in the USA was first reported in 1941-1943 (US Tariff Commission, 1945). Production reached a peak in 1969 when five US companies reported combined production of 1.1 million kg (US Tariff Commission, 1971). Separate production data have not been reported since 1970, but US consumption of lauroyl peroxide in 1982 was about 630 thousand kg (Anon., 1982). Three companies currently produce lauroyl peroxide in the USA. Separate data on US imports and exports are not published.

One company in Canada produces lauroyl peroxide, and usage in 1980 was estimated at 75 thousand kg (Anon., 1981). It is produced commercially by three companies in Germany, two companies in the UK and one in the Netherlands.

Commercial production in Japan of lauroyl peroxide started in 1953. Two Japanese companies currently manufacture it by the reaction of lauroyl chloride with sodium peroxide; their combined production in 1983 is estimated to have been 50-60 thousand kg.

(b) *Use*

Lauroyl peroxide is believed to be used almost exclusively as an initiator in the production of polymers. It has found extensive use in the suspension polymerization of vinyl chloride and was used in ethylene, styrene and acrylate polymerization, and also to some extent in the moulding of polyester resins (Mageli & Kolczynski, 1968). Lauroyl peroxide is now used (usually as part of a mixture with other initiators) in the suspension polymerization of vinyl chloride and in the elevated temperature curing of unsaturated polyester resins (Pastorino et al., 1983). It has also been recommended for use in the polymerization/curing of acrylates and styrenated alkyd resins and for drying and bleaching oils (Pennwalt Corporation, 1983).

In Japan, lauroyl peroxide is used almost exclusively as an initiator in the polymerization of vinyl chloride. It also finds limited use as a bleaching agent and as a drying agent for tung oil.

Lauroyl peroxide is approved by the US Food and Drug Administration (1980) for the following uses as an indirect food additive: as a component of adhesives used in articles in contact with foods; as a polymerization catalyst in components of paper and paperboard in contact with aqueous and fatty foods; and as a catalyst, at a level of 1.5% max, in crosslinked polyester resins used in articles for repeated contact with foods.

As part of the Hazardous Materials Regulations of the US Department of Transportation (1980), shipments of lauroyl peroxide are subject to a variety of labelling, packaging, quantity and shipping restrictions consistent with its designation as a hazardous material.

2.2 Occurrence

(a) *Natural occurrence*

Lauroyl peroxide is not known to occur as a natural product.

(b) *Occupational exposure*

On the basis of a 1974 National Occupational Hazard Survey, the National Institute for Occupational Safety and Health (1980, 1981) estimated that about 500 US workers in the industrial organic chemicals industry were exposed to lauroyl peroxide.

2.3 Analysis

General methods for the analysis of acyl peroxides (which include lauroyl peroxide) in various matrices have been reviewed (Mair & Hall, 1971). A spectrophotometric method for the analysis of lauroyl peroxide in workplace air has been described (Bianchi & Muccioli, 1978).

3. Biological Data Relevant to the Evaluation of Carcinogenic Risk to Humans

3.1 Carcinogenicity studies in animals

(a) *Subcutaneous and/or intramuscular administration*

Mouse: Groups of 50 and 30 female ICR mice/Ha Swiss mice, eight weeks of age, were given weekly subcutaneous injections of 0.1 and 10 mg lauroyl peroxide [purity unspecified] in tricaprylin, respectively. A total of 110 female mice (from three groups) received weekly injections of 0.05 ml tricaprylin only and served as vehicle controls. All mice were treated and observed throughout life span. Median survival times were 368-535 days in the control groups, 324 days in the low-dose group and 331 days in the high-dose group. No tumour was observed in the controls; in the low-dose group, 3/50 mice developed a fibrosarcoma at the injection site [$p = 0.03$]; no tumour was observed in the high-dose group (Van Duuren et al., 1966).

A group of 15 female BALB/c mice, aged two months, received twice-weekly subcutaneous injections of 0.05 mg lauroyl peroxide (purity, >99%) in 0.1 ml tricaprylin for 52 weeks. Ten mice that received tricaprylin alone served as controls. The experiment was terminated after 18 months, at which time 14/15 mice were still alive. Grossly, no tumour was observed either at the injection site or in internal organs of treated animals; one fibrosarcoma was found at the injection site in the control group (Swern et al., 1970).

Rat: A group of 20 female Sprague-Dawley rats, six weeks of age, received weekly subcutaneous injections of 11 mg lauroyl peroxide [purity unspecified] in 0.2 ml tricaprylin; further groups of 20 females received injections of 0.1 ml tricaprylin only or no injection and served as vehicle or untreated controls. The experiment was terminated on day 542. Median survival times were 537 days in the vehicle- and untreated control groups and 488 days in the lauroyl peroxide-treated group. No tumour was found at the injection site in any group, although local palpable masses associated with tissue necrosis, inflammatory reaction and vascular thrombosis were noted in the treated group (Van Duuren et al., 1967).

(b) Skin application

Mouse: A group of 30 male ICR/Ha Swiss mice, eight weeks of age, received thrice-weekly skin applications of approximately 100 mg of a solution of lauroyl peroxide [purity unspecified] in 5% benzene solution. Three groups of 30 and one group of 60 mice received applications of approximately 100 mg benzene alone. Median survival times were 262-412 days in the benzene-treated groups and 437 days in the lauroyl peroxide-treated group. One animal treated with lauroyl peroxide developed a skin papilloma at the site of application; 11 of the 150 animals treated with benzene only developed tumours at the treatment site, one of which was a carcinoma and the others papillomas (Van Duuren et al., 1963).

Groups of 30 female SENCAR mice, aged seven to nine weeks, were used to test the tumour-promoting (A), tumour-initiating (B) and complete carcinogenic (C) activities of lauroyl peroxide on the skin. Mice in experiment (A) received a single topical application of 10 nmol 7,12-dimethylbenz[a]anthracene (DMBA) in 0.2 ml acetone, followed one week later by applications of 1, 10 or 20 mg lauroyl peroxide in 0.2 ml acetone twice weekly for 25 weeks or acetone only (controls). The numbers of mice with papillomas at 25 weeks were 0/30 (controls), 7/28 (1 mg), 13/30 (10 mg) and 19/29 (20 mg). Mice in experiment (B) received a single topical application of 20 mg lauroyl peroxide in 0.2 ml acetone or acetone only (controls), followed one week later by twice-weekly applications of 2 μg 12-O-tetradecanoylphorbol 13-acetate in acetone for 25 weeks. The numbers of mice with papillomas at 25 weeks were: controls, 4/29; treated animals, 4/28. Mice in experiment (C) received twice-weekly topical applications of 20 mg lauroyl peroxide in 0.2 ml acetone or acetone only (controls) for 25 weeks; 1/28 treated animals and 0/29 controls had papillomas at that time. No squamous-cell carcinoma was found when these animals were observed up to 50 weeks (Klein-Szanto & Slaga, 1982). [The Working Group noted the absence of a DMBA-treated control group for the promotion experiment and the short duration of the experiment for complete carcinogenicity.]

3.2 Other relevant biological data

(a) Experimental systems

Toxic effects

A single application of either 20 or 40 mg lauroyl peroxide to the skin of mice induced mild hyperplasia and a temporary increase in the number of dark basal keratinocytes. No major inflammatory or vascular change was noted (Klein-Szanto & Slaga, 1982).

Effects on reproduction and prenatal toxicity

Doses of 0.25 and 0.50 µmol lauroyl peroxide (98%, purum) were dissolved in acetone and injected into the inner-shell membrane in the air chamber of three-day-old White Leghorn chicken eggs. Dosage was limited by the solubility of lauroyl peroxide in the acetone vehicle. In the high-dose group there was elevated embryonic mortality (10%) compared to that in controls (0%); malformations were seen in both treated groups, but the effect was not dose-related (25% at 0.25 µmol/egg compared to 13% at 0.50 µmol/egg) (Korhonen *et al.*, 1984).

Absorption, distribution, excretion and metabolism

No data were available to the Working Group.

Mutagenicity and other short-term tests (see also 'Appendix: Activity Profiles for Short-term Tests', p. 346)

Lauroyl peroxide was not mutagenic to *Salmonella typhimurium* TA98 or TA100 in the presence of an exogenous metabolic system from polychlorinated-biphenyl-induced rat liver (Yamaguchi & Yamashita, 1980). [The Working Group noted that lauroyl peroxide was tested at only one dose in each strain.]

(b) *Humans*

No data were available to the Working Group.

3.3 Case reports and epidemiological studies of carcinogenicity to humans

No data were available to the Working Group.

4. Summary of Data Reported and Evaluation

4.1 Exposure data

Lauroyl peroxide was first produced commercially in about 1941. It is used principally in the production of polymers; small amounts are employed in food packaging.

4.2 Experimental data

Lauroyl peroxide was tested by subcutaneous administration in mice and rats and by skin application in mice. In one study in mice by subcutaneous administration, the evidence concerning a carcinogenic effect was inconclusive; in two other studies, no increase in tumour incidence was observed. Two studies in mice by skin application were inadequate for an evaluation of complete carcinogenicity; one study indicated that lauroyl peroxide has promoting activity in mouse skin.

The available data are inadequate to evaluate the teratogenic potential of lauroyl peroxide in mammals.

The available data are inadequate to evaluate the activity in short-term tests of lauroyl peroxide.

4.3 Human data

No case report or epidemiological study of the carcinogenicity of lauroyl peroxide to humans was available to the Working Group.

4.4 Evaluation[1]

There is *inadequate evidence* for the carcinogenicity of lauroyl peroxide to experimental animals.

In the absence of epidemiological data, no evaluation could be made of the carcinogenicity of lauroyl peroxide to humans.

5. References

Anon. (1981) Organic peroxides. Lucidol still works at closing the loopholes. *Corpus Market Focus*, 9 February, 5691.1.1-5691.1.2

Anon. (1982) Organic peroxides. *Modern Plastics*, September, pp. 70, 71

Bianchi, A. & Muccioli, G. (1978) Determination of organic hydroperoxides present in some work atmospheres. Determination of total hydroperoxides in 4-vinylcyclohexene (Ital.). *Ann. Ist. super. Sanita*, *14*, 447-453 [*Chem. Abstr.*, *92*, 168296z]

Hawley, G.G., ed. (1981) *The Condensed Chemical Dictionary*, 10th ed., New York, Van Nostrand Reinhold Co., p. 603

Klein-Szanto, A.J.P. & Slaga, T.J. (1982) Effects of peroxides on rodent skin: Epidermal hyperplasia and tumor promotion. *J. invest. Dermatol.*, *79*, 30-34

Korhonen, A., Hemminki, K. & Vainio, H. (1984) Embryotoxic effects of eight organic peroxides and hydrogen peroxide on three-day chicken embryos. *Environ. Res.*, *33*, 54-61

Luperox GmbH (1979) *Organic Peroxides as Polymerization Initiators*, Günzburg

Mageli, O.L. & Kolczynski, J.R. (1968) *Peroxy compounds*. In: Bikales, N.M., ed., *Encyclopedia of Polymer Science and Technology, Plastics, Resins, Rubbers, Fibers*, Vol. 9, New York, Interscience, pp. 818, 819, 831, 841

Mair, R.D. & Hall, R.T. (1971) *Determination of organic peroxides by physical, chemical and colorimetric methods*. In: Swern, D., ed., *Organic Peroxides*, Vol. 2, New York, Wiley-Interscience, pp. 535-635

---

[1]For definitions of the italicized terms, see the Preamble, pp. 15-16.

National Institute for Occupational Safety and Health (1980) *Projected Number of Occupational Exposures to Chemical and Physical Hazards*, Cincinnati, OH, p. 81

National Institute for Occupational Safety and Health (1981) *National Occupational Hazard Survey Data* (microfiche), Cincinnati, OH, p. 5288 (generics excluded)

Pastorino, R.L., Lewis, R.N. & Halle, R. (1983) Organic peroxides. In: Agranoff, J., ed., *Modern Plastics Encyclopedia 1983-1984*, New York, McGraw-Hill, pp. 150, 152, 156

Pennwalt Corporation (1983) *Product Bulletin: Diacyl Peroxides*, Buffalo, NY, Lucidol Division

Sheppard, C.S. & Mageli, O.L. (1982) Peroxides and peroxy compounds, organic. In: Grayson, M., ed., *Kirk-Othmer Encyclopedia of Chemical Technology*, 3rd ed., Vol. 17, New York, John Wiley & Sons, pp. 65, 69, 90

Silbert, L.S. (1971) Physical properties of organic peroxides. In: Swern, D., ed., *Organic Peroxides*, Vol. 2, New York, Wiley-Interscience, pp. 637-639, 689, 704

Swern, D., Wieder, R., McDonough, M., Meranze, D.R. & Shimkin, M.B. (1970) Investigation of fatty acids and derivatives for carcinogenic activity. *Cancer Res.*, 30, 1037-1046

US Department of Transportation (1980) Identification numbers, hazardous wastes, hazardous substances, international descriptions, improved descriptions, forbidden materials, and organic peroxides. *US Code Fed. Regul.*, Title 49, Parts 171-174, 176, 177; *Fed. Regist.*, 45 (No. 101), pp. 34560, 34622

US Food and Drug Administration (1980) Food and drugs. *US Code Fed. Regul.*, Title 21, Parts 175.105, 176.170, 177.2420

US Tariff Commission (1945) *Synthetic Organic Chemicals, US Production and Sales, 1941-43 (Report No. 153, Second Series)*, Washington DC, US Government Printing Office, p. 127

US Tariff Commission (1971) *Synthetic Organic Chemicals, US Production and Sales, 1969 (TC Publication 412)*, Washington DC, US Government Printing Office, pp. 206, 231

Van Duuren, B.L., Nelson, N., Orris, L., Palmes, E.D. & Schmitt, F.L. (1963) Carcinogenicity of epoxides, lactones, and peroxy compounds. *J. natl Cancer Inst.*, 31, 41-55

Van Duuren, B.L., Langseth, L., Orris, L., Teebor, G., Nelson, N. & Kuschner, M. (1966) Carcinogenicity of epoxides, lactones, and peroxy compounds. IV. Tumor response in epithelial and connective tissue in mice and rats. *J. natl Cancer Inst.*, 37, 825-838

Van Duuren, B.L., Langseth, L., Orris, L., Baden, M. & Kuschner, M. (1967) Carcinogenicity of epoxides, lactones, and peroxy compounds. V. Subcutaneous injection in rats. *J. natl Cancer Inst.*, 39, 1213-1216

Yamaguchi, T. & Yamashita, Y. (1980) Mutagenicity of hydroperoxides of fatty acids and some hydrocarbons. *Agric. biol. Chem.*, 44, 1675-1678

APPENDIX

APPENDIX:
ACTIVITY PROFILES FOR SHORT-TERM TESTS

Introduction

Data from short-term tests are summarized and evaluated in sections 3.2 and 4.2 of the individual monographs in this volume. This Appendix serves to provide a graphical representation of those data and information on the doses tested.

As with the data in the monographs, the tests have been classified here according to the end-point detected (DNA damage, mutation, chromosomal effects, cell transformation) and to the test organism (arranged phylogenetically within each group, see the Preamble and Table 1). Either the minimal effective dose or the maximal dose tested that produced no effect has been recorded for each test system, and this information is presented as an activity profile (Garrett et al., 1984) for each compound considered by the Working Group. Only those data used in making the evaluation were incorporated in the profile. Results of studies described only in abstracts are cited in sections 3.2 but are not used in making the evaluation (sections 4.2) nor in the activity profiles. The profiles are therefore in accordance with the assessments made by the Working Group.

Table 1. Classification scheme for short-term tests for genetic activity and cell transformation

GENETIC ACTIVITY
  DNA damage
    Prokaryotes
    Fungi/green plants
    Mammalian cells (in vitro)
    Mammals (in vivo)
    Humans (in vivo)
  Mutation
    Prokaryotes/bacteriophages
    Fungi/green plants
    Insects
    Mammalian cells (in vitro)
    Mammals (in vivo)
    Humans (in vivo)
  Chromosomal effects
    Fungi/green plants
    Insects
    Mammalian cells (in vitro)
    Mammals (in vivo)
    Humans (in vivo)

CELL TRANSFORMATION
    Mammalian cells (in vitro)

Representation of an activity profile for short-term tests

A data set, consisting of a discrete number of assays and of the doses required to produce responses in those assays, is represented in a bar graph. Test systems (representative of the assays on the compounds discussed in these monographs) are shown on the x-axis, and values corresponding to qualitative responses and to the doses employed to obtain them for each test system are given on the y-axis. The codes on the x-axis are given in Table 2.

Table 2. Sequence of codes assigned to individual test systems

	Code	Definition
GENETIC ACTIVITY		
DNA damage		
Prokaryotes	PIN	Prophage induction test
		Escherichia coli Pol A (W3110-P3478)
	REP	Spot test
	RET	Liquid suspension
	REW	*Bacillus subtilis* rec (H17-M45), spot test
	REC	DNA repair-deficient bacteria
Fungi/green plants	DBY	DNA strand breaks, yeast
Mammalian cells (*in vitro*)		Unscheduled DNA synthesis
	UDP	Rat primary hepatocytes
	UDH	Human diploid fibroblasts
	UDS	Other mammalian cells
		Inhibition of DNA synthesis
	IDR	Rodent cells
		DNA strand breaks
	DBH	Human cells
	DBR	Rodent cells
Mammals (*in vivo*)	DBO	DNA strand breaks, rodents
Mutation		
Prokaryotes/bacteriophages		*Salmonella typhimurium*
	SA5	TA1535
	SA7	TA1537
	SA8	TA1538
	SA9	TA98
	SA0	TA100
	SAM	Miscellaneous strains
	THM	T_2 bacteriophage, mutation of *Escherichia coli* B
		Escherichia coli
	WP2	WP2
	WPU	WP2 uvrA
	ECO	Miscellaneous strains
	BSM	*Bacillus subtilis* sporulation test
	HIM	*Haemophilus influenzae*
	KPM	*Klebsiella pneumoniae*
	MAF	*Micrococcus aureus*
	BFA	Body fluids, mammals
	BFH	Body fluids, humans
	HMA	Host-mediated assays
Fungi/green plants	SGR	*Streptomyces griseoflavus*, reverse mutation
		Streptomyces coelicolor
	STF	Forward mutation
	STR	Reverse mutation
		Saccharomyces cerevisiae
	YEC	Gene conversion
	YEH	Homozygosis (through recombination or gene conversion)
	YEF	Forward mutation
	YER	Reverse mutation
	YEY	*Schizosaccharomyces pombe*, forward mutation
		Aspergillus nidulans
	ASF	Forward mutation
	ASR	Reverse mutation
	NER	*Neurospora crassa*, reverse mutation
	HOM	*Hordeum vulgare*
	PGM	Plant gene mutation

APPENDIX: ACTIVITY PROFILES

	Code	Definition
Insects		*Drosophila melanogaster*,
	SRL	Sex-linked recessive lethal test
Mammalian cells (*in vitro*)		Chinese hamster cells
	CHO	Ovary HGPRT or ATPase locus
	V7H	Lung (V79) HGPRT locus
	V70	Lung (V79) ATPase locus
		Mouse lymphoma (L5178Y cells)
	L5M	Methotrexate locus
	L5T	TK locus
Mammals (*in vivo*)	MST	Mouse spot test
	SLT	Mouse specific locus test
Humans (*in vivo*)	HVL	Human variant lymphocytes
Chromosomal effects		
Fungi/green plants	ALC	*Allium cepa*, cytogenetics
	HOC	*Hordeum vulgare*, cytogenetics
	VIC	*Vicia faba*, cytogenetics
	PYC	Plant chromosome studies
Insects		*Drosophila melanogaster*
	DAG	Sex chromosome gain
	DAP	Aneuploidy-partial sex chromosome loss
	DAN	Aneuploidy-all tests
	DHT	Heritable (reciprocal) translocation
	DMM	Mosaics
Mammalian cells (*in vitro*)		Sister chromatid exchanges
	SCC	Chinese hamster ovary (CHO)
	SCV	Chinese hamster lung fibroblasts (V79)
	SCL	Human lymphocytes
	SC1	Other human cells
		Chromosomal aberrations
	CYU	Chinese hamster
	CYV	Syrian golden hamster
	CYX	Rat
	CYY	Mouse
	CYH	Human lymphocytes
	CYZ	Other human cells
	CYM	Tumour cells
	CYC	Other cell types
	MNC	Micronuclei *in vitro*
Mammals (*in vivo*)	SC3	Sister chromatid exchanges, animals except humans
		Chromosomal aberrations, animals except humans
	CYB	Bone marrow
	OYL	Lymphocytes
	CYO	Oocyte or early embryo studies
	CYS	Spermatogonia treated, spermatogonia observed
	CYT	Spermatocytes treated, spermatocytes observed
	DLM	Dominant lethal test, mouse
	DLR	Dominant lethal test, rat
	MHT	Heritable translocation, mouse
	MNH	Micronucleus test, hamster
	MNM	Micronucleus test, mouse
	MNR	Micronucleus test, rat
Humans (*in vivo*)	SC4	Sister chromatid exchanges
	CYD	Chromosomal aberrations, lymphocytes
	MNS	Micronuclei
CELL TRANSFORMATION		
Mammalian cells (*in vitro*)	CTH	C3H10T1/2 cells
	CT7	SA-7/Syrian hamster embryo cells

The term 'dose', as used here, does not take into consideration length of treatment or exposure and may therefore be considered synonymous with concentration. Doses and concentrations are reported in the literature in various units (e.g., parts per million, percentage, mass per volume, volume per volume); for comparative assessment of liquids and solids, therefore, units are converted to mass per volume. Concentrations of gaseous compounds reported as parts per million on a volume per volume basis are converted to parts per million, mass per volume, according to the ideal gas laws. Dose units for in-vitro test systems are expressed as micrograms per millilitre; for microbial plate-incorporation tests, a volume of 2 ml is assumed for the top agar, and a 1-ml volume is taken for differential toxicity assays. Doses used in in-vivo bioassays are expressed as milligrams per kilogram body weight of the treated animal.

Only those results accompanied by a dose value were included in the profile. Line-heights were derived as follows: for negative test results, the highest dose tested is defined as the highest ineffective dose (HID). If there was evidence of extreme toxicity, the next highest dose was used. A single dose tested with a negative result was considered equivalent to the HID. Similarly, for positive results, the lowest effective dose (LED) was recorded. If the original data had been analysed statistically by the author, the dose that gave a positive effect was taken as that for which the results were significant ($p < 0.05$). If the original data had not been analysed statistically, the dose required to produce an effect was estimated as follows: when a dose-related positive response was observed with two or more doses, the lower of the doses was taken as the LED; a single dose resulting in a positive response was considered equivalent to the LED. When two or more studies using the same assay were available, an average of the logarithmic values of the data subset was calculated. When conflicting results were encountered with a given assay, the subset of data corresponding to the result judged valid by the Working Group was used.

Because of the wide range of doses encountered in the 100 assays, a logarithmic scale is used; the LED is plotted on an inverted scale [100 000-1 µg/ml (or mg/kg)] on the positive y-axis and the HID [1-100 000 µg/ml (or mg/kg)] on the negative y-axis. Negative results obtained with less than 1 µg/ml are arbitrarily assigned to a dose of 1 µg/ml.

The LED or HID for any given assay will depend on the characteristics of the performance and response of each test system. Because activity is plotted on a logarithmic scale, differences in molecular weights of compounds do not greatly influence comparisons of their overall responses.

It should be noted that in the *Salmonella*/microsome assay, multiple tester strains can be used, and these are represented independently in the profiles. However, a positive response in any one strain dictates an overall positive response for the assay.

References

Garrett, N.E., Stack, H.F., Gross, M.R. & Waters, M.D. (1984) An analysis of the spectra of genetic activity produced by known or suspected human carcinogens. *Mutat. Res.*, *134*, 89-111

APPENDIX: ACTIVITY PROFILES

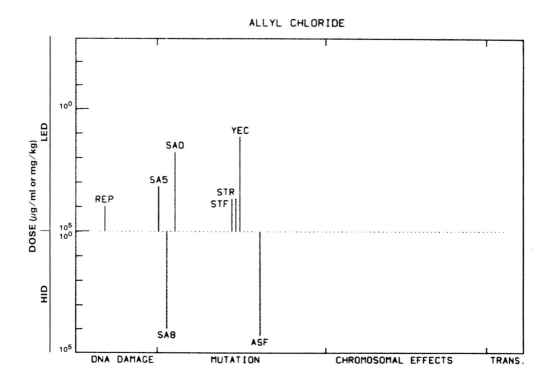

ALLYL CHLORIDE

TEST	RESULT -S9/+S9	DOSE (HID or LED)	REFERENCE
REP	+/0	9400	MCCOY ET AL., 1978
SA5	+/+	940	MCCOY ET AL., 1978
SA5	+/+	2350	BIGNAMI ET AL., 1980
SA8	-/-	9400	MCCOY ET AL., 1978
SA0	+/0		NORPOTH ET AL., 1980
SA0	+/0	0.02	SIMMON, 1981
SA0	-/-	9400	MCCOY ET AL., 1978
SA0	(+)/-	9400	BIGNAMI ET AL., 1980
SA0	+/(+)		EDER ET AL., 1982a
SA0	+/(+)		EDER ET AL., 1982b
SA0	+/(+)	1150	EDER ET AL., 1980
STF	+/0	4700	BIGNAMI ET AL., 1980
STR	+/0	4700	BIGNAMI ET AL., 1980
YEC	+/0	14	MCCOY ET AL., 1978
ASF	-/0	18800	BIGNAMI ET AL., 1980

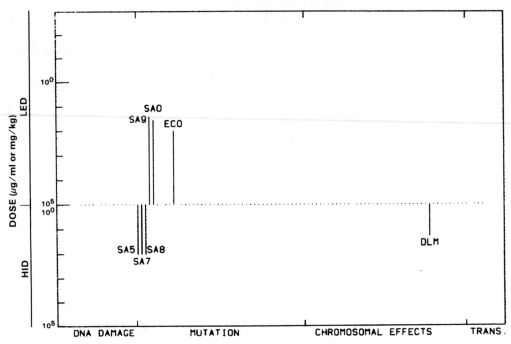

ALLYL ISOTHIOCYANATE

```
TEST RESULT      DOSE        REFERENCE
     -S9/+S9    (HID or LED)

     REW  -/0              ODA ET AL., 1978
     SA5  -/0       100    YAMAGUCHI, 1980
     SA7  -/0       100    YAMAGUCHI, 1980
     SA8  -/0       100    YAMAGUCHI, 1980
     SA9  +/0        25    YAMAGUCHI, 1980
     SA9  -/-       250    KASAMAKI ET AL., 1982
     SAO  +/0        25    YAMAGUCHI, 1980
     SAO  -/-       250    KASAMAKI ET AL., 1982
     SAO  (+)/-            EDER ET AL., 1982a
     SAO  (+)/-            EDER ET AL., 1982b
     SAO  (+)/-      50    EDER ET AL., 1980
     ECO  -/+        99    RIHOVA, 1982
     HMA  -/0              USFDA 1975
     SRL  +/0              AUERBACH & ROBSON, 1944
     SRL  -/0              SCHALET & HERSKOWITZ, 1954
     SRL  (+)/0            AUERBACH & ROBSON, 1947
     ALC  +/0              SHARMA & SHARMA, 1962
     PYC  +/0              SWAMINATHAN & NATARAJAN, 1959
     PYC  +/0              SWAMINATHAN & NATARAJAN, 1956
     DAP  -/0              AUERBACH & ROBSON, 1947
     CYU  +/+              KASAMAKI ET AL., 1982
     CYZ  -/0              USFDA 1975
     CYB  -/0              USFDA 1975
     DLM  -/0        19    EPSTEIN ET AL., 1972
     DLR  -/0              USFDA 1975
```

APPENDIX: ACTIVITY PROFILES

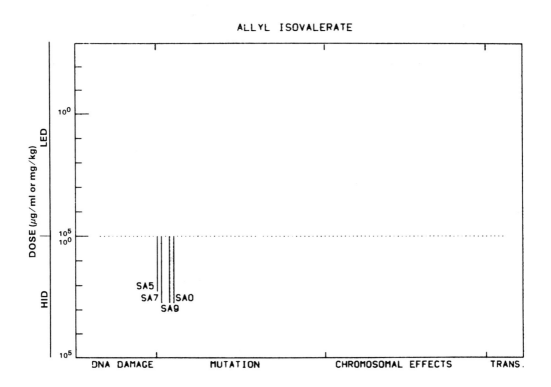

ALLYL ISOVALERATE

TEST	RESULT −S9/+S9	DOSE (HID or LED)	REFERENCE
SA5	−/−	167	NTP 1983
SA7	−/−	500	NTP 1983
SA9	−/−	500	NTP 1983
SA0	−/−	500	NTP 1983

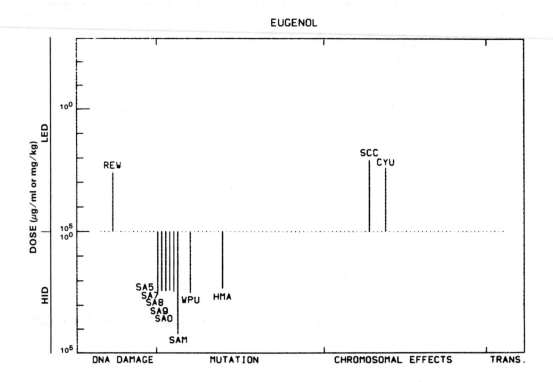

APPENDIX: ACTIVITY PROFILES 333

EUGENOL

TEST	RESULT -S9/+S9	DOSE (HID or LED)	REFERENCE		TEST	RESULT -S9/+S9	DOSE (HID or LED)	REFERENCE
REW	+/0	400	SEKIZAWA & SHIBAMOTO, 1982		SA9	-/-	250	POOL & LIN, 1982
REW	-/0	100000	YOSHIMURA ET AL., 1981		SA9	-/-	250	YOSHIMURA ET AL., 1981
SA5	-/-	167	HAWORTH ET AL., 1983		SA9	-/-	300	SEKIZAWA & SHIBAMOTO, 1982
SA5	-/-	167	NTP 1983		SA9	-/-	410	SWANSON ET AL., 1979
SA5	-/-	250	TO ET AL., 1982		SA9	-/-	492	FLORIN ET AL., 1980
SA5	-/-	250	POOL & LIN, 1982		SA9	-/-	1000	NESTMANN ET AL., 1980
SA5	-/-	250	YOSHIMURA ET AL., 1981		SA9	-/0	32.8	DORANGE ET AL., 1977
SA5	-/-	300	SEKIZAWA & SHIBAMOTO, 1982		SA9	0/-		ROCKWELL & RAW, 1979
SA5	-/-	410	SWANSON ET AL., 1979		SA0	-/-		EDER ET AL., 1982a
SA5	-/-	492	FLORIN ET AL., 1980		SA0	-/-		EDER ET AL., 1982b
SA5	-/-	492	DELAFORGE ET AL., 1977		SA0	-/-	164	SWANSON ET AL., 1979
SA5	-/-	1000	NESTMANN ET AL., 1980		SA0	-/-	164	MILLER ET AL., 1979
SA5	-/0	32.5	DORANGE ET AL., 1977		SA0	-/-	167	HAWORTH ET AL., 1983
SA7	-/-	167	HAWORTH ET AL., 1983		SA0	-/-	167	NTP 1983
SA7	-/-	167	NTP 1983		SA0	-/-	250	TO ET AL., 1982
SA7	-/-	250	TO ET AL., 1982		SA0	-/-	250	POOL & LIN, 1982
SA7	-/-	250	POOL & LIN, 1982		SA0	-/-	250	YOSHIMURA ET AL., 1981
SA7	-/-	250	YOSHIMURA ET AL., 1981		SA0	-/-	300	SEKIZAWA & SHIBAMOTO, 1982
SA7	-/-	300	SEKIZAWA & SHIBAMOTO, 1982		SA0	-/-	492	FLORIN ET AL., 1980
SA7	-/-	492	FLORIN ET AL., 1980		SA0	-/-	1000	NESTMANN ET AL., 1980
SA7	-/-	492	DELAFORGE ET AL., 1977		SA0	-/-	1500	EDER ET AL., 1980
SA7	-/-	1000	NESTMANN ET AL., 1980		SA0	-/-	500	RAPSON ET AL., 1980
SA7	-/0	32.8	DORANGE ET AL., 1977		SA0	-/0	32.8	DORANGE ET AL., 1977
SA8	-/-	250	TO ET AL., 1982		SA0	0/-		ROCKWELL & RAW, 1979
SA8	-/-	250	POOL & LIN, 1982		SAM	-/-	16000	GREEN & SAVAGE, 1978
SA8	-/-	250	YOSHIMURA ET AL., 1981		WPU	-/-	300	SEKIZAWA & SHIBAMOTO, 1982
SA8	-/-	300	SEKIZAWA & SHIBAMOTO, 1982		BFA	0/-		ROCKWELL & RAW, 1979
SA8	-/-	492	DELAFORGE ET AL., 1977		HMA	-/0	200	GREEN & SAVAGE, 1978
SA8	-/-	1000	NESTMANN ET AL., 1980		YEC	-/0		NESTMANN & LEE, 1983
SA8	-/0	32.8	DORANGE ET AL., 1977		YER	-/0		NESTMANN & LEE, 1983
SA9	-/-	167	HAWORTH ET AL., 1983		SCC	(+/+)	123	NTP 1983
SA9	-/-	167	NTP 1983		CYU	-/+	324	NTP 1983
SA9	-/-	250	TO ET AL., 1982		CYU	(+)/0	200	STICH ET AL., 1981

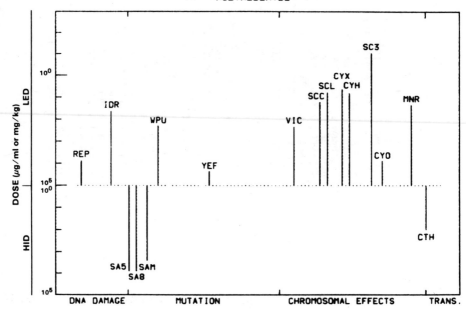

ACETALDEHYDE

```
TEST RESULT    DOSE        REFERENCE
     -S9/+S9  (HID or LED)

     REP  (+)/0    7800     ROSENKRANZ, 1977
     IDR  (+)/0      44     BIRD & DRAPER, 1980
     SA5   -/0     7800     ROSENKRANZ, 1977
     SA8   -/0     7800     ROSENKRANZ, 1977
     SA9   -/-              SASAKI & ENDO, 1978
     SAO   -/-              SASAKI & ENDO, 1978
     SAM   -/0     2500     MARNETT ET AL., 1984
     WPU   +/0       39     VEGHELYI ET AL., 1978
     WPU   +/0     1000     IGALI & GAZSO, 1980
     YEF  (+)/0   23400     BANDAS, 1982
     VIC   +/0      220     RIEGER & MICHAELIS, 1960
     SCC   +/+       39     DE RAAT ET AL., 1983
     SCC   +/0       31     OBE & RISTOW, 1977
     SCC   +/0      3.9     OBE & BEEK, 1979
     SCL   +/0              VEGHELYI & OSZTOVICS, 1978
     SCL   +/0        6     JANSSON, 1982
     SCL   +/0       16     BOHLKE ET AL., 1983
     SCL   +/0      1.8     VEGHELYI ET AL., 1978
     SCL   +/0      7.8     RISTOW & OBE, 1978
     CYU   +/0              AU & BADR, 1979
     CYX   +/0      4.4     BIRD ET AL., 1982
     CYH   +/0              VEGHELYI & OSZTOVICS, 1978
     CYH   +/0       16     BOHLKE ET AL., 1983
     CYH   +/0       20     BADR & HUSSAIN, 1977
     CYH   +/0      7.8     OBE ET AL., 1984
     CYH   +/0              OBE ET AL., 1979
     CYH  (+)/0     0.78    OBE ET AL., 1978
     SC3   +/0      0.5     KORTE ET AL., 1981
     SC3   +/0     0.016    OBE ET AL., 1979
     CYO   +/0     7800     BARILIAK & KOZACHUK, 1983
     MNR   +/0       22     BIRD ET AL., 1982
     CTH   -/0      100     ABERNETHY ET AL., 1982
```

APPENDIX: ACTIVITY PROFILES

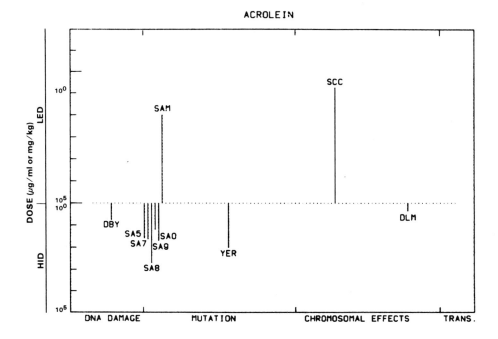

ACROLEIN

```
TEST RESULT    DOSE       REFERENCE
     -S9/+S9  (HID or LED)
     REP  +/0              BILIMORIA, 1975
     DBY  -/0      5.6     FLEER & BRENDEL, 1982
     SAE  ./       17      HAWORTH ET AL., 1983
     SA5  -/-      28      LOQUET ET AL., 1981
     SA5  -/-     500      LIJINSKY & ANDREWS, 1980
     SA5  -/-       8.4    FLORIN ET AL., 1980
     SA5  -/(+)    0.005   HALES, 1982
     SA7  -/-      17      HAWORTH ET AL., 1983
     SA7  -/-     500      LIJINSKY & ANDREWS, 1980
     SA7  -/-       8.4    FLORIN ET AL., 1980
     SA8  -/-     500      LIJINSKY & ANDREWS, 1980
     SA9  +/-       8.4    LIJINSKY & ANDREWS, 1980
     SA9  -/-      17      HAWORTH ET AL., 1983
     SA9  -/-      28      LOQUET ET AL., 1981
     SA9  -/-       8.4    FLORIN ET AL., 1980
     SAO  +/-       2.1    LUTZ ET AL., 1982
     SAO  -/-      28      LOQUET ET AL., 1981
     SAO  -/-      50      HAWORTH ET AL., 1983
     SAO  -/-     500      LIJINSKY & ANDREWS, 1980
     SAO  -/-       8.4    FLORIN ET AL., 1980
     SAM  +/0      10      MARNETT ET AL., 1984
     WPU  (+)/0            HEMMINKI ET AL., 1980
     STF  +/0              ORTALI ET AL., 1977
     YER  -/0     100      IZARD, 1973
     ASF  -/0              BIGNAMI ET AL., 1977
     SCC  +/-       0.56   AU ET AL., 1980
     DLM  -/0       2.2    EPSTEIN ET AL., 1972
     CTH  -/0              ABERNETHY ET AL., 1983
```

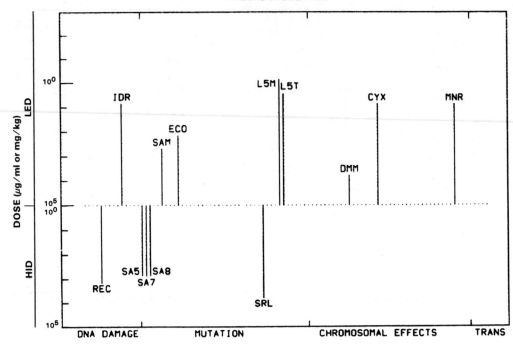

MALONDIALDEHYDE

TEST	RESULT -S9/+S9	DOSE (HID or LED)	REFERENCE
REC	-/0	1441	YONEI & FURUI, 1981
IDR	+/0	7.2	BIRD & DRAPER, 1980
SA5	-/0		MARNETT & TUTTLE, 1980
SA5	-/0	720	MUKAI & GOLDSTEIN, 1976
SA7	-/0	720	MUKAI & GOLDSTEIN, 1976
SA8	-/0		MARNETT & TUTTLE, 1980
SA8	-/0	720	MUKAI & GOLDSTEIN, 1976
SAM	+/0	288	MUKAI & GOLDSTEIN, 1976
SAM	+/0	500	SHAMBERGER ET AL., 1979
SAM	+/0	2500	MARNETT ET AL., 1984
SAM	(+)/0		BASU & MARNETT, 1983
SAM	(+)/0	90	MARNETT & TUTTLE, 1980
SAM	(+)/0	1000	LEVIN ET AL., 1982
ECO	+/0	144	YONEI & FURUI, 1981
SRL	-/0	6125	SZABAD ET AL., 1983
L5M	+/0	0.72	YAU, 1979
L5T	+/0	2.9	YAU, 1979
DMM	+/0	6125	SZABAD ET AL., 1983
CYX	+/0	7.2	BIRD ET AL., 1982b
MNR	+/0	7.2	BIRD ET AL., 1982b

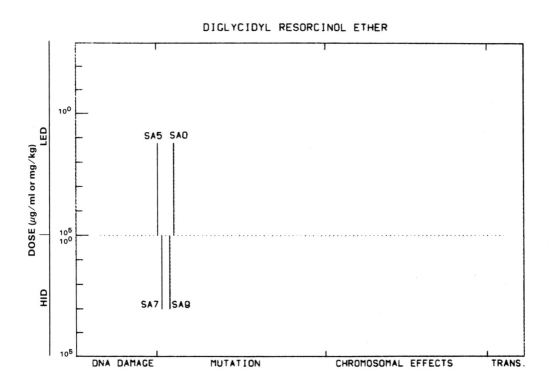

DIGLYCIDYL RESORCINOL ETHER

```
TEST RESULT     DOSE      REFERENCE
     -S9/+S9  (HID or LED)

 SA5   +/+      16.5     NTP, 1985
 SA7   -/-      1000     NTP, 1985
 SA9   -/-      1000     NTP, 1985
 SA0   +/+      16.5     NTP, 1985
```

338 IARC MONOGRAPHS VOLUME 36

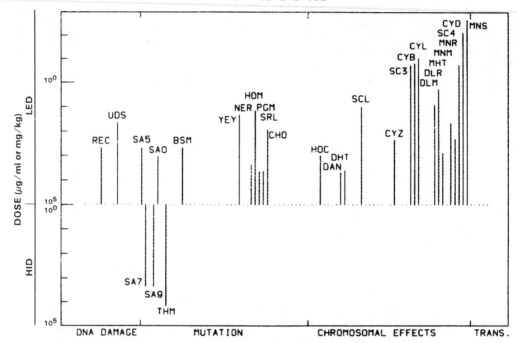

APPENDIX: ACTIVITY PROFILES 339

ETHYLENE OXIDE

TEST	RESULT -S9/+S9	DOSE (HID or LED)	REFERENCE		TEST	RESULT -S9/+S9	DOSE (HID or LED)	REFERENCE
REC	+/0	480	TANOOKA, 1979		CYZ	+/0	220	POIRIER & PAPADOPOULO, 1982
UDS	+/+	44	PERO ET AL., 1981		SC3	+/0		KLIGERMAN ET AL., 1984
SA5	+/+		DE FLORA, 1981		SC3	+/0	0.8	KLIGERMAN ET AL., 1983
SA5	+/0	210	RANNUG ET AL., 1976		SC3	+/0	0.09	YAGER & BENZ, 1982
SA5	+/0	1100	PFEIFFER & DUNKELBERG, 1980		SC3	+/0	0.09	LYNCH ET AL., 1984b
SA7	-/-		DE FLORA, 1981		CYB	+/0	9	STREKALOVA, 1971
SA7	-/0	2200	PFEIFFER & DUNKELBERG, 1980		CYB	+/0	0.03	FOMENKO & STREKALOVA, 1973
SA8	-/-		DE FLORA, 1981		CYB	+/0	0.45	EMBREE & HINE, 1975
SA9	-/-		DE FLORA, 1981		CYB	+/0	0.004	STREKALOVA ET AL., 1975
SA9	-/0	2200	PFEIFFER & DUNKELBERG, 1980		CYL	+/0	0.09	LYNCH ET AL., 1984b
SA0	+/+		DE FLORA, 1981		CYL	-/0	0.8	KLIGERMAN ET AL., 1983
SA0	+/0	1100	PFEIFFER & DUNKELBERG, 1980		DLM	+/0	150	GENEROSO ET AL., 1980
THM	-/0	14500	COOKSON ET AL., 1971		DLM	+/0	0.45	GENEROSO ET AL., 1983
BSM	+/0	480	TANOOKA, 1979		DLR	+/0	1.8	EMBREE ET AL., 1977
YEY	+/+	22	MIGLIORE ET AL., 1982		MHT	+/0	750	GENEROSO ET AL., 1980
NER	+/0		DE SERRES, 1983		MNM	+/0	10	CONAN ET AL., 1979
NER	+/0	1100	KOLMARK & WESTERGAARD, 1953		MNM	+/0	200	APPELGREN ET AL., 1978
NER	+/0	2200	KOLMARK & KILBEY, 1968		MNR	+/0	200	APPELGREN ET AL., 1978
NER	+/0	6170	KILBEY & KOLMARK, 1968		SC4	+/0		LAURENT ET AL., 1983
HOM	+/0		KUCERA ET AL., 1975		SC4	+/0		GARRY ET AL., 1984
HOM	+/0	0.18	SULOVSKA ET AL., 1969		SC4	+/0	50	LAURENT ET AL., 1984
HOM	+/0	1200	EHRENBERG & GUSTAFSSON, 1957		SC4	+/0	1.4	YAGER ET AL., 1983
PGM	+/0	4400	JANA & ROY, 1975		SC4	+/0	0.06	GARRY ET AL., 1979
SRL	+/0	3970	NAKAO & AUERBACH, 1961		SC4	+/0	0.002	SARTO ET AL., 1984
SRL	+/0	4000	WATSON, 1966		SC4	+/0	0.018	STOLLEY ET AL., 1984
SRL	+/0	5000	BIRD, 1952		SC4	-/0		HUSGAFVEL-PURSIAINEN ET AL., 1980
CHO	+/+	88	TAN ET AL., 1981		SC4	-/0		HEINER ET AL., 1982
DVH	+/0		HATCH ET AL., 1983		SC4	-/0	0.002	HOGSTEDT ET AL., 1983
DVD	+/0		HATCH ET AL., 1983		SC4	-/0	0.009	HANSEN ET AL., 1984
HOC	+/0	1000	MOUTSCHEN-DAHMEN ET AL., 1968		CYD	+/0	0.02	PERO ET AL., 1981
DAN	+/0	5010	FAHMY & FAHMY, 1970		CYD	+/0	0.002	HOGSTEDT ET AL., 1983
DHT	+/0	3970	NAKAO & AUERBACH, 1961		CYD	+/0	0.009	THIESS ET AL., 1981a
DHT	+/0	4000	WATSON, 1966		CYD	-/0	0.0002	VAN SITTERT ET AL., 1984
SCL	+/0		GARRY ET AL., 1984		MNS	(+)/0	0.002	HOGSTEDT ET AL., 1983
SCL	+/0	10	GARRY ET AL., 1982		CT7	+/0		HATCH ET AL., 1982

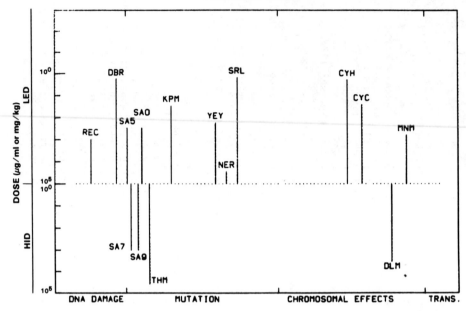

PROPYLENE OXIDE

TEST	RESULT -S9/+S9	DOSE (HID or LED)	REFERENCE
REC	+/0	1000	BOOTMAN ET AL., 1979
DBR	+/0	1.7	SINA ET AL., 1983
SA5	+/+	50	BOOTMAN ET AL., 1979
SA5	+/0	1000	PFEIFFER & DUNKELBERG, 1980
SA5	(+)/0	500	WADE ET AL., 1978
SA7	-/-	350	BOOTMAN ET AL., 1979
SA7	-/0		WADE ET AL., 1978
SA7	-/0	2900	PFEIFFER & DUNKELBERG, 1980
SA9	-/-	350	BOOTMAN ET AL., 1979
SA9	-/0		WADE ET AL., 1978
SA9	-/0	2900	PFEIFFER & DUNKELBERG, 1980
SA0	+/+	50	BOOTMAN ET AL., 1979
SA0	+/0	200	YAMAGUCHI, 1982
SA0	+/0	1000	PFEIFFER & DUNKELBERG, 1980
SA0	(+)/0	750	WADE ET AL., 1978
THM	-/0	39000	COOKSON ET AL., 1971
WP2	+/+		BOOTMAN ET AL., 1979
KPM	+/0	29	VOOGD ET AL., 1981
YEY	+/+	174	MIGLIORE ET AL., 1982
NER	+/0	29000	KOLMARK & GILES, 1955
SRL	+/0	1.5	HARDIN ET AL., 1983
CYH	+/0	1.85	BOOTMAN ET AL., 1979
CYC	+/0	25	DEAN & HODSON-WALKER, 1979
DLM	-/0	3500	BOOTMAN ET AL., 1979
DLR	-/0	0.7	HARDIN ET AL., 1983
MNM	+/0	600	BOOTMAN ET AL., 1979

APPENDIX: ACTIVITY PROFILES

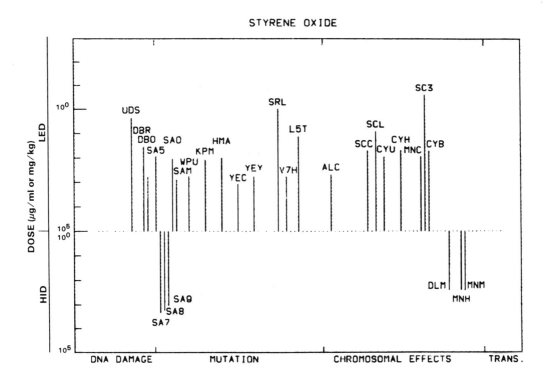

STYRENE OXIDE

TEST RESULT -S9/+S9		DOSE (HID or LED)	REFERENCE	TEST RESULT -S9/+S9		DOSE (HID or LED)	REFERENCE
UDS	+/0	528	LOFRIENO ET AL., 1978	SAO	+/+	768	DE MEESTER ET AL., 1981
UDS	+/0	0.01	AUDETTE ET AL., 1979	SAO	+/0	48	PAGANO ET AL., 1982
DBR	+/0	36	SINA ET AL., 1983	SAO	+/0	60	GLATT ET AL., 1983
DEO	+/0	600	WALLES & ORSEN, 1983	SAO	+/0	120	TURCHI ET AL., 1981
SA5	+/+		DE FLORA, 1981	SAO	+/0	144	SUGIURA & GOTO, 1981
SA5	+/+	24	DE MEESTER ET AL., 1978	SAO	+/0	146	SUGIURA ET AL., 1978a
SA5	+/+	60	BUSK, 1979a	SAO	+/0	200	MILVY & GARRO, 1976
SA5	+/+	60	LOFRIENO ET AL., 1978	SAO	+/0	250	WATABE ET AL., 1978
SA5	+/+	0.6	VAINIO ET AL., 1976	SAO	+/0	250	WADE ET AL., 1978
SA5	+/+	125	STOLTZ & WITHEY, 1977	SAO	+/0	600	WATABE ET AL., 1978
SA5	+/+	768	DE MEESTER ET AL., 1981	SAO	+/+	768	DE MEESTER ET AL., 1981
SA5	+/0	50	WATABE ET AL., 1978	SAM	+/+	480	SUGIURA & GOTO, 1981
SA5	+/0	5000	MILVY & GARRO, 1976	WFU	+/0	720	SUGIURA ET AL., 1978b
SA5	(+)/0	250	WADE ET AL., 1978	WFU	+/0	120	VOOGD ET AL., 1981
SA7			DE FLORA, 1981	KPM	(+)/0	100	LOFRIENO ET AL., 1976
SA7	-/-	600	VAINIO ET AL., 1976	HMA	0/+	1200	LOFRIENO ET AL., 1976
SA7	-/-	1150	DE MEESTER ET AL., 1981	YEC	0/+	600	LOFRIENO ET AL., 1976
SA7	-/-	6000	DE MEESTER ET AL., 1978	YEY	+/0	1.0	DONNER ET AL., 1979
SA7	-/0		WADE ET AL., 1978	SRL	+/-	240	BEIJE & JENSSEN, 1982
SA7	-/0	5000	MILVY & GARRO, 1976	V7H	+/0	504	LOFRIENO ET AL., 1978
SA7	(+)/0	25	WATABE ET AL., 1978	V7H	+/0	1020	BONATTI ET AL., 1978
SA8	-/+	6	VAINIO ET AL., 1976	V7H	+/0	1020	LOFRIENO ET AL., 1976
SA8	-/-		DE FLORA, 1981	LST	+/(+)	13.8	AMACHER & TURNER, 1982
SA8	-/-	1150	DE MEESTER ET AL., 1981	ALC	+/0	500	LINNAINMAA ET AL., 1978a,b
SA8	-/-	6000	DE MEESTER ET AL., 1978	SCC	+/+	50	DE RAAT, 1978
SA8	-/0	250	WATABE ET AL., 1978	SCL	+/0	8.4	NORPPA ET AL., 1981
SA8	-/0	5000	MILVY & GARRO, 1976	CYU	+/0	90	TURCHI ET AL., 1981
SA9	-/-		DE FLORA, 1981	CYH	+/0	24	NORPPA ET AL., 1981
SA9	-/-	250	UENO ET AL., 1978	CYH	+/0	60	FABRY ET AL., 1978
SA9	-/-	600	VAINIO ET AL., 1976	CYH	+/0	80	LINNAINMAA ET AL., 1978a,b
SA9	-/-	1150	DE MEESTER ET AL., 1981	MNC	+/0	80	LINNAINMAA ET AL., 1978a,b
SA9	-/-	6000	DE MEESTER ET AL., 1978	MNC	+/0	90	TURCHI ET AL., 1981
SA9	-/0	250	WADE ET AL., 1978	SC3	-/0	0.48	NORPPA ET AL., 1979
SA9	-/0	250	WATABE ET AL., 1978	SC3	(+)/0	0.24	CONNER ET AL., 1982
SA9	-/0	5000	MILVY & GARRO, 1976	CYB	(+)/0	50	LOFRIENO ET AL., 1978
SAO	+/+		DE FLORA, 1981	CYB	-/0	250	FABRY ET AL., 1978
SAO	+/+	60	DE MEESTER ET AL., 1978	CYB	-/0	0.48	NORPPA ET AL., 1979
SAO	+/+	0.6	VAINIO ET AL., 1976	DLM	-/0	250	FABRY ET AL., 1978
SAO	+/+	120	BUSK, 1979a	MNH	-/0	250	PENTTILA ET AL., 1980
SAO	+/+	240	YOSHIKAWA ET AL., 1980	MNM	-/0	250	FABRY ET AL., 1978

APPENDIX: ACTIVITY PROFILES

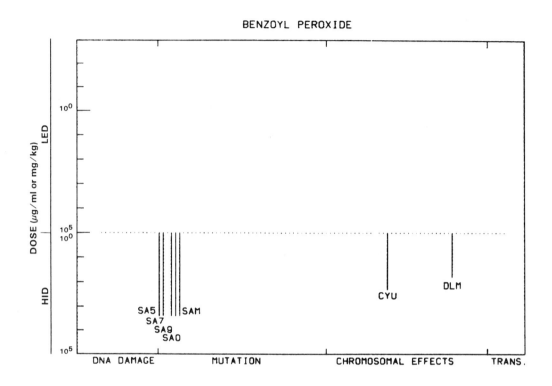

BENZOYL PEROXIDE

TEST	RESULT -S9/+S9	DOSE (HID or LED)	REFERENCE
SA5	-/-	2500	ISHIDATE ET AL., 1980
SA7	-/-	2500	ISHIDATE ET AL., 1980
SA9	-/-	2500	ISHIDATE ET AL., 1980
SAO	-/-	2500	ISHIDATE ET AL., 1980
SAM	-/-	2500	ISHIDATE ET AL., 1980
SCC	-/+		JARVENTAUS ET AL., 1984
CYU	-/0	200	ISHIDATE ET AL., 1980
DLM	-/0	62	EPSTEIN ET AL., 1972

344 IARC MONOGRAPHS VOLUME 36

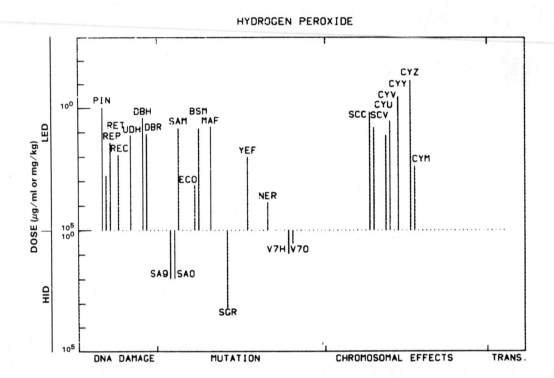

HYDROGEN PEROXIDE

TEST	RESULT -S9/+S9	DOSE (HID or LED)	REFERENCE
PIN	+/0	1.0	NORTHROP, 1958
REP	+/0	600	ROSENKRANZ, 1973
RET	+/0	340	ANANTHASWAMY & EISENSTARK, 1977
RET	(+)/0	2.0	HARTMAN & EISENSTARK, 1978
REC	+/0	20	HARTMAN & EISENSTARK, 1978
REC	+/0	340	ANANTHASWAMY & EISENSTARK, 1977
UDH	+/0	20	STICH ET AL., 1978
UDH	+/0	9.0	COPPINGER ET AL., 1983
DBH	+/0		TAYLOR ET AL., 1979
DBH	+/0	2.0	WANG ET AL., 1980
DBH	+/0	3.4	HOFFMANN & MENEGHINI, 1979
DBR	+/0	12	BRADLEY ET AL., 1979
SA9	-/-	0.9	XU ET AL., 1984
SA9	-/-	340	STICH ET AL., 1978
SA9	0/-	25	YAMAGUCHI & YAMASHITA, 1980
SA0	+/0	136	NORKUS ET AL., 1983
SA0	-/-	0.9	XU ET AL., 1984
SA0	-/0	340	STICH ET AL., 1978
SA0	0/-	25	YAMAGUCHI & YAMASHITA, 1980
SAM	+/0	30	AMES ET AL., 1981
SAM	+/0	50	LEVIN ET AL., 1982
SAM	+/0	0.22	XU ET AL., 1984
ECO	+/0	1500	DE MEREC ET AL., 1951
BSM	+/0	7.2	SACKS & MACGREGOR, 1982
HIM	(+)/0		KIMBALL & HIRSCH, 1975
MAF	+/0	6	CLARK, 1953
SGR	-/0	1440	MASHIMA & IKEDA, 1958
YEF	+/0	100	THACKER & PARKER, 1976
YEF	+/0	100	THACKER, 1976
ASR	+/0		NANDA ET AL., 1975
NER	+/0	6800	JENSEN ET AL., 1951
NER	+/0	7140	DICKEY ET AL., 1949
SRL	-/0	0.3	DI PAOLO, 1952
V7H	-/0	12	BRADLEY ET AL., 1979
V7H	-/0	17	BRADLEY & ERICKSON, 1981
V7H	-/0	3.4	TSUDA, 1981
V70	-/0	3.4	TSUDA, 1981
SCC	+/0	0.13	MACRAE & STICH, 1979
SCC	(+)/0	17	WILMER & NATARAJAN, 1981
SCV	+/0	12	BRADLEY ET AL., 1979
SCV	+/0	3.4	SPEIT ET AL., 1982
SCI	+/0		ESTERVIG & WANG, 1979
CYU	+/0	34	SASAKI ET AL., 1980
CYU	+/0	1.0	HANHAM ET AL., 1983
CYU	+/0	3.4	TSUDA, 1981
CYU	(+)/0	10	STICH ET AL., 1978
CYU	(+)/0	340	WILMER & NATARAJAN, 1981
CYV	+/0	3.4	TSUDA, 1981
CYY	+/0	0.34	TSUDA, 1981
CYZ	+/0	0.07	PARSHAD ET AL., 1980
CYM	+/0	170	SCHONEICH ET AL., 1970
CYM	+/0	340	SCHONEICH, 1967
CYB	-/0		KAWACHI ET AL., 1980

346 IARC MONOGRAPHS VOLUME 36

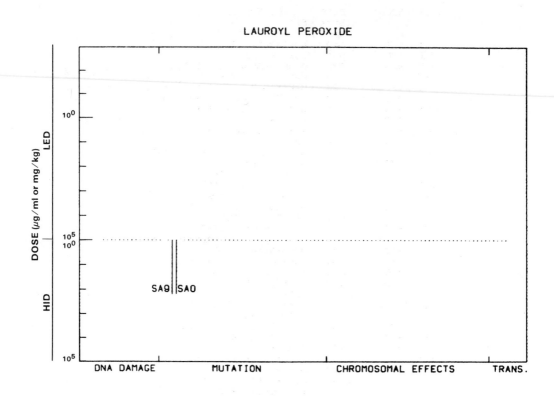

SUPPLEMENTARY CORRIGENDA TO VOLUMES 1-35

Corrigenda covering volumes 1-6 appeared in volume 7; others appeared in volumes 8, 10-13 and 15-35.

Volume 8

p. 336   *After* Ponceau 3R *delete* see Ponceau MX 189 and see Ponceau 3R *and* Ponceau 4R see Ponceau MX 189

Supplement 2

p. 364	Table 11, line 2	*replace* 1.245 *by* 1.061
	Table 11, last line	*replace* 4.80 *by* 4.62
p. 365	line 2	*replace* 1.48 *by* 1.49 *and replace* 4.80 *by* 4.62
	line 3	*replace* 1.48 *by* 1.49, *replace* 4.80 *by* 4.62 *and replace* 0.68 *by* 0.69

SUPPLEMENTARY CORRIGENDA TO VOLUME

CUMULATIVE INDEX TO IARC MONOGRAPHS ON THE EVALUATION OF THE CARCINOGENIC RISK OF CHEMICALS TO HUMANS

Numbers in italics indicate volume, and other numbers indicate page. References to corrigenda are given in parentheses. Compounds marked with an asterisk(*) were considered by the working groups in the year indicated, but monographs were not prepared because adequate data on carcinogenicity were not available.

A

Acetaldehyde	*36*, 101
Acetaldehyde formylmethylhydrazone	*31*, 163
Acetamide	*7*, 197
Acetylsalicyclic acid (1976)*	
Acridine orange	*16*, 145
Acriflavinium chloride	*13*, 31
Acrolein	*19*, 479
	36, 133
Acrylic acid	*19*, 47
Acrylic fibres	*19*, 86
Acrylonitrile	*19*, 73
	Suppl. 4, 25
Acrylonitrile-butadiene-styrene copolymers	*19*, 91
Actinomycins	*10*, 29 (corr. *29*, 399; *34*, 197)
	Suppl. 4, 27
Adipic acid (1978)*	
Adriamycin	*10*, 43
	Suppl. 4, 29
AF-2	*31*, 47
Aflatoxins	*1*, 145 (corr. *7*, 319)
	(corr. *8*, 349)
	10, 51
	Suppl. 4, 31
Agaritine	*31*, 63
Aldrin	*5*, 25
	Suppl. 4, 35
Allyl chloride	*36*, 39
Allyl isothiocyanate	*36*, 55
Allyl isovalerate	*36*, 69
Aluminium production	*34*, 37
Amaranth	*8*, 41
5-Aminoacenaphthene	*16*, 243
2-Aminoanthraquinone	*27*, 191
para-Aminoazobenzene	*8*, 53
ortho-Aminoazotoluene	*8*, 61 (corr. *11*, 295)
para-Aminobenzoic acid	*16*, 249
4-Aminobiphenyl	*1*, 74 (corr. *10*, 343)
	Suppl. 4, 37
3-Amino-1,4-dimethyl-5*H*-pyrido-[4,3-*b*] – indole and its acetate	*31*, 247
1-Amino-2-methylanthraquinone	*27*, 199

3-Amino-1-methyl-5H-pyrido-[4,3-b] —
   indole and its acetate   *31*, 255
2-Amino-5-(5-nitro-2-furyl)-1,3,4-thiadiazole   *7*, 143
4-Amino-2-nitrophenol   *16*, 43
2-Amino-4-nitrophenol (1977)*
2-Amino-5-nitrophenol (1977)*
2-Amino-5-nitrothiazole   *31*, 71
6-Aminopenicillanic acid (1975)*
Amitrole   *7*, 31
    Suppl. 4, 38
Amobarbital sodium (1976)*
Anaesthetics, volatile   *11*, 285
    Suppl. 4, 41
Aniline   *4*, 27 (corr. *7*, 320)
    27, 39
    Suppl. 4, 49
Aniline hydrochloride   *27*, 40
ortho-Anisidine and its hydrochloride   *27*, 63
para-Anisidine and its hydrochloride   *27*, 65
Anthanthrene   *32*, 95
Anthracene   *32*, 105
Anthranilic acid   *16*, 265
Apholate   *9*, 31
Aramite®   *5*, 39
Arsenic and arsenic compounds   *1*, 41
    2, 48
    23, 39
    Suppl. 4, 50
   Arsanilic acid
   Arsenic pentoxide
   Arsenic sulphide
   Arsenic trioxide
   Arsine
   Calcium arsenate
   Dimethylarsinic acid
   Lead arsenate
   Methanearsonic acid, disodium salt
   Methanearsonic acid, monosodium salt
   Potassium arsenate
   Potassium arsenite
   Sodium arsenate
   Sodium arsenite
   Sodium cacodylate
Asbestos   *2*, 17 (corr. *7*, 319)
    14 (corr. *15*, 341)
      (corr. *17*, 351)
    Suppl. 4, 52
   Actinolite
   Amosite
   Anthophyllite
   Chrysotile
   Crocidolite
   Tremolite

Asiaticoside (1975)*	
Auramine	*1*, 69 (corr. *7*, 319)
	Suppl. 4, 53 (corr. *33*, 223)
Aurothioglucose	*13*, 39
5-Azacytidine	*26*, 37
Azaserine	*10*, 73 (corr. *12*, 271)
Azathioprine	*26*, 47
	Suppl. 4, 55
Aziridine	*9*, 37
2-(1-Aziridinyl)ethanol	*9*, 47
Aziridyl benzoquinone	*9*, 51
Azobenzene	*8*, 75
B	
Benz[a] acridine	*32*, 123
Benz[c]acridine	*3*, 241
	32, 129
Benzal chloride	*29*, 65
	Suppl. 4, 84
Benz[a]anthracene	*3*, 45
	32, 135
Benzene	*7*, 203 (corr. *11*, 295)
	29, 93, 391
	Suppl. 4, 56 (corr. *35*, 249)
Benzidine and its salts	*1*, 80
	29, 149, 391
	Suppl. 4, 57
Benzo[b]fluoranthene	*3*, 69
	32, 147
Benzo[j]fluoranthene	*3*, 82
	32, 155
Benzo [k]fluoranthene	*32*, 163
Benzo[ghi]fluoranthene	*32*, 171
Benzo[a]fluorono	*32*, 177
Benzo[b]fluorene	*32*, 183
Benzo[c]fluorene	*32*, 189
Benzo[ghi]perylene	*32*,195
Benzo[c]phenanthrene	*32*, 205
Benzo[a]pyrene	*3*, 91
	Suppl. 4, 227
	32, 211
Benzo[e]pyrene	*3*, 137
	32, 225
para-Benzoquinone dioxime	*29*, 185
Benzotrichloride	*29*, 73
	Suppl. 4, 84
Benzoyl peroxide	*36*, 267
Benzoyl chloride	*29*, 83
	Suppl. 4, 84
Benzyl chloride	*11*, 217 (corr. *13*, 243)
	29, 49 (corr. *30*, 407)
	Suppl. 4, 84

Benzyl violet 4B	*16*, 153
Beryllium and beryllium compounds	*1*, 17
	23, 143 (corr. *25*, 392)
Bertrandite	*Suppl. 4*, 60
Beryllium acetate	
Beryllium acetate, basic	
Beryllium-aluminium alloy	
Beryllium carbonate	
Beryllium chloride	
Beryllium-copper alloy	
Beryllium-copper-cobalt alloy	
Beryllium fluoride	
Beryllium hydroxide	
Beryllium-nickel alloy	
Beryllium oxide	
Beryllium phosphate	
Beryllium silicate	
Beryllium sulphate and its tetrahydrate	
Beryl ore	
Zinc beryllium silicate	
Bis(1-aziridinyl)morpholinophosphine sulphide	*9*, 55
Bis(2-chloroethyl)ether	*9*, 117
N,N-Bis(2-chloroethyl)-2-naphthylamine (chlornaphazine)	*4*, 119 (corr. *30*, 407)
	Suppl. 4, 62
Bischloroethyl nitrosourea (BCNU)	*26*, 79
	Suppl. 4, 63
Bis-(2-chloroisopropyl)ether (1976)*	
1,2-Bis(chloromethoxy)ethane	*15*, 31
1,4-Bis(chloromethoxymethyl)benzene	*15*, 37
Bis(chloromethyl)ether	*4*, 231 (corr. *13*, 243)
	Suppl. 4, 64
Bitumens	*35*, 39
Bleomycins	*26*, 97
	Suppl. 4, 66
Blue VRS	*16*, 163
Boot and shoe manufacture and repair	*25*, 249
	Suppl. 4, 138
Brilliant blue FCF diammonium and disodium salts	*16*, 171 (corr. *30*, 407)
1,4-Butanediol dimethanesulphonate (Myleran)	*4*, 247
	Suppl. 4, 68
Butyl benzyl phthalate	*29*, 194 (corr. *32*, 455)
Butyl-*cis*-9,10-epoxystearate (1976)*	
β-Butyrolactone	*11*, 225
γ-Butyrolactone	*11*, 231

C

Cadmium and cadmium compounds	*2*, 74
	11, 39 (corr. *27*, 320)
Cadmium acetate	*Suppl. 4*, 71
Cadmium chloride	
Cadmium oxide	
Cadmium sulphate	
Cadmium sulphide	

CUMULATIVE INDEX

Calcium cyclamate	22, 58 (corr. 25, 391)
	Suppl. 4, 97
Calcium saccharin	22, 120 (corr. 25, 391)
	Suppl. 4, 225
Cantharidin	10, 79
Caprolactam	19, 115 (corr. 31, 293)
Captan	30, 295
Carbaryl	12, 37
Carbazole	32, 239
Carbon blacks	3, 22
	33, 35
Carbon tetrachloride	1, 53
	20, 371
	Suppl. 4, 74
Carmoisine	8, 83
Carpentry and joinery	25, 139
	Suppl. 4, 139
Carrageenans (native)	10, 181 (corr. 11, 295)
	31, 79
Catechol	15, 155
Chloramben (1982)*	
Chlorambucil	9, 125
	26, 115
	Suppl. 4, 77
Chloramphenicol	10, 85
	Suppl. 4, 79
Chlordane	20, 45 (corr. 25, 391)
	Suppl. 4, 80
Chlordecone (Kepone)	20, 67
Chlordimeform	30, 61
Chlorinated dibenzodioxins	15, 41
	Suppl. 4, 211, 238
Chlormadinone acetate	6, 149
	21, 365
	Suppl. 4, 192
Chlorobenzilate	5, 75
	30, 73
1-(2-Chloroethyl)-3-cyclohexyl-1-nitrosourea (CCNU)	26, 137 (corr. 35, 249)
	Suppl. 4, 83
Chloroform	1, 61
	20, 401
	Suppl. 4, 87
Chloromethyl methyl ether	4, 239
	Suppl. 4, 64
4-Chloro-*ortho*-phenylenediamine	27, 81
4-Chloro-*meta*-phenylenediamine	27, 82
Chloroprene	19, 131
	Suppl. 4, 89
Chloropropham	12, 55
Chloroquine	13, 47
Chlorothalonil	30, 319
para-Chloro-*ortho*-toluidine and its hydrochloride	16, 277
	30, 61

5-Chloro-*ortho*-toluidine (1977)*
Chlorotrianisene   *21*, 139
Chlorpromazine (1976)*
Cholesterol   *10*, 99
                                                     31, 95
Chromium and chromium compounds   *2*, 100
                                                     23, 205
                                                     Suppl. *4*, 91
   Barium chromate
   Basic chromic sulphate
   Calcium chromate
   Chromic acetate
   Chromic chloride
   Chromic oxide
   Chromic phosphate
   Chromite ore
   Chromium carbonyl
   Chromium potassium sulphate
   Chromium sulphate
   Chromium trioxide
   Cobalt-chromium alloy
   Ferrochromium
   Lead chromate
   Lead chromate oxide
   Potassium chromate
   Potassium dichromate
   Sodium chromate
   Sodium dichromate
   Strontium chromate
   Zinc chromate
   Zinc chromate hydroxide
   Zinc potassium chromate
   Zinc yellow
Chrysene   *3*, 159
                                                     32, 247
Chrysoidine   *8*, 91
C.I. Disperse Yellow 3   *8*, 97
Cinnamyl anthranilate   *16*, 287
                                                     31, 133
Cisplatin   *26*, 151
                                                     Suppl. *4*, 93
Citrus Red No. 2   *8*, 101 (corr. *19*, 495)
Clofibrate   *24*, 39
                                                     Suppl. *4*, 95
Clomiphene and its citrate   *21*, 551
                                                     Suppl. *4*, 96
Coal gasification   *34*, 65
Coal-tars and derived products   *35*, 83
Coke production   *34*, 101
Conjugated œstrogens   *21*, 147
                                                     Suppl. *4*, 179
Copper 8-hydroxyquinoline   *15*, 103
Coronene   *32*, 263

Coumarin	*10*, 113
meta-Cresidine	*27*, 91
para-Cresidine	*27*, 92
Cycasin	*1*, 157 (corr. *7*, 319)
	10, 121
Cyclamic acid	*22*, 55 (corr. *25*, 391)
Cyclochlorotine	*10*, 139
Cyclohexylamine	*22*, 59 (corr. *25*, 391)
	Suppl. 4, 97
Cyclopenta[*cd*]pyrene	*32*, 269
Cyclophosphamide	*9*, 135
	26, 165
	Suppl. 4, 99
D	
2,4-D and esters	*15*, 111
	Suppl. 4, 101, 211
Dacarbazine	*26*, 203
	Suppl. 4, 103
D and C Red No. 9	*8*, 107
Dapsone	*24*, 59
	Suppl. 4, 104
Daunomycin	*10*, 145
DDT and associated substances	*5*, 83 (corr. *7*, 320)
	Suppl. 4, 105
DDD (TDE)	
DDE	
Diacetylaminoazotoluene	*8*, 113
N,N'-Diacetylbenzidine	*16*, 293
Diallate	*12*, 69
	30, 235
2,4-Diaminoanisole and its sulphate	*16*, 51
	27, 103
2,5-Diaminoanisole (1977)*	
4,4'-Diaminodiphenyl ether	*16*, 301
	29, 203
1,2-Diamino-4-nitrobenzene	*16*, 63
1,4-Diamino-2-nitrobenzene	*16*, 73
2,4-Diaminotoluene	*16*, 83
2,5-Diaminotoluene and its sulphate	*16*, 97
Diazepam	*13*, 57
Diazomethane	*7*, 223
Dibenz[*a,h*]acridine	*3*, 247
	32, 277
Dibenz[*a,j*]acridine	*3*, 254
	32, 283
Dibenz[*a,c*]anthracene	*32*, 289 (corr. *34*, 197)
Dibenz[*a,h*]anthracene	*3*, 178
	32, 299
Dibenz[*a, j*]anthracene	*32*, 309
7H-Dibenzo[*c,g*]carbazole	*3*, 260
	32, 315
Dibenzo[*a,e*]fluoranthene	*32*, 321

Dibenzo[*h,rst*]pentaphene	*3*, 197
Dibenzo[*a,e*]pyrene	*3*, 201
	32, 327
Dibenzo[*a,h*]pyrene	*3*, 207
	32, 331
Dibenzo[*a,i*]pyrene	*3*, 215
	32, 337
Dibenzo[*a,l*]pyrene	*3*, 224
	32, 343
1,2-Dibromo-3-chloropropane	*15*, 139
	20, 83
ortho-Dichlorobenzene	*7*, 231
	29, 213
	Suppl. 4, 108
para-Dichlorobenzene	*7*, 231
	29, 215
	Suppl. 4, 108
3,3'-Dichlorobenzidine and its dihydrochloride	*4*, 49
	29, 239
	Suppl. 4, 110
trans-1,4-Dichlorobutene	*15*, 149
3,3'-Dichloro-4,4'-diaminodiphenyl ether	*16*, 309
1,2-Dichloroethane	*20*, 429
Dichloromethane	*20*, 449
	Suppl. 4, 111
Dichlorvos	*20*, 97
Dicofol	*30*, 87
Dicyclohexylamine	*22*, 60 (corr. *25*, 391)
Dieldrin	*5*, 125
	Suppl. 4, 112
Dienoestrol	*21*, 161
	Suppl. 4, 183
Diepoxybutane	*11*, 115 (corr. *12*, 271)
Di-(2-ethylhexyl) adipate	*29*, 257
Di-(2-ethylhexyl) phthalate	*29*, 269 (corr. *32*, 455)
1,2-Diethylhydrazine	*4*, 153
Diethylstilboestrol	*6*, 55
	21, 173 (corr. *23*, 417)
	Suppl. 4, 184
Diethylstilboestrol dipropionate	*21*, 175
Diethyl sulphate	*4*, 277
	Suppl. 4, 115
Diglycidyl resorcinol ether	*11*, 125
	36, 181
Dihydrosafrole	*1*, 170
	10, 233
Dihydroxybenzenes	*15*, 155
Dihydroxymethylfuratrizine	*24*, 77
Dimethisterone	*6*, 167
	21, 377
	Suppl. 4, 193
Dimethoate (1977)*	

Dimethoxane	*15*, 177
3,3'-Dimethoxybenzidine (*ortho*-Dianisidine)	*4*, 41
	Suppl. 4, 116
para-Dimethylaminoazobenzene	*8*, 125 (corr. *31*, 293)
para-Dimethylaminobenzenediazo sodium sulphonate	*8*, 147
trans-2[(Dimethylamino)methylimino]-5-[2-(5-nitro-2-furyl)vinyl]-1,3,4-oxadiazole	*7*, 147 (corr. *30*, 407)
3,3'-Dimethylbenzidine (*ortho*-Tolidine)	*1*, 87
Dimethylcarbamoyl chloride	*12*, 77
	Suppl. 4, 118
1,1-Dimethylhydrazine	*4*, 137
1,2-Dimethylhydrazine	*4*, 145 (corr. *7*, 320)
1,4-Dimethylphenanthrene	*32*, 349
Dimethyl sulphate	*4*, 271
	Suppl. 4, 119
Dimethylterephthalate (1978)*	
1,8-Dinitropyrene	*33*, 171
Dinitrosopentamethylenetetramine	*11*, 241
1,4-Dioxane	*11*, 247
	Suppl. 4, 121
2,4'-Diphenyldiamine	*16*, 313
Diphenylthiohydantoin (1976)*	
Direct Black 38	*29*, 295 (corr. *32*, 455)
	Suppl. 4, 59
Direct Blue 6	*29*, 311
	Suppl. 4, 59
Direct Brown 95	*29*, 321
	Suppl. 4, 59
Disulfiram	*12*, 85
Dithranol	*13*, 75
Dulcin	*12*, 97
E	
Endrin	*5*, 157
Enflurane (1976)*	
Eosin and its disodium salt	*15*, 183
Epichlorohydrin	*11*, 131 (corr. *18*, 125)
	(corr. *26*, 387)
	Suppl. 4, 122
	(corr. *33*, 223)
1-Epoxyethyl-3,4-epoxycyclohexane	*11*, 141
3,4-Epoxy-6-methylcyclohexylmethyl-3,4-epoxy-6-methyl-cyclohexane carboxylate	*11*, 147
cis-9,10-Epoxystearic acid	*11*, 153
Ethinyloestradiol	*6*, 77
	21, 233
	Suppl. 4, 186
Ethionamide	*13*, 83
Ethyl acrylate	*19*, 57
Ethylene	*19*, 157
Ethylene dibromide	*15*, 195
	Suppl. 4, 124

358   IARC MONOGRAPHS VOLUME 36

Ethylene oxide	*11*, 157
	Suppl. *4*, 126
	36, 189
Ethylene sulphide	*11*, 257
Ethylenethiourea	*7*, 45
	Suppl. *4*, 128
Ethyl methanesulphonate	*7*, 245
Ethyl selenac	*12*, 107
Ethyl tellurac	*12*, 115
Ethynodiol diacetate	*6*, 173
	21, 387
	Suppl. *4*, 194
Eugenol	*36*, 75
Evans blue	*8*, 151

F
Fast green FCF	*16*, 187
Ferbam	*12*, 121 (corr. *13*, 243)
Fluometuron	*30*, 245
Fluoranthene	*32*, 355
Fluorene	*32*, 365
Fluorescein and its disodium salt (1977)*	
Fluorides (inorganic, used in drinking-water and dental preparations)	*27*, 237
Fluorspar	
Fluosilicic acid	
Sodium fluoride	
Sodium monofluorophosphate	
Sodium silicofluoride	
Stannous fluoride	
5-Fluorouracil	*26*, 217
	Suppl. *4*, 130
Formaldehyde	*29*, 345
	Suppl. *4*, 131
2-(2-Formylhydrazino)-4-(5-nitro-2-furyl)thiazole	*7*, 151 (corr. *11*, 295)
Furazolidone	*31*, 141
The furniture and cabinet-making industry	*25*, 99
	Suppl. *4*, 140
2-(2-Furyl)-3-(5-nitro-2-furyl)acrylamide	*31*, 47
Fusarenon-X	*11*, 169
	31, 153

G
L-Glutamic acid-5-[2-(4-Hydroxymethyl) phenylhydrazide)	*31*, 63
Glycidaldehyde	*11*, 175
Glycidyl oleate	*11*, 183
Glycidyl stearate	*11*, 187
Griseofulvin	*10*, 153
Guinea green B	*16*, 199
Gyromitrin	*31*, 163

H

Haematite	*1*, 29
	Suppl. 4, 254
Haematoxylin (1977)*	
Hair dyes, epidemiology of	*16*, 29
	27, 307
Halothane (1976)*	
Heptachlor and its epoxide	*5*, 173
	20, 129
	Suppl. 4, 80
Hexachlorobenzene	*20*, 155
Hexachlorobutadiene	*20*, 179
Hexachlorocyclohexane (α-,β-,δ-,ε-,technical HCH and lindane)	*5*, 47
	20, 195 (corr. *32*, 455)
	Suppl. 4, 133
Hexachloroethane	*20*, 467
Hexachlorophene	*20*, 241
Hexamethylenediamine (1978)*	
Hexamethylphosphoramide	*15*, 211
Hycanthone and its mesylate	*13*, 91
Hydralazine and its hydrochloride	*24*, 85
	Suppl. 4, 135
Hydrazine	*4*, 127
	Suppl. 4, 136
Hydrogen peroxide	*36*, 285
Hydroquinone	*15*, 155
4-Hydroxyazobenzene	*8*, 157
17α-Hydroxyprogesterone caproate	*21*, 399 (corr. *31*, 293)
	Suppl. 4, 195
8-Hydroxyquinoline	*13*, 101
Hydroxysenkirkine	*10*, 265

I

Indeno[1,2,3-cd]pyrene	*3*, 229
	32, 373
Iron and steel founding	*34*, 133
Iron-dextran complex	*2*, 161
	Suppl. 4, 145
Iron-dextrin complex	*2*, 161 (corr. *7*, 319)
Iron oxide	*1*, 29
Iron sorbitol-citric acid complex	*2*, 161
Isatidine	*10*, 269
Isoflurane (1976)*	
Isonicotinic acid hydrazide	*4*, 159
	Suppl. 4, 146
Isophosphamide	*26*, 237
Isoprene (1978)*	
Isopropyl alcohol	*15*, 223
	Suppl. 4, 151
Isopropyl oils	*15*, 223
	Suppl. 4, 151
Isosafrole	*1*, 169
	10, 232

J
Jacobine   *10*, 275

K
Kaempferol   *31*, 171

L
Lasiocarpine   *10*, 281
Lauroyl peroxide   *36*, 315
Lead and lead compounds   *1*, 40 (corr. *7*, 319)
                                                 2, 52 (corr. *8*, 349)
                                                 2, 150
                                                 23, 39, 205, 325
                                                 Suppl. 4, 149
    Lead acetate and its trihydrate
    Lead carbonate
    Lead chloride
    Lead naphthenate
    Lead nitrate
    Lead oxide
    Lead phosphate
    Lead subacetate
    Lead tetroxide
    Tetraethyllead
    Tetramethyllead
The leather goods manufacturing industry (other than   *25*, 279
    boot and shoe manufacture and tanning)
                                                 Suppl. 4, 142
The leather tanning and processing industries   *25*, 201
                                                 Suppl. 4, 142
Ledate   *12*, 131
Light green SF   *16*, 209
Lindane   *5*, 47
                                                 20, 196
The lumber and sawmill industries (including logging)   *25*, 49
                                                 Suppl. 4, 143
Luteoskyrin   *10*, 163
Lynoestrenol   *21*, 407
                                                 Suppl. 4, 195
Lysergide (1976)*

M
Magenta   *4*, 57 (corr. *7*, 320)
                                                 Suppl. 4, 152
Malathion   *30*, 103
Maleic hydrazide   *4*, 173 (corr. *18*, 125)
Malonaldehyde   *36*, 163
Maneb   *12*, 137
Mannomustine and its dihydrochloride   *9*, 157
MCPA   *Suppl. 4*, 211
                                                 30, 255
Medphalan   *9*, 168

Medroxyprogesterone acetate	6, 157
	21, 417 (corr. 25, 391)
	Suppl. 4, 196
Megestrol acetate	21, 431
	Suppl. 4, 198
Melphalan	9, 167
	Suppl. 4, 154
6-Mercaptopurine	26, 249
	Suppl. 4, 155
Merphalan	9, 169
Mestranol	6, 87
	21, 257 (corr. 25, 391)
	Suppl. 4, 188
Methacrylic acid (1978)*	
Methallenoestril (1978)*	
Methotrexate	26, 267
	Suppl. 4, 157
Methoxsalen	24, 101
	Suppl. 4, 158
Methoxychlor	5, 193
	20, 259
Methoxyflurane (1976)*	
Methylacrylate	19, 52
2-Methylaziridine	9, 61
Methylazoxymethanol	10, 121
Methylazoxymethanol acetate	1, 164
	10, 131
Methyl bromide (1978)*	
Methyl carbamate	12, 151
1-,2-,3-,4-,5-and 6-Methylchrysenes	32, 379
N-Methyl-N,4-dinitrosoaniline	1, 141
4,4'-Methylene bis(2-chloroaniline)	4, 65 (corr. 7, 320)
4,4'-Methylene bis(N,N-dimethyl)benzenamine	27, 119
4,4'-Methylene bis(2-methylaniline)	4, 73
4,4'-Methylenedianiline	4, 79 (corr. 7, 320)
4,4'-Methylenediphenyl diisocyanate	19, 314
2-and 3-Methylfluoranthenes	32, 399
Methyl iodide	15, 245
Methyl methacrylate	19, 187
Methyl methanesulphonate	7, 253
2-Methyl-1-nitroanthraquinone	27, 205
N-Methyl-N'-nitro-N-nitrosoguanidine	4, 183
Methyl parathion	30, 131
1-Methylphenanthrene	32, 405
Methyl protoanemonin (1975)*	
Methyl red	8, 161
Methyl selenac	12, 161
Methylthiouracil	7, 53
Metronidazole	13, 113
	Suppl. 4, 160
Mineral oils	3, 30
	Suppl. 4, 227
	33, 87

Mirex	5, 203
	20, 283 (corr. 30, 407)
Miristicin (1982)*	
Mitomycin C	10, 171
Modacrylic fibres	19, 86
Monocrotaline	10, 291
Monuron	12, 167
5-(Morpholinomethyl)-3-[(5-nitrofurfurylidene)amino]-2-oxazolidinone	7, 161
Mustard gas	9, 181 (corr. 13, 243)
	Suppl. 4, 163

N

Nafenopin	24, 125
1,5-Naphthalenediamine	27, 127
1,5-Naphthalene diisocyanate	19, 311
1-Naphthylamine	4, 87 (corr. 8, 349)
	(corr. 22, 187)
	Suppl. 4, 164
2-Naphthylamine	4, 97
	Suppl. 4, 166
1-Naphthylthiourea (ANTU)	30, 347
Nickel and nickel compounds	2, 126 (corr. 7, 319)
	11, 75
	Suppl. 4, 167
Nickel acetate and its tetrahydrate	
Nickel ammonium sulphate	
Nickel carbonate	
Nickel carbonyl	
Nickel chloride	
Nickel-gallium alloy	
Nickel hydroxide	
Nickelocene	
Nickel oxide	
Nickel subsulphide	
Nickel sulphate	
Nihydrazone (1982)*	
Niridazole	13, 123
Nithiazide	31, 179
5-Nitroacenaphthene	16, 319
5-Nitro-ortho-anisidine	27, 133
9-Nitroanthracene	33, 179
6-Nitrobenzo[a]pyrene	33, 187
4-Nitrobiphenyl	4, 113
6-Nitrochrysene	33, 195
Nitrofen	30, 271
3-Nitrofluoranthene	33, 201
5-Nitro-2-furaldehyde semicarbazone	7, 171
1[(5-Nitrofurfurylidene)amino]-2-imidazolidinone	7, 181
N-[4-(5-Nitro-2-furyl)-2-thiazolyl]acetamide	1, 181
	7, 185

Nitrogen mustard and its hydrochloride	9, 193
	Suppl. 4, 170
Nitrogen mustard N-oxide and its hydrochloride	9, 209
2-Nitropropane	29, 331
1-Nitropyrene	33, 209
N-Nitrosatable drugs	24, 297 (corr. 30, 407)
N-Nitrosatable pesticides	30, 359
N-Nitrosodi-n-butylamine	4, 197
	17, 51
N-Nitrosodiethanolamine	17, 77
N-Nitrosodiethylamine	1, 107 (corr. 11, 295)
	17, 83 (corr. 23, 417)
N-Nitrosodimethylamine	1, 95
	17, 125 (corr. 25, 391)
N-Nitrosodiphenylamine	27, 213
para-Nitrosodiphenylamine	27, 227 (corr. 31, 293)
N-Nitrosodi-n-propylamine	17, 177
N-Nitroso-N-ethylurea	1, 135
	17, 191
N-Nitrosofolic acid	17, 217
N-Nitrosohydroxyproline	17, 304
N-Nitrosomethylethylamine	17, 221
N-Nitroso-N-methylurea	1, 125
	17, 227
N-Nitroso-N-methylurethane	4, 211
N-Nitrosomethylvinylamine	17, 257
N-Nitrosomorpholine	17, 263
N'-Nitrosonornicotine	17, 281
N-Nitrosopiperidine	17, 287
N-Nitrosoproline	17, 303
N-Nitrosopyrrolidine	17, 313
N-Nitrososarcosine	17, 327
N-Nitrososarcosine ethyl ester (1977)*	
Nitrovin	31, 185
Nitroxolne (1976)*	
Nivalenol (1976)*	
Norethisterone and its acetate	6, 179
	21, 441
	Suppl. 4, 199
Norethynodrel	6, 191
	21, 461 (corr. 25, 391)
	Suppl. 4, 201
Norgestrel	6, 201
	21, 479
	Suppl. 4, 202
Nylon 6	19, 120
Nylon 6/6 (1978)*	

O
Ochratoxin A	10, 191
	31, 191 (corr. 34, 197)

Oestradiol-17β   6, 99
                                                     21, 279
                                                     Suppl. 4, 190
Oestradiol 3-benzoate   21, 281
Oestradiol dipropionate   21, 283
Oestradiol mustard   9, 217
Oestradiol-17β-valerate   21, 284
Oestriol   6, 117
                                                     21, 327
Oestrone   6, 123
                                                     21, 343 (corr. 25, 391)
                                                     Suppl. 4, 191
Oestrone benzoate   21, 345
                                                     Suppl. 4, 191
Oil Orange SS   8, 165
Orange I   8, 173
Orange G   8, 181
Oxazepam   13, 58
Oxymetholone   13, 131
                                                     Suppl. 4, 203
Oxyphenbutazone   13, 185

P
Panfuran S (Dihydroxymethylfuratrizine)   24, 77
Parasorbic acid   10, 199 (corr. 12, 271)
Parathion   30, 153
Patulin   10, 205
Penicillic acid   10, 211
Pentachlorophenol   20, 303
                                                     Suppl. 4, 88, 205
Pentobarbital sodium (1976)*
Perylene   32, 411
Petasitenine   31, 207
Phenacetin   13, 141
                                                     24, 135
                                                     Suppl. 4, 47
Phenanthrene   32, 419
Phenazopyridine [2,6-Diamino-3-(phenylazo)pyridine] 8, 117
  and its hydrochloride
                                                     24, 163 (corr. 29, 399)
                                                     Suppl. 4, 207
Phenelzine and its sulphate   24, 175
                                                     Suppl. 4, 207
Phenicarbazide   12, 177
Phenobarbital and its sodium salt   13, 157
                                                     Suppl. 4, 208
Phenoxybenzamine and its hydrochloride   9, 223
                                                     24, 185
Phenylbutazone   13, 183
                                                     Suppl. 4, 212
ortho-Phenylenediamine (1977)*

meta-Phenylenediamine and its hydrochloride	*16*, 111
para-Phenylenediamine and its hydrochloride	*16*, 125
N-Phenyl-2-naphthylamine	*16*, 325 (corr. *25*, 391)
	Suppl. 4, 213
ortho-Phenylphenol and its sodium salt	*30*, 329
N-Phenyl-*para*-phenylenediamine (1977)*	
Phenytoin and its sodium salt	*13*, 201
	Suppl. 4, 215
Piperazine oestrone sulphate	*21*, 148
Piperonyl butoxide	*30*, 183
Polyacrylic acid	*19*, 62
Polybrominated biphenyls	*18*, 107
Polychlorinated biphenyls	*7*, 261
	18, 43
	Suppl. 4, 217
Polychloroprene	*19*, 141
Polyethylene (low-density and high-density)	*19*, 164
Polyethylene terephthalate (1978)*	
Polyisoprene (1978)*	
Polymethylene polyphenyl isocyanate	*19*, 314
Polymethyl methacrylate	*19*, 195
Polyoestradiol phosphate	*21*, 286
Polypropylene	*19*, 218
Polystyrene	*19*, 245
Polytetrafluoroethylene	*19*, 288
Polyurethane foams (flexible and rigid)	*19*, 320
Polyvinyl acetate	*19*, 346
Polyvinyl alcohol	*19*, 351
Polyvinyl chloride	*7*, 306
	19, 402
Polyvinylidene fluoride (1978)*	
Polyvinyl pyrrolidone	*19*, 463
Ponceau MX	*8*, 189
Ponceau 3R	*8*, 199
Ponceau SX	*8*, 207
Potassium bis (2-hydroxyethyl)dithiocarbamate	*12*, 183
Prednisone	*26*, 293
	Suppl. 4, 219
Procarbazine hydrochloride	*26*, 311
	Suppl. 4, 220
Proflavine and its salts	*24*, 195
Progesterone	*6*, 135
	21, 491 (corr. *25*, 391)
	Suppl. 4, 202
Pronetalol hydrochloride	*13*, 227 (corr. *16*, 387)
1,3-Propane sultone	*4*, 253 (corr. *13*, 243)
	(corr. *20*, 591)
Propham	*12*, 189
β-Propiolactone	*4*, 259 (corr. *15*, 341)
n-Propyl carbamate	*12*, 201
Propylene	*19*, 213
Propylene oxide	*11*, 191
	36, 227

Propylthiouracil   7, 67
                                                   Suppl. 4, 222
The pulp and paper industry   25, 157
                                                   Suppl. 4, 144

Pyrazinamide (1976)*
Pyrene   32, 431
Pyrimethamine   13, 233
Pyrrolizidine alkaloids   10, 333

Q
Quercitin   31, 213
Quinoestradol (1978)*
Quinoestrol (1978)*
para-Quinone   15, 255
Quintozene (Pentachloronitrobenzene)   5, 211

R
Reserpine   10, 217
                                                   24, 211 (corr. 26, 387)
                                                           (corr. 30, 407)
                                                   Suppl. 4, 222
Resorcinol   15, 155
Retrorsine   10, 303
Rhodamine B   16, 221
Rhodamine 6G   16, 233
Riddelliine   10, 313
Rifampicin   24, 243
Rotenone (1982)*
The rubber industry   28 (corr. 30, 407)
                                                   Suppl. 4, 144
Rugulosin (1975)*

S
Saccharated iron oxide   2, 161
Saccharin   22, 111 (corr. 25, 391)
                                                   Suppl. 4, 224
Safrole   1, 169
                                                   10, 231
Scarlet red   8, 217
Selenium and selenium compounds   9, 245 (corr. 12, 271)
                                                          (corr. 30, 407)
Semicarbazide hydrochloride   12, 209 (corr. 16, 387)
Seneciphylline   10, 319
Senkirkine   10, 327
                                                   31, 231
Shale-oils   35, 161
Simazine (1982)*
Sodium cyclamate   22, 56 (corr. 25, 391)
                                                   Suppl. 4, 97
Sodium diethyldithiocarbamate   12, 217
Sodium equilin sulphate   21, 148

Sodium oestrone sulphate	21, 147
Sodium saccharin	22, 113 (corr. 25, 391)
	Suppl. 4, 224
Soot and tars	3, 22
	Suppl. 4, 227
Soots	35, 219
Spironolactone	24, 259
	Suppl. 4, 229
Sterigmatocystin	1, 175
	10, 245
Streptozotocin	4, 221
	17, 337
Styrene	19, 231
	Suppl. 4, 229
Styrene-acrylonitrile copolymers	19, 97
Styrene-butadiene copolymers	19, 252
Styrene oxide	11, 201
	19, 275
	Suppl. 4, 229
	36, 245
Succinic anhydride	15, 265
Sudan I	8, 225
Sudan II	8, 233
Sudan III	8, 241
Sudan brown RR	8, 249
Sudan red 7B	8, 253
Sulfafurazole (Sulphisoxazole)	24, 275
	Suppl. 4, 233
Sulfallate	30, 283
Sulfamethoxazole	24, 285
	Suppl. 4, 234
Sulphamethazine (1982)*	
Sunset yellow FCF	8, 257
Symphytine	31, 239
T	
2,4,5-T and esters	15, 273
	Suppl. 4, 211, 235
Tannic acid	10, 253 (corr. 16, 387)
Tannins	10, 254
Terephthalic acid (1978)*	
Terpene polychlorinates (Strobane[R])	5, 219
Testosterone	6, 209
	21, 519
Testosterone oenanthate	21, 521
Testosterone propionate	21, 522
2,2′,5,5′-Tetrachlorobenzidine	27, 141
Tetrachlorodibenzo-*para*-dioxin (TCDD)	15, 41
	Suppl. 4, 211, 238
1,1,2,2-Tetrachloroethane	20, 477
Tetrachloroethylene	20, 491
	Suppl. 4, 243
Tetrachlorvinphos	30, 197

Tetrafluoroethylene	*19*, 285
Thioacetamide	*7*, 77
4,4'-Thiodianiline	*16*, 343
	27, 147
Thiouracil	*7*, 85
Thiourea	*7*, 95
Thiram	*12*, 225
2,4-Toluene diisocyanate	*19*, 303
2,6-Toluene diisocyanate	*19*, 303
ortho-Toluenesulphonamide	*22*, 121
	Suppl. *4*, 224
ortho-Toluidine and its hydrochloride	*16*, 349
	27, 155
	Suppl. *4*, 245
Toxaphene (Polychlorinated camphenes)	*20*, 327
Treosulphan	*26*, 341
	Suppl. *4*, 246
Trichlorphon	*30*, 207
1,1,1-Trichloroethane	*20*, 515
1,1,2-Trichloroethane	*20*, 533
Trichloroethylene	*11*, 263
	20, 545
	Suppl. *4*, 247
2,4,5-and 2,4,6-Trichlorophenols	*20*, 349
	Suppl. *4*, 88, 249
Trichlorotriethylamine hydrochloride	*9*, 229
Trichlorphon	*30*, 207
T$_2$-Trichothecene	*31*, 265
Triethylene glycol diglycidyl ether	*11*, 209
Trifluralin (1982)*	
2,4,5-Trimethylaniline and its hydrochloride	*27*, 177
2,4,6-Trimethylaniline and its hydrochloride	*27*, 178
Triphenylene	*32*, 447
Tris(aziridinyl)-*para*-benzoquinone (Triaziquone)	*9*, 67
	Suppl. *4*, 251
Tris(1-aziridinyl)phosphine oxide	*9*, 75
Tris(1-aziridinyl)phosphine sulphide (Thiotepa)	*9*, 85
	Suppl. *4*, 252
2,4,6-Tris(1-aziridinyl)-*s*-triazine	*9*, 95
1,2,3-Tris(chloromethoxy)propane	*15*, 301
Tris(2,3-dibromopropyl)phosphate	*20*, 575
Tris(2-methyl-1-aziridinyl)phosphine oxide	*9*, 107
Trp-P-1	*31*, 247
Trp-P-2	*31*, 255
Trypan blue	*8*, 267

U
Uracil mustard	*9*, 235
	Suppl. *4*, 256
Urethane	*7*, 111

V

Vinblastine sulphate	*26*, 349 (corr. *34*, 197)
	Suppl. 4, 257
Vincristine sulphate	*26*, 365
	Suppl. 4, 259
Vinyl acetate	*19*, 341
Vinyl bromide	*19*, 367
Vinyl chloride	*7*, 291
	19, 377
	Suppl. 4, 260
Vinyl chloride-vinyl acetate copolymers	*7*, 311
	19, 412
4-Vinylcyclohexene	*11*, 277
Vinylidene chloride	*19*, 439
	Suppl. 4, 262 (corr. *31*, 293)
Vinylidene chloride-vinylchloride copolymers	*19*, 448
Vinylidene fluoride (1978)*	
N-Vinyl-2-pyrrolidone	*19*, 461

X

2,4-Xylidine and its hydrochloride	*16*, 367
2,5-Xylidine and its hydrochloride	*16*, 377
2,6-Xylidine (1977)*	

Y

Yellow AB	*8*, 279
Yellow OB	*8*, 287

Z

Zearalenone	*31*, 279
Zectran	*12*, 237
Zineb	*12*, 245
Ziram	*12*, 259

IARC MONOGRAPHS ON THE EVALUATION OF THE CARCINOGENIC RISK OF CHEMICALS TO HUMANS

Available from WHO Sales Agents. See addresses on back cover

Some Inorganic Substances, Chlorinated Hydrocarbons, Aromatic Amines, N-Nitroso Compounds, and Natural Products	Volume 1, 1972; 184 pages (out of print)	Some Halogenated Hydrocarbons	Volume 20, 1979; 609 pages US$ 35.00; Sw. fr. 60.--
Some Inorganic and Organometallic Compounds	Volume 2, 1973; 181 pages US$ 3.60; Sw. fr. 12.-- (out of print)	Sex Hormones (II)	Volume 21, 1979; 583 pages US$ 35.00; Sw. fr. 60.--
		Some Non-nutritive Sweetening Agents	Volume 22, 1980; 208 pages US$ 15.00; Sw. fr. 25.--
Certain Polycyclic Aromatic Hydrocarbons and Heterocyclic Compounds	Volume 3, 1973; 271 pages (out of print)	Some Metals and Metallic Compounds	Volume 23, 1980; 438 pages US$ 30.00; Sw. fr. 50.--
Some Aromatic Amines, Hydrazine and Related Substances, N-Nitroso Compounds and Miscellaneous Alkylating Agents	Volume 4, 1974; 286 pages US$ 7.20; Sw. fr. 18.--	Some Pharmaceutical Drugs	Volume 24, 1980; 337 pages US$ 25.00; Sw. fr. 40.--
		Wood, Leather and Some Associated Industries	Volume 25, 1980; 412 pages US$ 30.00; Sw. fr. 60.--
Some Organochlorine Pesticides	Volume 5, 1974; 241 pages US$ 7.20; Sw. fr. 18.-- (out of print)	Some Anticancer and Immunosuppressive Drugs	Volume 26, 1981; 411 pages US$ 30.00; Sw. fr. 62.--
Sex Hormones	Volume 6, 1974; 243 pages US$ 7.20; Sw. fr. 18.--	Some Aromatic Amines, Anthraquinones and Nitroso Compounds and Inorganic Fluorides Used in Drinking-Water and Dental Preparations	Volume 27, 1982; 341 pages US$ 25.00; Sw. fr. 40.--
Some Anti-thyroid and Related Substances, Nitrofurans and Industrial Chemicals	Volume 7, 1974; 326 pages US$ 12.80; Sw. fr. 32.--		
Some Aromatic Azo Compounds	Volume 8, 1975; 357 pages US$ 14.40; Sw. fr. 36.--	The Rubber Industry	Volume 28, 1982; 486 pages US$ 35.00; Sw. fr. 70.--
Some Aziridines, N-, S- and O-Mustards and Selenium	Volume 9, 1975; 268 pages US$ 10.80; Sw. fr. 27.--	Some Industrial Chemicals and Dyestuffs	Volume 29, 1982; 416 pages US$ 30.00; Sw. fr. 60.--
Some Naturally Occurring Substances	Volume 10, 1976; 353 pages US$ 15.00; Sw. fr. 38.--	Miscellaneous Pesticides	Volume 30, 1983; 424 pages US$ 30.00; Sw. fr. 60.--
Cadmium, Nickel, Some Epoxides, Miscellaneous Industrial Chemicals and General Considerations on Volatile Anaesthetics	Volume 11, 1976; 306 pages US$ 14.00; Sw. fr. 34.--	Some Feed Additives, Food Additives and Naturally Occurring Substances	Volume 31, 1983; 314 pages US$ 30.00; Sw. fr. 60. --
Some Carbamates, Thiocarbamates and Carbazides	Volume 12, 1976; 282 pages US$ 14.00; Sw. fr. 34.--	Chemicals and Industrial Processes Associated with Cancer in Humans (IARC Monographs 1-20)	Supplement 1, 1979; 71 pages (out of print)
Some Miscellaneous Pharmaceutical Substances	Volume 13, 1977; 255 pages US$ 12.00; Sw. fr. 30.--	Long-term and Short-term Screening Assays for Carcinogens: A Critical Appraisal	Supplement 2, 1980; 426 pages US$ 25.00; Sw. fr. 40.--
Asbestos	Volume 14, 1977; 106 pages US$ 6.00; Sw. fr. 14.--	Cross Index of Synonyms and Trade Names in Volumes 1 to 26	Supplement 3, 1982; 199 pages US$ 30.00; Sw. fr. 60.--
Some Fumigants, the Herbicides 2,4-D and 2,4,5-T, Chlorinated Dibenzodioxins and Miscellaneous Industrial Chemicals	Volume 15, 1977; 354 pages US$ 20.00; Sw. fr. 50.--	Chemicals, Industrial Processes and Industries Associated with Cancer in Humans (IARC Monographs Volumes 1 to 29)	Supplement 4, 1982; 292 pages US$ 30.00; Sw. fr. 60.--
Some Aromatic Amines and Related Nitro Compounds - Hair Dyes, Colouring Agents and Miscellaneous Industrial Chemicals	Volume 16, 1978; 400 pages US$ 20.00; Sw. fr. 50.--	Polynuclear Aromatic Compounds, Part 1, Chemical, Environmental and Experimental Data	Volume 32, 1983 ; 477 pages US $ 35.00 ; Sw.fr. 70
Some N-Nitroso Compounds	Volume 17, 1978; 365 pages US$ 25.00; Sw. fr. 50.--	Polynuclear Aromatic Compounds, Part 2, Carbon Blacks, Mineral Oils and Some Nitroarenes	Volume 33, 1984; 245 pages US$ 25.00; Sw. fr. 50.--
Polychlorinated Biphenyls and Polybrominated Biphenyls	Volume 18, 1978; 140 pages US$ 13.00; Sw. fr. 20.--	Polynuclear Aromatic Compounds, Part 3, Industrial Exposures in Aluminium Production, Coal Gasification, Coke Production, and Iron and Steel Founding	Volume 34, 1984 ; 219 pages US$ 20.00 ; Sw. fr. 48.—
Some Monomers, Plastics and Synthetic Elastomers, and Acrolein	Volume 19, 1979; 513 pages US$ 35.00; Sw. fr. 60.--		
		Polynuclear Aromatic Compounds, Part 4, Bitumens, Coal-tars and Derived Products, Shale-oils and Soots	Volume 35, 1985 ; 271 pages US$ 25.00 ; Sw. fr. 70.—
		Allyl Compounds, Aldehydes, Epoxides and Peroxides	Volume 36, 1985 ; 369 pages US$ 25.00 ; Sw. fr. 70.—

IARC SCIENTIFIC PUBLICATIONS

Available from Oxford University Press, Walton Street, Oxford OX2 6DP, UK and in London, New York, Toronto, Delhi, Bombay, Calcutta, Madras, Karachi, Kuala Lumpur, Singapore, Hong Kong, Tokyo, Nairobi, Dar es Salaam, Cape Town, Melbourne, Auckland and associated companies in Beirut, Berlin, Ibadan, Mexico City, Nicosia.

Liver Cancer	No. 1, 1971; 176 pages $10.—	Nasopharyngeal Carcinoma: Etiology and Control	No. 20, 1978; 610 pages £35.—
Oncogenesis and Herpesviruses	No. 2, 1972; 515 pages £30.—	Cancer Registration and Its Techniques	No. 21, 1978; 235 pages £11.95
N-Nitroso Compounds, Analysis and Formation	No. 3, 1972; 140 pages £8.50	Environmental Carcinogens—Selected Methods of Analysis, Vol. 2: Methods for the Measurement of Vinyl Chloride in Poly(vinyl chloride), Air, Water and Foodstuffs	No. 22, 1978; 142 pages £35.—
Transplacental Carcinogenesis	No. 4, 1973; 181 pages £11.95		
Pathology of Tumours in Laboratory Animals—Volume I—Tumours of the Rat, Part 1	No. 5, 1973; 214 pages £17.50	Pathology of Tumours in Laboratory Animals—Volume II—Tumours of the Mouse	No. 23, 1979; 669 pages £35.—
Pathology of Tumours in Laboratory Animals—Volume I—Tumours of the Rat, Part 2	No. 6, 1976; 319 pages £17.50	Oncogenesis and Herpesviruses III	No. 24, 1978; Part 1, 580 pages £20.— Part 2, 522 pages $20.—
Host Environment Interactions in the Etiology of Cancer in Man	No. 7, 1973; 464 pages £30.—	Carcinogenic Risks—Strategies for Intervention	No. 25, 1979; 283 pages £20.—
Biological Effects of Asbestos	No. 8, 1973; 346 pages £25.—	Directory of On-Going Research in Cancer Epidemiology 1978	No. 26, 1978; 550 pages (OUT OF PRINT)
N-Nitroso Compounds in the Environment	No. 9, 1974; 243 pages £15.—	Molecular and Cellular Aspects of Carcinogen Screening Tests	No. 27, 1980; 371 pages £20.—
Chemical Carcinogenesis Essays	No. 10, 1974; 230 pages £15.—	Directory of On-Going Research in Cancer Epidemiology 1979	No. 28, 1979; 672 pages (OUT OF PRINT)
Oncogenesis and Herpesviruses II	No. 11, 1975; Part 1, 511 pages £30.— Part 2, 403 pages £30.—	Environmental Carcinogens—Selected Methods of Analysis, Vol. 3: Analysis of Polycyclic Aromatic Hydrocarbons in Environmental Samples	No. 29, 1979; 240 pages £17.50
Screening Tests in Chemical Carcinogenesis	No. 12, 1976; 666 pages £30.—	Biological Effects of Mineral Fibres	No. 30, 1980; Volume 1, 494 pages £25.— Volume 2, 513 pages £25.—
Environmental Pollution and Carcinogenic Risks	No. 13, 1976; 454 pages £17.50		
Environmental N-Nitroso Compounds—Analysis and Formation	No. 14, 1976; 512 pages £35.—	N-Nitroso Compounds: Analysis, Formation and Occurrence	No. 31, 1980; 841 pages £30.—
Cancer Incidence in Five Continents—Volume III	No. 15, 1976; 584 pages £35.—	Statistical Methods in Cancer Research, Vol. 1: The Analysis of Case-Control Studies	No. 32, 1980; 338 pages £17.50
Air Pollution and Cancer in Man	No. 16, 1977; 331 pages £30.—		
Directory of On-Going Research in Cancer Epidemiology 1977	No. 17, 1977; 599 pages (OUT OF PRINT)	Handling Chemical Carcinogens in the Laboratory—Problems of Safety	No. 33, 1979; 32 pages £3.95
Environmental Carcinogens—Selected Methods of Analysis, Vol. 1: Analysis of Volatile Nitrosamines in Food	No. 18, 1978; 212 pages £30.—	Pathology of Tumours in Laboratory Animals—Volume III—Tumours of the Hamster	No. 34, 1982; 461 pages £30.—
		Directory of On-Going Research in Cancer Epidemiology 1980	No. 35, 1980; 660 pages (OUT OF PRINT)
Environmental Aspects of N-Nitroso Compounds	No. 19, 1978; 566 pages £35.—	Cancer Mortality by Occupation and Social Class 1851-1971	No. 36, 1982, 253 pages £20.—

Laboratory Decontamination and Destruction of Aflatoxins B_1, B_2, G_1, G_2 in Laboratory Wates	No. 37, 1980; 59 pages £ 5.95
Directory of On-Going Research in Cancer Epidemiology 1981	No 38, 1981; 696 pages (OUT OF PRINT)
Host Factors in Human Carcinogenesis	No. 39, 1982; 583 pages £ 35.—
Environmental Carcinogens—Selected Methods of Analysis, Vol. 4: Some Aromatic Amines and Azo Dyes in the General and Industrial Environment	No. 40, 1981; 347 pages £ 20.—
N-Nitroso Compounds: Occurrence and Biological Effect	No. 41, 1982; 755 pages £ 35.—
Cancer Incidence in Five Continents—Volume IV	No. 42, 1982; 811 pages £ 35.—
Laboratory Decontamination and Destruction of Carcinogens in Laboratory Wastes: Some N-Nitrosamines	No. 43, 1982; 73 pages £ 6.50
Environmental Carcinogens—Selected Methods of Analysis, Vol. 5: Mycotoxins	No. 44, 1983; 455 pages £ 20.—
Environmental Carcinogens—Selected Methods of Analysis, Vol. 6: N-Nitroso Compounds	No. 45, 1983; 508 pages £ 20.—
Directory of On-Coing Research in Cancer Epidemiology 1982	No. 46, 1982; 722 pages (OUT OF PRINT)
Cancer Incidence in Singapore	No. 47, 1982; 174 pages £ 10.—
Cancer Incidence in the USSR Second Revised Edition	No. 48, 1982; 75 pages £ 10.—
Laboratory Decontamination and Destruction of Carcinogens in Laboratory Wastes: Some Polycyclic Aromatic Hydrocarbons	No. 49, 1983; 81 pages £ 7.95
Directory of On-Going Research in Cancer Epidemiology 1983	No. 50, 1983: 740 pages (OUT OF PRINT
Modulators of Experimental Carcinogenesis	No. 51, 1983; 307 pages £ 25.—
Second Cancers Following Radiation Treatment for Cancer of the Uterine Cervix: Results of a Cancer Registry Collaborative Study	No. 52, 1984; 207 pages £ 17.50
Nickel in the Human Environment	No. 53, 1984; 529 pages £ 30.—
Laboratory Decontamination and Destruction of Carcinogens in Laboratory Wastes: Some Hydrazine	No. 54, 1983; 87 pages £ 6.95
Laboratory Decontamination and Destruction of Carcinogens in Laboratory Wastes: Some N-Nitrosamides	No. 55, 1984; 65 pages £ 6.95
Models, Mechanisms and Etiology of Tumour Promotion	No. 56, 1985; 532 pages £ 30.—
N-Nitroso Compounds: Occurence, Biological Effects and Relevance to Human Cancer	No. 57, 1984; 1013 pages £ 75.—
Age-related Factors in Carcinogenesis	No. 58, 1985 (in press)
Monitoring Human Exposure to Carcinogenic and Mutagenic Agents	No. 59, 1985; 457 pages £ 25.—
Burkitt's Lymphoma: A Human Cancer Model	No. 60, 1985; 484 pages £ 25.—
Laboratory Decontamination and Destruction of Carcinogens in Laboratory Wastes: Some Haloethers	No. 61, 1984; 55 pages £ 5.95
Directory of On-Going Research in Cancer Epidemiology 1984	No. 62, 1984; 728 pages £ 18.—
Virus-associated Cancers in Africa	No. 63, 1984; 773 pages £ 20.—
Laboratory Decontamination and Destruction of Carcinogens in Laboratory Wastes: Some Aromatic Amines and 4-Nitrobiphenyl	No. 64, 1985; 85 pages £ 5.95.—
Interpretation of Negative Epidemiological Evidence for Carcinogenicity	No. 65, 1985 (in press)
The Role of the Registry in Cancer Control	No. 66, 1985 (in press)
Transformation Assay of Established Cell Lines: Mechanisms and Application	No. 67, 1985 (in press)
Environmental Carcinogens-Selected Methods of Analysis, Vol. 7: Some Volatile Halogenated Hydrocarbons	No. 68, 1985 (in press)
Directory of On-going Research in Cancer Epidemiology 1985	No. 69, 1985 (in press)

NON-SERIAL PUBLICATIONS

Available from IARC

Alcool et Cancer	1978; 42 pages Fr. fr. 35-; Sw. fr. 14.-
Cancer Morbidity and Causes of Death Among Danish Brewery Workers	1980, 145 pages US$ 25.00; Sw. fr. 45.-